Flora of the Voynich Codex

Arthur O. Tucker • Jules Janick

Flora of the Voynich Codex

An Exploration of Aztec Plants

Springer

Arthur O. Tucker
Department of Agriculture and Natural Resources
Delaware State University
Dover, DE, USA

Jules Janick
Department of Horticulture and Landscape Architecture
Purdue University
West Lafayette, IN, USA

ISBN 978-3-030-19376-8 ISBN 978-3-030-19377-5 (eBook)
https://doi.org/10.1007/978-3-030-19377-5

This Springer imprint is published by the registered company Springer Nature Switzerland AG
The registered company address is: Gewerbestrasse 11, 6330 Cham, Switzerland

For Sharon L. Tucker, Patricia M. Ryan, and our children: Angelica, Arthur IV, Melissa, Peter, and Robin.

This book is dedicated to the memory of Dr. Arthur O. Tucker who sadly died on August 5, 2019 when this book was in press.

Cover image: The cover image, folio 94r of the *Voynich Codex*, is identified as *Viola bicolor* Pursh (*V. refinesquii* Greene). The phytomorph clearly shows linear terminal stipular lobes as in the North American native *V. bicolor*, not spatulate as in the Eurasian *V. tricolor* L. Its blue flowers match *V. bicolor* in contrast to the tri-colored ones of *V. tricolor*. The delineation of *V. bicolor* as native to the New World and not introduced from elsewhere, was only elucidated in 1961.

Reverend Dr. Hugh O'Neill (1894–1969)

Preface

The mysterious *Voynich Codex* has been shown to be a sixteenth-century Mexican manuscript, on the basis of the presence of New World plants and animals and the presence of Mexican landmarks. The evidence dates to a seminal 1944 note by the Reverend Dr. Hugh O'Neill, who reported the presence of sunflower and capsicum pepper in the *Voynich Codex,* clear evidence that the *Voynich Codex* must be post-Columbus because these two species are indigenous to the New World. This counterevidence to the almost universally accepted dogma that the codex was a European fifteenth-century manuscript was ignored, discounted, or explained away by most of the Voynich community, all non-botanists. The number of identified phytomorphs was increased to 37 (Tucker and Talbert 2013) and then to 58 (Tucker and Janick 2016). In our book, *Unravelling the Voynich Codex,* the count went up to 60 (Janick and Tucker 2018). This new work raises the number of phytomorphs identified as New World plants to 169 and provides evidence that the *Voynich Codex* must be included as one of the seminal contributions to sixteenth-century Mexican botany. Thus, we have titled this book *Flora of the Voynich Codex: An Exploration of Aztec Plants*. It is too early to know the reception of this work by an exuberant, albeit seemingly fanatical, community unable or unwilling to accept botanical evidence.

We dedicate this work to the memory of Hugh O'Neill, who has been disregarded and vilified despite his credentials as a distinguished botanical taxonomist with outstanding work on tropical American flora.

Dover, DE, USA — Arthur O. Tucker
West Lafayette, IN, USA — Jules Janick

Literature Cited

Janick, J., and A. O. Tucker, eds. 2018. *Unraveling the Voynich Codex*. Cham, Switzerland: Springer Nature Switzerland.

O'Neill, H. 1944. Botanical observations on the Voynich MS. *Speculum* 19: 126.

Tucker, A. O., and J. Janick. 2016. Identification of phytomorphs in the Voynich Codex. *Horticultural Reviews* 44: 1–64.

Tucker, A. O., and R. H. Talbert. 2013. A preliminary analysis of the botany, zoology, and mineralogy of the Voynich manuscript. *HerbalGram* 100: 70–85.

Contents

Chapter 1
An Introduction to the *Voynich Codex*

The Mysterious Codex

The *Voynich Codex* is a bizarre, profusely illustrated manuscript that has been considered the most mysterious manuscript ever written. The work is encyclopedic and has been divided by convention as follows:

Herbal, 127 pages containing 131 plant images plus text (Fig. 1.1)
Pharmaceutical, 20 pages with 228 small images of plants or plant parts, plus apothecary jars, many labeled with symbolic script (Fig. 1.2)
Balneological, 19 pages with nude nymphs, singly or in groups, many cavorting in pools with strange plumbing or vascular systems (Fig. 1.3)
Astrological, 12 pages containing signs of the zodiac surrounded by nymphs, mostly nude (Fig. 1.4)
Cosmological, 16 pages including combinations of the sun, moon, planets, and stars (Fig. 1.5)
Rosette, a large insert the size of 6 pages, with 11 circles or rosettes resembling a kabbalah tree-of-life symbol that is a map of central Mexico cities and volcanos (Fig. 1.6)
Recipe, 23 pages of text, probably medicinal prescriptions, or possibly poems or incantations, with each phrase highlighted by a six- to eight-pointed star in the left margin (Fig. 1.7).

The *Voynich Codex* includes symbolic writing, referred to as *Voynichese*, that has evaded decipherment by eminent cryptologists, and its original language has been under dispute. The symbolic script follows rules of languages, and the suggestion that the *Voynich Codex* is a modern hoax has been universally discredited. Robert Brumbaugh (1978) assumed it was a forgery, whereas Gordon Rugg (2004a, b) declared it was gibberish. These theories have been tested by modern cryptographic analysis based on the frequency of letters or words. Antoine Casanova (1999) concluded that the *Voynich Codex* had the qualities of a synthetic language whose

A. O. Tucker, J. Janick, *Flora of the Voynich Codex*,
https://doi.org/10.1007/978-3-030-19377-5_1

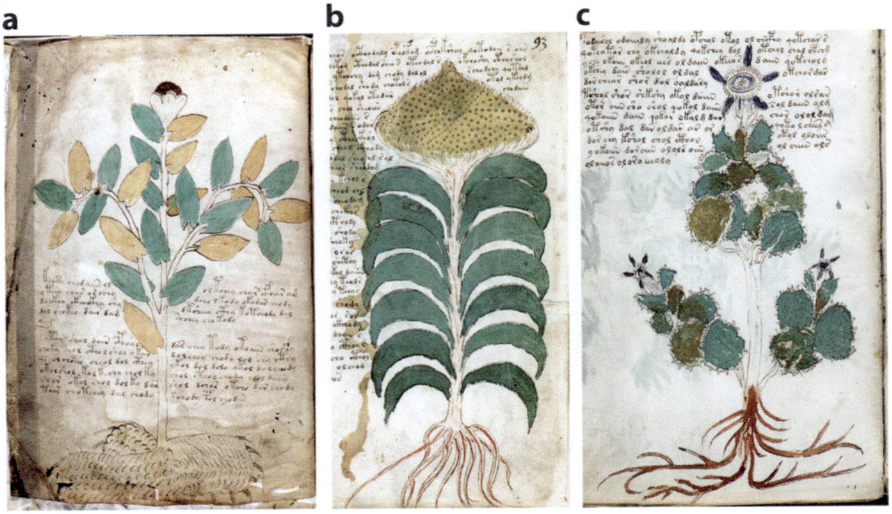

Fig. 1.1 Herbal section phytomorphs: (**a**) folio 1v, *Ipomoea arborescens*; (**b**) folio 93r, *Helianthus annuus*; (**c**) folio 23v, *Passiflora sp.* section *Decaloba*; cf. *P. morifolia*

alphabet was subject to transformations. M. A. Montemurro and D. H. Zanette (2013) demonstrated that the language organization is complex, but the distribution of words is compatible with real language sequences and the distribution of letters follows Zipf's law, as in other languages. D. R. Amancio et al. (2013) concluded that words display compatibility with natural languages and are incompatible with random texts.

Provenance

An untitled manuscript was purchased by a Polish book dealer, Wilfred Voynich (Fig. 1.8), in the Villa Mondragone at Frascati, Italy. This patrician complex, constructed in 1573 on the site of a Roman villa, had been transformed into a Jesuit college for young aristocrats in 1865 and was discreetly selling off parts of its library.

The work, now known as the *Voynich Codex*, has long bedeviled historians of science. It has been traced to the Emperor Rudolf II (1576–1612), Emperor of the Holy Roman Empire (Fig. 1.9) and a collector and connoisseur of art. A 1667 letter to the Jesuit scholar Athanasius Kircher from Joannes Marci, Rector of Prague, noted that the manuscript was sold to Rudolf II for 600 gold ducats. Its association with the court of Rudolf II was confirmed by the embedded signature on the first page of the manuscript of Jacobi à Tepeanecz, the 1607-ennobled name of Jaobus

Fig. 1.2 A page in the Pharmaceutical section (folio 100r) with 16 images of phytomorphs labeled in Voynichese symbols and two apothecary jars

Horĉicky (1575–1622), Imperial Chemist and personal physician of Rudolf II. However, the origin and the author of the work has been a mystery.

Wilfrid Voynich considered the manuscript to be the work of Roger Bacon (ca. 1219–1292), but his attempts to sell it at a high price ($160,000) after he moved to the United States were unsuccessful. After the death of Wilfrid Voynich's wife, Ethel Lilian Boole, in 1960, the manuscript was sold to the book dealer H.P. Kraus, who also failed to sell it. In 1969 it was donated it to the Beinecke Rare Book & Manuscript Library at Yale University, where it now resides as Manuscript 408.

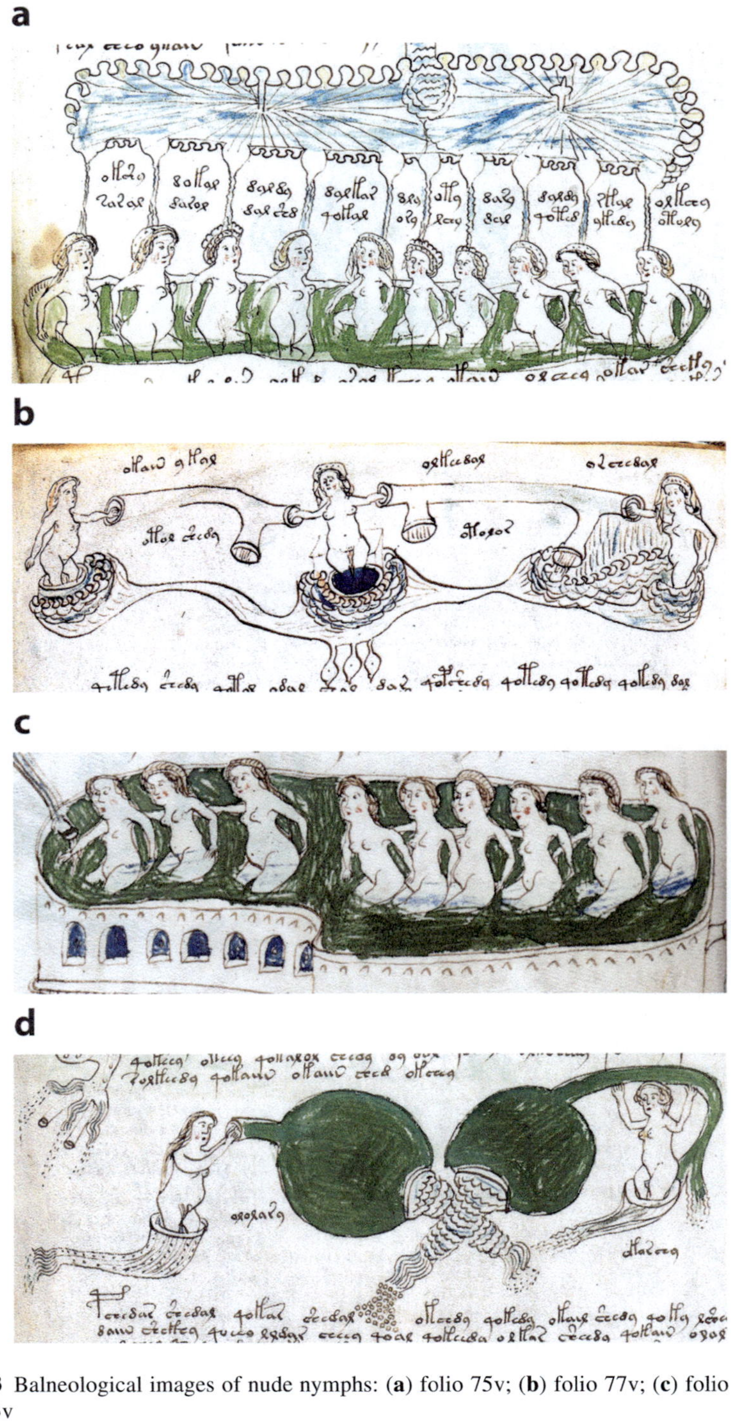

Fig. 1.3 Balneological images of nude nymphs: (**a**) folio 75v; (**b**) folio 77v; (**c**) folio 78v; (**d**) folio 83v

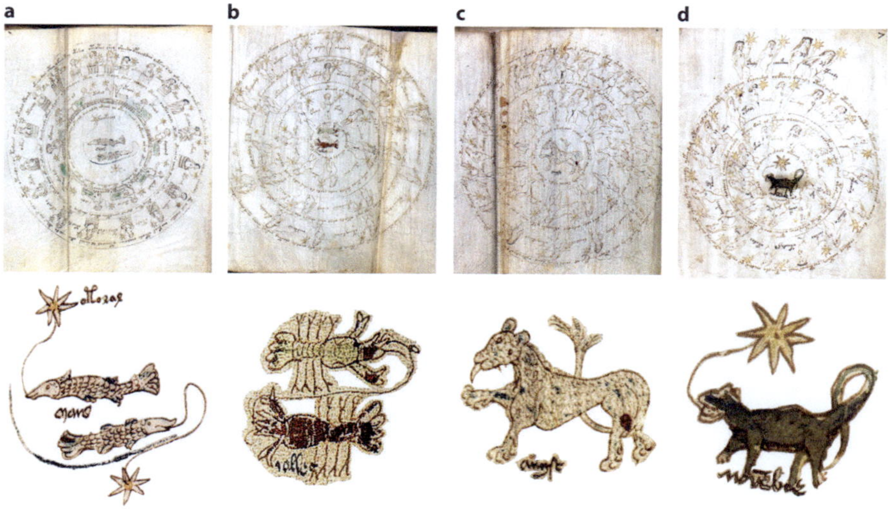

Fig. 1.4 Astrological signs with Mexican animals: (**a**) folio 70r, Pisces represented by an alligator gar; (**b**) folio 71v, Cancer represented by Mexican crayfish; (**c**) folio 72r(1), Leo represented by on ocelot; and (**d**) folio 73r, Scorpio represented by a jaguarundi

Dating and Origin of the Codex

In 2011, the University of Arizona issued a press release on the age of the *Voynich Codex* vellum using carbon dating (Stolte 2011). Dr. Gregory Hodgins of the university's Department of Physics performed the analysis in 2009 using an accelerator mass spectrometer to measure the ratio of ^{14}C (carbon-14) to ^{12}C (carbon-12) in four snippets of the manuscript. The samples dated from 1404 to 1438. Following this, Hodgins made several oral presentations and announcements with additional data, but the information has not been published (Zandbergen 2012).

This fact, as well as its provenance in Italy, has led to the widely held view and current dogma that it was a fifteenth century European work. However, botanical evidence indicates that the work contains New World plants (O'Neill 1944; Guy 1991; Tucker and Talbert 2013; Tucker and Janick 2016; Janick and Tucker 2018), which contravenes the conventional wisdom based on the dating of the vellum. A Mexican connection to the *Voynich Codex* had been hinted at by Jacques B.M. Guy (1991), proposed by James C. Comegys (2001), published in a peer–reviewed journal by Arthur O. Tucker and Rexford T. Talbert (2013), and reported in an unpublished manuscript by John D. Comegys that appeared online in 2014 (copyrighted 2013). This hypothesis was fully expanded in *Unraveling the Voynich Codex* by Jules Janick and Arthur Tucker (2018). A signature in the first botanical image provides evidence that the author was Gaspar de Torres, a Spaniard of Jewish heritage born in Santo Domingo, and Master of Students of the *Colegio Imperial de Santa Cruz* at Tlatelolco 1568–1572. Ligated initials on the same folio suggests that the Illustrator was Juan Gerson, an indigenous artist (*tlacuilo*) born in Mexico (Janick and Tucker, 2017).

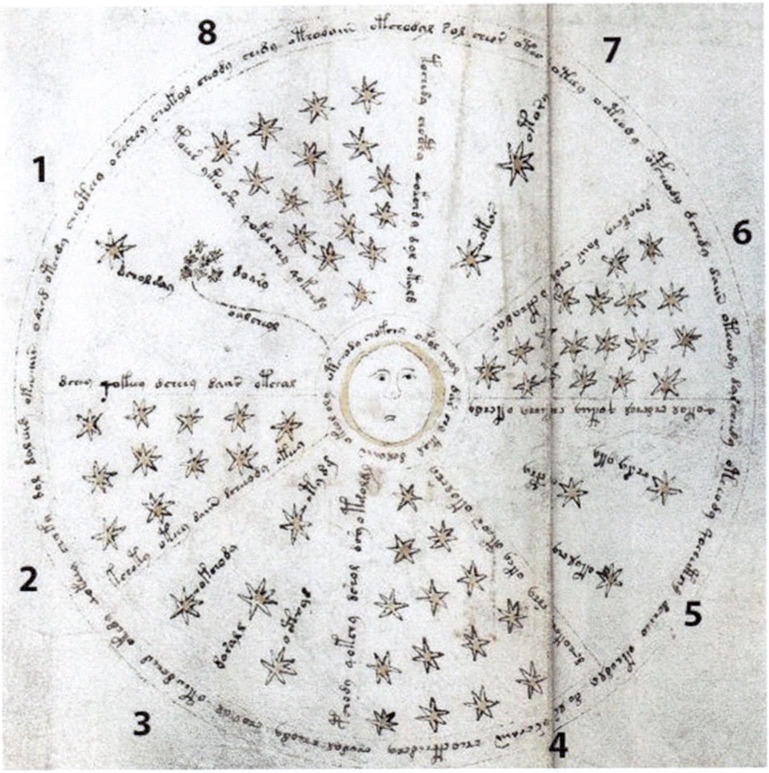

Moon, and stars, fol. 68r(3)

Fig. 1.5 Cosmological image, folio 68r(3), showing sun and labeled stars, including the Pleiades star cluster in section 1

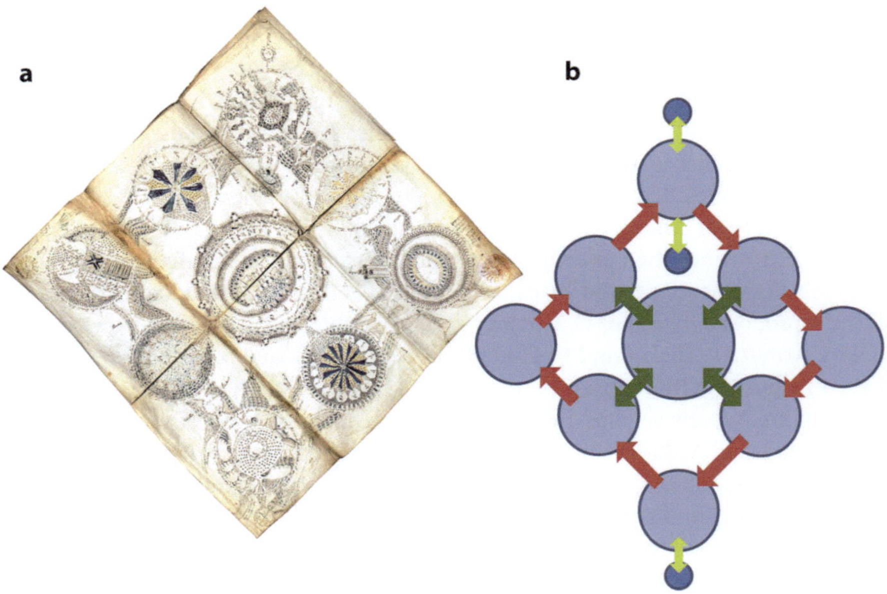

Fig. 1.6 The Rosette page, folio 86v: (**a**) oriented with suns in an East-West position, representing a kabbalah-like map. (**b**) Schematic diagram of folio 86v

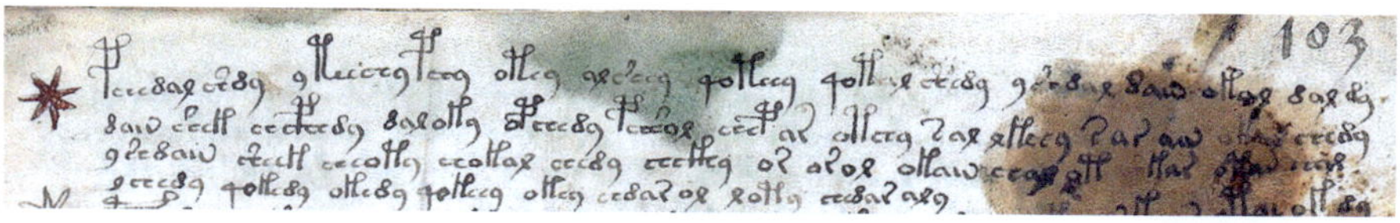

Fig. 1.7 The first entry of folio 103r in the Recipe section, showing symbolic script known as Voynichese

Fig. 1.8 Wilfrid Voynich (1865–1930)

The connection of the *Voynich Codex* to sixteenth century New Spain suggests that the book was either a palimpsest written over the vellum of an old book, or simply written on reused vellum that was known to be imported to the *Colegio Imperial de Santa Cruz*. In New Spain in times of scarcity, use was made of already printed paper, which was sold as a waste product (Reyes 1948, cited in Romero Ramírez 2013:39 Vol. 1), and new Spanish paper was available in the large market at Tlatelolco from paper sellers called *amanamac* or *amaoztomecatl* (Hirth 2016). Reused parchment was available at the scriptorium of *El Colegio de Santa Cruz de Tlatelolco* in New Spain (Gravier 2011). Researchers from the Bodleian Library at the University of Oxford have demonstrated that the *Codex Selden*, also known as the *Codex Añute* (1560) and written on a deer hide, was a palimpsest written over a pre-colonial manuscript (Snijders et al. 2016).

Evidence of a New World origin is supported by the following:

1. The plants are indigenous to the New World or circumboreal, with the exception of one weedy Spanish introduction.
2. The animals identified are indigenous to the New World, including alligator gar, armadillo, coati mundi, caecilian, jaguarundi, horned lizard, Mexican crayfish, and ocelot, or are Spanish introductions.

Fig. 1.9 Emperor Rudolf II (1552–1612), portrait by Hans von Aachen

3. The landmarks in the Rosette page (folio 86v) include cities and volcanoes of central Mexico.
4. The first botanical illustration includes the ligated initials of Juan Gerson, Tlacuilo ("JGT"), an indigenous artist of Mexico, and the name "Gasp.Torres" (Gaspar de Torres), born in Santo Domingo and master of students at the *Colegio Imperial de Santa Cruz.*

Description of the Work

The *Voynich Codex* is composed of folded sheets called *folios*, each composed of two pages numbered *recto* (r) (right side) and *verso* (v) (reverse side). Folios are referred to by their individual numbers as 1r, 1v, 2r, 2v, and so on, up to 116v. However, 14 folios are missing (12, 59, 60, 61, 62, 63, 64, 74, 91, 92, 97, 98, 109, 110). Adjusting for missing folios and foldouts of two to four pages, the original work is equivalent to 262 pages, but only 234 remain. The folios are bound into set quires of four folios (8 pages each) and sewn together, but improper binding has altered the original sequence so that some of the sections are not contiguous.

Illustrations

With the exception of the recipe section, almost every page contains tinted illustrations drawn with a pen and most tinted with a limited palette of four colors: blue, red, green, and red-brown. Pigment analysis in 2009 by McCrone Associates Inc. of Westmont, Illinois, indicated that symbols are made with iron gall ink made by reaction of iron sulfate with oak galls. The green pigment was a copper and copper-chlorine resinate, probably derived from atacamite. The blue pigment was derived from azurite, and the red-brown pigment was red ochre derived from hematite. The binder was unidentified but was not gum arabic. The blue and red-brown pigments are compatible with a New World origin; the green pigment is unlikely to be of European origin because European greens were only rarely derived from a source of copper-chlorine pigment such as decayed bronze (which provides paratacamite, not atacamite).

Decipherment

Attempts to decipher the *Voynich Codex* have failed gloriously. The fact that William Friedman, the world's most eminent cryptologist and breaker of the Japanese Purple cipher, has been unsuccessful in deciphering Voynichese suggests that ciphers or secret writing is not involved. Friedman's conclusion, published in a cipher code revealed in 1970, indicates that Voynichese is an artificial or universal language of the *a priori* type (*i.e.,* an invented or synthetic language).

Decipherment of the plain text of the *Voynich Codex* requires both that the symbols be decoded phonetically (transliteration) and that the appropriate language be identified (translation). The simplest interpretation of Voynichese, consisting of about 37 symbols, is that they are derived from scribal shorthand of sixteenth century New Spain; these transliterate to either a Nahuatl *lingua franca* or a phonetic alphabet or syllabary of either a real and/or an invented language.

The labeled plants in the Pharmaceutical section provided a path to a decipherment of the symbols. Tucker and Talbert (2013) decoded the Voynichese symbols based on the identification of two plants derived from primary names, *nāshtli,* a phonetic derivative from Nahuatl, *nochtli* (*Opuntia ficus indica*) and *māgueoy,* derived via Taino from *maguey* (*Agave atrovirens*). Other plant names expanded the symbol decoding, and although none were previously recorded Nahuatl names, many were descriptive in the manner of Nahuatl names. This decoding system allowed decipherment of Nahuatl cognates and some cities and words, including dialects of Nahuatl, Latin American Spanish, Mixtec, Huastec, and Taino,; that gave us confidence that the decoding was on the right track, but much to our chagrin, it failed strict Classical Nahuatl. It might be a synthetic language (as suggested by William Friedman) composed of various indigenous languages. We are convinced that a Nahuatl dialect is involved, because the decoded symbols identified the city of Huejotzingo, confirming a previous independent identification. We agree with William Friedman that Voynichese may be a mixed synthetic language. It probably was invented from various contemporary languages such as Latin American Spanish (including Nahuatl, Taino, and other languages of central Mexico) and perhaps a North African Arabic dialect for astronomical names extant in sixteenth century New Spain. Another possibility was a Nahuatl *lingua franca* used in commerce by the *pochteca,* Aztec traders (Dakin, 1981; Janick and Tucker, 2018).

Objectives

It is our belief that the inability to decipher the symbols and translate *Voynich Codex* stems from the erroneous belief that the manuscript must be a European work. This has led to plant misidentification, wrong linguistic analysis, and circular reasoning. We are convinced that the original sin of failure to correctly identify the plants of the *Voynich Codex* as New World plants and accept the possibility of New World languages is a serious error in *Voynich Codex* scholarship. The objective of this work has three interconnected objectives:

1. Identification of all plants in the Herbal section of the *Voynich Codex* and many in the Pharmaceutical section
2. Association of the plants with Aztec botany and medicine
3. Relation of the plants of the *Voynich Codex* to period herbals of New Spain and Mexico, as well as 16th to early seventeenth century herbals that describe plants of the New World

Literature Cited

Amancio, D.R., E.G. Altmann, D. Rybski, O.N. Oliveira, and L. da F. Costa. 2013. Probing the statistical properties of unknown texts: Application to the Voynich Manuscript. *PLoS One* 8 (7): e67310. https://doi.org/10.1371/journal.pone.0067310.

Brumbaugh, R.S. 1978. *The most mysterious manuscript.* Carbondale: Southern Illinois Press.

Casanova, A. 1999. Méthodes d'analyse du langage crypté: Une contribution à l'étude du manuscript du Voynich. Docteur thesis. Université de Paris. http://voynich.free.fr/a_casanova_these_19mars1999.pdf. Accessed 18 Jan 2014.

Comegys, J.C. 2001. *Keys for the Voynich scholar*, 2006. Madera: Library of Congress.

Comegys, J.D. 2014 [2013]. The Voynich Manuscript: Aztec herbal from New Spain. [Unpublished pdf file created March 22, 2014 but title page says ©2013]. http://voynichms.com/wp-content/uploads/2014/03/VMSAztecHerbal-JDComegys.pdf. Accessed 23 Aug 2014.

Dakin, K. 1981. The characteristics of a Nahuatl lingua franca. In: *Nahuatl studies in memory of Fernando Horcasitas*, ed. Frances Karttunen. *Texas Linguistic Forum* 18: 55–67.

Gravier, M.G. 2011. Sahagún's Codex and book design in the indigenous context. In *Colors between two worlds. The Florentine Codex of Bernardino de Sahagún*, ed. G. Wolf and J. Connors, 156–197. Milan: Officina Libraria.

Guy, J. 1991. Voynich revisited. *Cryptologia* 15: 161–166.

Hirth, K.G. 2016. *The Aztec economic world: Merchants and markets in ancient Mesoamerica*. New York: Cambridge University Press.

Janick, J., and A.O. Tucker. 2017. Were Juan Gerson the illustrator and Gaspar de Torres the author of the Voynich Codex? *Notulae Botanicae Horti Agrobotanici Cluj-Napoca* 45 (2): 343–352.

Janick, J., and A.O. Tucker, eds. 2018. *Unraveling the Voynich Codex*. Cham: Springer Nature.

Montemurro, M.A., and D.H. Zanette. 2013. Keywords and co-occurrence patterns in the Voynich manuscript: An information-theoretic analysis. *PLoS One* 8 (6): e66344. https://doi.org/10.1371/journal.pone.0066344.

O'Neill, H. 1944. Botanical observations on the Voynich MS. *Speculum* 19: 126.

Romero Ramírez, M.E. 2013. *Limp, laced-case binding in parchment on sixteenth-century Mexican printed books. Ph.D. Thesis*. Vol. 2. London: Camberwell College of Arts, University of the Arts.

Rugg, G. 2004a. The mystery of the Voynich Manuscript: New analysis of a famously cryptic medieval document suggests that it contains nothing but gibberish. *Scientific American* 291 (1): 104–109.

Rugg, G. 2004b. An elegant hoax? A possible solution to the Voynich Manuscript. *Cryptologia* 28: 31–46.

Snijders, L., T. Zaman, and D. Howell. 2016. Using hyperspectral imaging to reveal a hidden pre-colonial Mesoamerican codex. *Journal of Archaeological Science: Reports* 9: 143–149.

Stolte, D. 2011. UA experts determine age of book 'nobody can read'. https://uanews.arizona.edu/story/ua-experts-determine-age-of-book-nobody-can-read. Accessed 13 Mar 2018.

Tucker, A.O., and R.H. Talbert. 2013. A preliminary analysis of the botany, zoology, and mineralogy of the Voynich Manuscript. *HerbalGram* 100: 70–85.

Tucker, A.O., and J. Janick. 2016. Identification of phytomorphs in the Voynich Codex. *Horticultural Reviews* 44: 1–64.

Zandbergen, R. 2012. The Voynich manuscript. http://www.voynich.nu/index.html. Accessed 29 Dec 2012.

Chapter 2
Aztec Botany, Agriculture, Trade, and Medicine

Mexico and the Aztec Empire

Mexico, a botanical paradise located in southern North America, is the third largest country of Latin America and one-fifth the size of the United States. Mexico's climate varies from temperate to tropical and from wetlands to desert. Mountain ranges as high as 7000 feet are responsible for the huge variation in microclimates and ecosystems.

With 23,324 native plant species, Mexico has the fourth largest floristic richness in the world; its endemic species (50%) are surpassed only by South Africa (Villaseñor 2016). Mexico is one of the centers of origin for many cultivated plants. The native food crops include allspice, amaranth, avocado, cacao, chayote, chia (*Saliva hispanica*), guava, chili pepper, lima bean, maize, papaya, *Phaseolus* beans, pumpkin, sunflower, sweet potato, tomato, and vanilla. It is also the source for many horticultural ornamentals, including cosmos, dahlia, frangipani, jacaranda, Jacobean lily (*Sprekelia*), marigold, Mexican poppy (*Argemone*)¸ Mexican sunflower (*Tithonia*), *Laelia* orchids, morning glory, sages, shell flower (*Tigridia*), tuberose, and zinnia. The wide variety of plants in Mexico selected over several millennia provides the knowledge base for the extensive medicinal lore of its indigenous people.

The present population of about 123 million persons now constitutes the world's largest population of Spanish speakers, 80% of whom live in Mexico City. This enormous city is located on the remains of the Aztec capital, Tenochtitlan, centered on Lake Texcoco, now drained. The Aztec empire (Fig. 2.1) extended to much of Mesoamerica and touched the Gulf of Mexico and the Pacific Ocean (Moreira 2018). Tenochtitlan, the magnificent, sumptuous, island city known for its *chinampas* ("floating gardens") was founded in 1325 and was destroyed by the conquistador Hernán Cortes in 1521 and decimated by diseases. The name *Aztecs* was coined by Francisco Clavijero in 1780 (Clavijero and de Mora 1844) based on their legendary home, *Aztlán*, but they referred to themselves as *Mexica*, and this word is

A. O. Tucker, J. Janick, *Flora of the Voynich Codex*,
https://doi.org/10.1007/978-3-030-19377-5_2

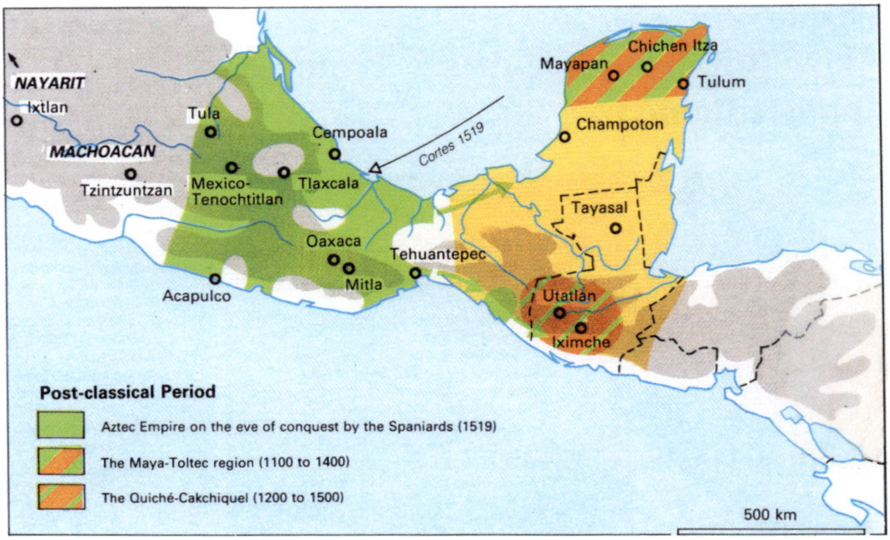

Fig. 2.1 Aztec empire in 1521. (Source: The Harper Atlas of World History, rev. 1992)

enshrined in the country's present name. They were also known as *Tenochca*, the people of Tenochtitlan. The Mexica spoke Nahuatl, part of the Uto-Aztecan language family group that extends from Oregon to Panama and still is spoken by more than a million people. Many Nahuatl-derived words from Mexico now are common in English, such as avocado, chili, chocolate, coyote, and tomato.

The Aztecs, latecomers to central Mexico, were the dominant rulers at the time of the Spanish encounter with the New World. At the time of the conquest, the Aztec population was estimated at about 11 million, but smallpox introduced by the Spanish in 1519–1520 killed five to eight million people. C*ocoliztli,* probably an indigenous hemorrhagic fever transmitted by rodent hosts and associated with a bacterial pathogen, killed another seven million in the epidemics of 1545 and 1576 (Acuña-Soto et al. 2002; Vágene et al. 2018). The devastation brought about by a combination of warfare and megadeath from introduced disease drastically reduced their population, but the remnants of the Aztec population interbred with Spanish conquistadors and immigrants, and their descendants now are an integral part of the Mexican population.

The Aztecs were a complex, advanced civilization. They practiced a bewildering state religion with stone pyramid temples and created a complex, interconnected religious and secular calendar. Their art was sophisticated and included feather weaving, painting, and sculpture, as well as illustrated codices. Priests and nobles studied astronomy, for they believed cyclical phenomena of the sun, stars, and planets were linked to earthly events including the agricultural calendar and ever-impending doom, ameliorated only by captive sacrifice. They utilized a logophonetic writing system, yet in many ways they were a Stone Age culture lacking beasts of burden and even the wheel.

Formed by a triple alliance between three city states (Tenochtitla, Texcoco, and Tlacopan), the Aztec Empire was a tribute system organized into various ethnic political units of city states (*altepetl*). At the time of the conquest, they were ruled by an emperor king, Moctezuma II. The Aztecs were militarily aggressive, with hegemony over 450 city states that were mostly in Central Mexico but reached as far as Guatemala. It was a trading society, with large markets carrying products from all over the empire.

Trade

Marketplaces existed at every major city, but the largest was at Tlatelolco, the sister city of Tenochititlan and not part of the Triple Alliance. However, this great market declined during the sixteenth century, with the trade passing to Tenochititlan. Anthony Pagden remarked on the letters of Cortés sent to the emperor: "*A new Indian market at San Hipólito came into being in the 1540s. Cervantes de Salazar described it as a 'square of such enormous size that it is wide enough for building a city'. It was flanked by a Franciscan monastery—containing the Colegio de Santiago de Tlatelolco—the residence of the Indian governor and an Indian prison. Cervantes de Salazar estimated that the number of Indians using this market amounted of twenty thousand or more.*" (Cortés 1986:507). At Tenochititlan, Cortés noted in his second letter to the emperor in 1519: "*There are streets of herbalists where all the medicines, herbs, and roots found in the land are sold. There are shops like apothecaries', where they sell ready-made medicines as well as liquid ointments and plasters.*" (Cortés 1986:103). Native *amatl* (amate) and maguey fiber papers used by Aztecs were available from paper vendors (*amanamacac, amaoztomecatl*) (Hirth 2016). Feldman (1978) provides a rather detailed map of a typical Mexica market, while Durand-Forest (1971) provides a map of the market at Tenochtitlan (Fig. 2.2).

The herb seller was the *quilnamacac*, "*a producer of herbs, a field worker, a plucker of herbs.*" (Sahagún 1961:92). The apothecary, the medicine collector, was the *pachichiuhqui* (Molina 1571). The medicine dealer apothecary was the *panamacac*: "*He sells all things, medicines, herbs, wood, stones, milk, alum…on a reed mat*" (Sahagún 1961:85–6). Hirth (2016:164) remarked, "*Medicine was another retail product. The apothecary (panamacac) sold a wide variety of herbal and natural remedies. These included fourteen kinds of different herbs together with types of wood, stones, milk, and alum. While he may have collected some of these, it is more likely that he relied on a selection of forager-collectors to provision him with a number of natural remedies. Some of the concoctions sold were undoubtedly prepared by the apothecary himself since Sahagún* [1961:86] *indicates that he sold things cooked in pots like skunk excretion. In this regard the medicine seller was probably part retailer and part producer-seller.*"

The herb sellers themselves were supplied by a special class of Aztec merchants called the *pochteca*. This class, including both men and women and nobility and commoners, often provided military and political aid to the nobility. The *naualoztomeca*

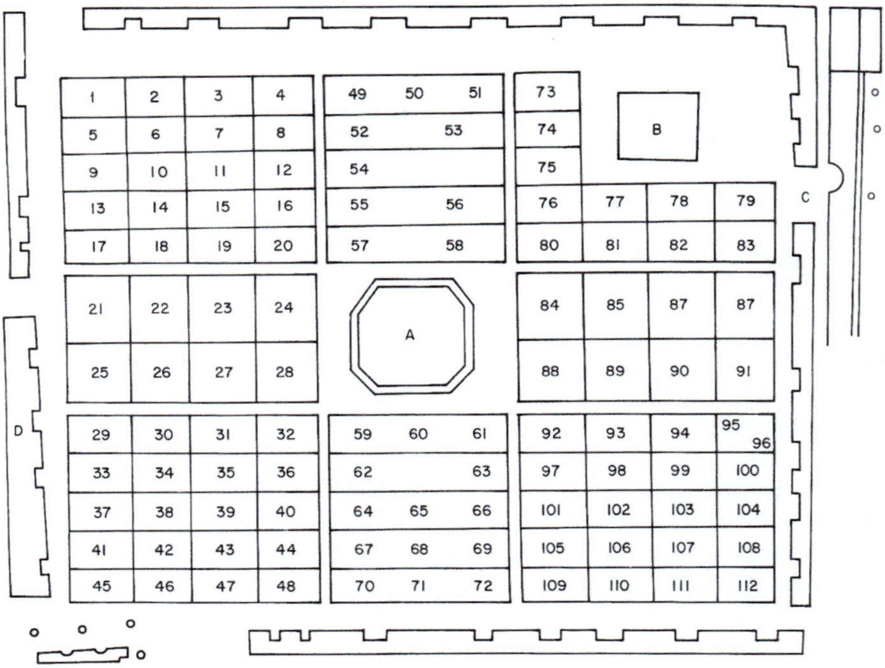

Fig. 2.2 Reconstruction of a pre-Hispanic Aztec market. (*Adapted from* Feldman 1978)

of the *pochteca* were "disguised traders" who were multilingual and acted as spies. The *pochteca* were wealthy and extremely secretive. Book 9 of Sahagún's *Florentine Codex* (Sahagún 1959) is devoted to these merchants. Their influence, and thus their trading routes, extended throughout the southeastern and southwestern United States to Georgia and Utah and as far south as the Andes in Peru (Berdan 1975, 1980, 1985, 1986, 1988; Nichols 2013; Oka and Kusimba 2008; Salomón 1978; Washburn et al. 2014; White and Weinstein 2008). In summary, the long-distance travels of the *pochteca* imply that the herbs (and the knowledge of their uses) sold at the markets did not have to come from Mexico alone but rather from the extent of the trading routes of these merchants. This fact also has implications for the language of the *Voynich Codex*, as the *pochteca* would have spoken a *lingua franca* based upon Nahuatl, such as discovered by Dakin (1981), a Pidgin- or Creole-like language now extinct and supplanted by Spanish.

Agriculture

Food production was vital in feeding the capital population and the one million others who lived in the central basin. The chief foods were agave, amaranth, chia, capsicum peppers, maize, and squash. The diverse agricultural production

The Parts of a Mexica Market[1]
Section I: Restaurants, Produce, Meat, and Fish

1., 2. *Tlaxcalnamacac*: The Tortilla Seller
3. *Aioachnamacac*: The Seller of Gourd Seeds and cakes of gourd seeds
4. *Tlaquetzalnamacac*: One who provides people with the chocolate drink of the rulers, of the nobles
5., 6. *Tlaxcalnamacac*: see above No. 1
7. *Atolnamacac*: The Atole Seller, sells hot atole drink
8. *Cacaoanamac*: The Cacao Seller, sells cacao beans
9. *Totoltenamac*: The Egg Seller
10. *Picienamacac*: The Seller of Fine Tobacco
11. *Chiennamacac*: The Chía Seller; The Oil Seller
12. *Henamac*: The Bean Seller
13. *Totolnamacac*: The Turkey Seller
14. *Vauhnamacac*: The Amaranth Seed Seller
15. *Tlaolnamacac*: The Seller of Maize Grains
16. *Chiquippantlacatl*: The Displayer of Wares on a Large Basket, the seller of colors, of various colors, of dyes
17. *Nacanamacac*: The Meat Seller
18., 19. *Xoquilacanamacac*: The Stench Seller (Fish Seller)
20. *Tochominamacac*: The Seller of Rabbit Hair

Section II: Meat and Produce

21. *Nacanamacac*: see above No. 17
22., 23. *Suchiqualpantlacatl*: The Man with Fruit; seller of maize stalks, green maize, tortillas, cooked gourds, tree fruit, cactus fruit, sweet potatoes, manioc
24. (thread)
25. *Nacanamacac*: see above No. 17
26., 27. *Suchiqualpantlacatl*: see above No. 22
28. (thread)

Section III: Hardware, Hides, Pharmacy, and Wine

29. *Ocnamacac*: Wine or Pulque Seller
30. *Amolnamacac*: Soap Seller
31. *Quinamaca Cuetlascactli*: Seller of Cured Leather Sandals
32. (rope)
33. *Quilnamac*: Herb Seller
34. *Hihujnamacac*: Feather Seller; the Spinner of Feathers into Thread
35. *Euanamacac*: Hide Seller (based upon Nahua rules of compound word construction)
36. *Siuhquilnamac*: Turquoise Herb Seller; sells black clay mixed with *uixachin* leaves for coloring things

(continued)

[1] Unless otherwise stated, all Nahua terms are from Sahagún (1961).

37. *Panamacac*: The Medicine Seller; sells medicines of herbs, wood, stones, milk, alum, and things cooked in ollas
38. *Tzacunamacac*: The Glue Seller
39. *Suchiocutzonamacac*: The Seller of Liquidambar; sells pine resin, possessor of pine resin trees
40. *Acaquauhnamacac*: The Smoking Tube Seller
41. *Ocutzonamac*: The Seller of Pine Resin; sells pine and other resin, a woodsman, a collector of pine resin
42. *Poponamacac*: The Broom Seller
43. *Vitzmallonamac*: The Needle Seller; sells needles, awls, punches, bells, axes, adzes, fish hooks, chisels
44. *Olnamacac*: The Rubber Seller
45. (cochineal)
46. not known
47. *Pinolnamacac*: The Pinole Seller (based upon Durand-Forest 1971)
48. not known

Section IV: Cloth, Minerals, Produce, and Services

49. *Tequixquinamacac*: The Seller of Saltpeter
50. *Tizanamacac*: The Chalk Seller
51. *Monamacac*: Self Seller (prostitute)
52. *Tomanamacac*: The Tomato Seller
53. *Chilnamacac*: The Chile Seller
54. *Chientzotzolnamacac*: Wrinkled Chia Seller
55., 56. *Iczotilmanamac*: The Seller of Palm Leaf Fiber Capes
57., 58. *Veicapantlacatl*: "The Principal Merchant;" sells worked capes, worked shifts, fine skirts

Section V: Maguey Goods and Building Material

59. *Necunamacac*: The Maguey Syrup Seller
60. *Aianamacac*: The Seller of Coarse Maguey Fiber Capes
61. (hats)
62. *Cacnamacac*: The Sandal Seller; sells sandals of maguey fiber
63. (hats)
64. *Cacnamacac*: see above No. 62
65. *Xiquipilnamacac*: The Bag Seller
66. *Neilpilonamocac*: The Sash Seller; sells narrow strips of cloth
67. *Quauhnanamaca*: The Wood Seller; sells lumber
68. *Quruhxinquil* "The Carpenter:" sells beams, wooden pillars, lintels, roofing, wooden columns, boards, planks
69. *Tenexnamacac*: Seller of Lime; adjacent to seller of building stone
70. *Quathnanamaca*: see above No. 67
71. *Quauhxinqui*: see above No. 68
72. *Tenexnamacac*: see above No. 69

(continued)

Section VI: Produce, Salt, Cloth, and Money

73. not known
74. *Chilnamacac*: see above No. 53
75. *Iztanannacau*: The Seller of Salt; sells salt balls and salt ollas
76. (pieces of *cueitl* or skirt cloth sold) (Durand-Forest 1971)
77. (pieces of *tilmatli* or *manta* cloth sold) (Durand-Forest 1971)
78. (selling of gold-filled quills)
79. (*ocotl* torches sold) (Durand-Forest 1971)
80.–82. *Tilmapan Tlacatl, Quachnamacac*: The Man With the Capes, The Seller of Large Cotton Capes
83. (seller of *tomitilmatli* or woolen *mantas*) (Durand-Forest 1971)

Section VII: High-Value Goods

84. *Tilmapan Tlacatl, Quachnamacac*: see above No. 80
85. *Xicalnamacac*: The Seller of Gourd Bowls
86. *Tlapitzalnamacac*: The Seller of Cast Metal Objects; sells objects made out of precious metal, either gold or silver
87. *Ilruinamacac*: The Feather Seller
88. *Tecoani*: The Slave Dealer
89. *Amanamacac*: The Paper Seller
90. *Chalchiuhnamacac*: The Seller of Green Stones
91. (sellers of bees' honey)

Section VIII: Woven Goods, Pots, Trinkets, and Services

92., 93. *Chiquiuhnamacac*: The Seller of Large Baskets
94. *Cozcatetecpanqui*: The Displayer of Necklaces
95. *Zoquichiahqui*: The Clay Worker; sells all pottery except griddles
96. *Comalnamacac*: The Comal Seller
97., 98. *Chiquiuhnamacac*: see above No. 92
99. not known
100. *Tananamacac*: The Seller of Small Baskets
101. (sellers of bird skins)
102. (sellers of shells)
103. *Tezcanamacac*: The Mirror-Stone Seller
104. *Petlanamacac*: The Reed Mat Seller
105. (barbers)
106. *ltznamacac*: The Obsidian Seller; sells blades on demand
107. *Otlachiquiuhnamacac*: The Seller of Stout Cane Carrying Baskets
108. *Petlanamacac*: see above No. 104
109.–112. Not known

Section IX: Structural Features

A. Stage
B. Court House
C. Entrance to the Courts of the Great Temple
D. Arcades

included *chinampa*, artificial gardens based on layered soil and organic matter on lake shores; *tlacolol*, a slash-and-burn system based on rotating crops along foothills; terrace farming on hills; and complex irrigation along river beds (Carmack et al. 2006:91). Chinampas would produce crops some seven times a year and provided 50% of the dietary needs of Tenochtitlan. The rest were provided via tributes from the provinces. Protein was supplemented mainly by turkeys, domesticated dogs, and game meat (such as tapir, deer, rabbit, and peccary, a New World pig). Secondary food sources were locusts, fish, crustaceans, water birds, grubs, and land lizards (Harris 1998:230). The noble class ate the flesh of victims of human sacrifice.

Medical Knowledge

Aztec medicine involved a variety of curing techniques from religious rituals to herbal remedies (von Hagen 1961:109–14; Aguilar-Moreno 2007:359; Schendel et al. 2014). Two types of doctors existed: the *tepati*, who used herbal medicine, and the *ticitl*, the sorcerers (Schendel et al. 2014).

Aztecs used and named hundreds of herbs, many shown to have pharmaceutical value (Appendix A). The medical knowledge of the Aztecs so impressed the Spanish rulers that they sent emissaries to New Spain to study the herbs, stones, trees, and roots used by Aztec physicians (Schendel et al. 2014). Francisco Hernández, a Spanish naturalist and court physician to Phillip II, was appointed "General protomedico of our Indies, islands and mainland of the Ocean Sea" in 1570 and was sent on a seminal scientific expedition to the New World to study medicinal plants. He visited Aztec botanical gardens.

Botanical Gardens

Although Mexicans were associated with gruesome rituals and warfare, they also were pioneering horticulturists and lovers of aesthetics long before their encounter with Spain. In 1457, a massive garden named Texcotzingo was constructed on a hill by the ruler Nezahualcoyotl, the ruler (*tlatoani*) of Texcoco, who also was responsible for the garden retreat at Chapultepec. Texcotzingo was more than just a botanical garden cultivating various flora from the empire; it represented a map of the political domain and was a hydrological engineering accomplishment. It contained paths, bas relief sculptures, and shrines. Aqueducts transferred water from Mount Tlaloc into pools and irrigations systems. Waterfalls, fountains, and baths adorned the park. Because of their southwestern placement and the mist produced from pools, certain sectors created an adequate micro-environment to grow tropical plants at high altitudes. The glories of Nezahualcotyl's garden were extolled by Granziera (2001):

> Nezahualcoyotl also laid out his garden on the sacred hill of Tezcotzinco and used the canalized waters flowing down from the springs of Mount Tlaloc to create ponds with panoramic views from the hillside. The hilltop was defined by a spiralling [sic] path reached by rock-cut stairs rising from the probable site of Nezahualcoyotl's villa on the southern slope below. Another path traverses the hill along the east-west axis. There were baths and pools placed at the cardinal points around the circumambulating path. On the summit were found the remains of a Tlaloc shrine. Below, to the east, monumental sculptures were dedicated to agricultural divinities. Other natural shrines in Tezcozinco include a cave at the transition-point between Nezahualcoyotl's residence and the botanical collections on the lower south slope of the hill.

This garden was occupied in the 1520s for 3 years by Pedro de Gante, one of the 12 priests who accompanied Cortés. Aztec botanic gardens such as Nezahualcoyotl facilitated the acquisition of fresh herbs, dried herbs, ointments, salves, wood, stones, sap, roots, and leaves for apothecaries and doctors, so they were readily available in the extensive market at Tlatelolco.

The glories of Nezahualcoyotl's garden can be found in the writings of his great grandson, Juan Bautista de Pomar, who claimed rights to the legacy of the palace with extensive water features, botanic gardens, and extensive baths (Douglas 2010; García Icazbalceta and Pomar 1891; Granziera 2001). In 1592, Pomar wrote *Relación de Tezcoco*, an account of the Aztecs and Tlatelolcas that included a discussion of Aztec foodstuffs, animals, trees, shrubs, and herbs and their medicinal uses, with Nahuatl names, and mentions familiarity with Francisco Hernández, chief physician to Phillip II. Pomar is also credited with one of the most important compilations of poetry, *Romances de los señores de Nueva España*.

Status rivalry among Aztec rulers was not limited to conquest, but also to beauty. In response to Nezahualcoyotl's creation, Moctezuma I reconstructed the ancient Huaxtepec, a semitropical botanical garden. It was a large, orderly garden of six miles circumference, constructed at a much lower altitude and hydraulically fed by a crossing river. The garden was filled with ornamental, aromatic, and medicinal plants, with coastal imports such as vanilla, cacao, and magnolia trees. According to the conquistador Díaz del Castillo, it was the best garden he and Cortés had seen. It served as a place of visual and olfactory pleasure. Recreational gardens and nurseries were commonplace in the Aztec world. Some served purely for pleasure, and nurseries supplied seeds and tree saplings for parks, as well as flowers and greens to decorate temples, which could be sold at central markets. The value placed on beautiful plants was so extreme that commoners were banned by decree from planting some of them (Evans 2004:477).

The comprehensive botanical gardens of the Aztecs in Tenochtitlan, and Chapultepec later inspired those in Europe. Some in Mexico included water features for ritualistic bathing combined with sweat lodges (*temezcalli*) or sweatbaths (Nuttall 1925; Granziera 2001, 2005). The large gardens of Moctezuma within Tenochtitlan and the gardens of the nobility elsewhere contained plants and herbs from many parts of Mexico. These extensive gardens indicate the wide knowledge and resources of native plants. The flower pleasure gardens of Moctezuma did not

Fig. 2.3 Torching of books by Spanish friars at Tlaxcala. (Image from Diego Muñoz Camargo, *Descripción de la ciudad y provincia de Tlaxcala*, Glasgow University Library, MS Hunter 242, folio 242r, courtesy of the Special Collections Department, Library, University of Glasgow)

contain food plants; the plants were solely for his personal enjoyment. Cortés specifically referred to the beautiful gardens of Moctezuma II, which contained channels for irrigation.

Conclusions

The diverse flora of Mexico and the great interest by the Aztecs in plants and gardens represent a rich source of botanical and historical information that has been neglected. Although Aztec botanical knowledge has been lost due to the decimation of Aztec culture and the destruction of their pre-Columbian codices (Fig. 2.3), their knowledge can be reconstructed based on sixteenth century works of New Spain in which indigenous Nahuans participated, and on European herbals as discussed in subsequent chapters. Identification of 131 images of plants from the Herbal section of the *Voynich Codex* and 41 plants from the Pharmaceutical section is the subject of Chap. 4. Chapter 5 analyzes which plants are common between these sources, and especially which plants of the *Voynich Codex* are duplicated. The results confirm that the *Voynich Codex* is a sixteenth century work based on flora known and named in Aztec culture.

Literature Cited

Acuña-Soto, R., D.W. Stahle, M.K. Cleaveland, and M.D. Therrell. 2002. Megadrought and mega-death in 16th Century Mexico. *Emerging Infectious Diseases* 8: 360–362.

Aguilar-Moreno, M. 2007. *Handbook to life in the Aztec world*. New York: Oxford University Press.

Berdan, F.F. 1975. *Trade, tribute, and market in the Aztec empire. Ph.D. Thesis. Department of Anthropology*. Austin: University of Texas.

Berdan, F.F. 1980. Aztec merchants and markets: Local-level economic activity in a non-industrial empire. *Mexicon* 2: 37–41.

Berdan, F.F. 1985. Markets in the economy of Aztec Mexico. In *Markets and marketing. Monographs in Economic Anthropology*, ed. S. Plattner, vol. 4, 339–367. Lanham: University Press of America.

Berdan, F.F. 1986. Enterprise and empire in Aztec and Early Colonial Mexico. In *Economic aspects of Prehispanic Highland Mexico. Research in Economic Anthropology, Supplement*, ed. B.L. Isaac, vol. 2, 281–302. Greenwich: JAI Press.

Berdan, F.F. 1988. Principles of regional and long-distance trade in the Aztec empire. In *Smoke and mist: Mesoamerican studies in memory of Thelma D. Sullivan*, ed. J.K. Josserand and K. Dakin, 639–656. Oxford: BAR International Series.

Carmack, R., J.L. Gasco, and G.H. Gossen. 2006. *The legacy of Mesoamerica: History and culture of a Native American civilization*. 2nd ed. Upper Saddle River: Prentice Hall.

Clavijero, F., and J. Joaquin de Mora. 1844. *Historia antigua de México y de su Conquista*. México: Lara.

Cortés, H. 1986. Letters from Mexico. Trans. and ed. A. Pagden, introd. J. H. Elliott. New Haven: Yale University Press.

Dakin, K. 1981. The characteristics of a Nahuatl *lingua franca*. In *Nahuatl studies in memory of Fernando Horcasitas*, ed. F. Karttunen and Texas Linguistic Forum, vol. 18, 55–67.

Douglas, E. de J. 2010. *In the Palace of Nezahualcoyotl: Painting manuscripts, writing the pre-Hispanic past in colonial period. Tetzcoco, Mexico*. Austin: University of Texas Press.

Durand-Forest, J. 1971. Cambios económicos y moneda entre los Aztecas. *Estudio de la Cultura Náhuatl (México: UNAM Instituto de Investigaciones Históricas)* 9: 105–124.

Evans, S.T. 2004. Aztec palaces. In *Palaces of the ancient New World*, ed. S.T. Evans and J. Pillsbury, 7–58. Washington, DC: Dumbarton Oaks Research Library and Collection.

Feldman, L.H. 1978. Inside a Mexican market. In *Mesoamerican communication routes and cultural contacts. New World Archeological Foundation*, ed. T.A. Lee and C. Navarette, 219–222. Provo: Brigham Young University.

García Icazbalceta, J., and J.B. Pomar. 1891. *Pomar y Zurita: Pomar, Relación de Tezcoco; Zurita, Breve relación de los señores de la Nueva España. Varias relaciones antiguas. (Siglo XVI)*. México: Francisco Diaz de Leon.

Granziera, P. 2001. Concept of the garden in pre-Hispanic Mexico. *Garden History* 29: 185–213.

Granziera, P. 2005. Huaxtepec: The sacred garden of an Aztec emperor. *Landscape Research* 30: 81–107.

Harris, M. 1998. *Good to eat: Riddles of food and culture*. Long Grove: Waveland Press, Inc.

Hirth, K.G. 2016. *The Aztec economic world: Merchants and markets in ancient Mesoamerica*. Cambridge, UK: Cambridge University Press.

Molina, A. de. 1571. *Vocabulario en lengua mexicana y castellana*. Mexico: Casa de Antonio de Spinosa.

Moreira, F. 2018. An historical context for the Voynich Codex: Aztec Mexico and Catholic Spain. Chapter 3. In *Unraveling the Voynich Codex: a 16th century Mexican manuscript*, ed. J. Janick and A.O. Tucker. Cham: Springer Nature.

Nichols, D.L. 2013. Merchants and merchandise: The archaeology of Aztec commerce at Otumba, Mexico. In *Merchants, markets, and exchange in the pre-Columbian world*, ed. K.G. Hirth and J. Pillsbury, 49–63. Washington, DC: Dumbarton Oaks Research.

Nuttall, Z. 1925. The gardens of ancient Mexico. *Annual Report of the Smithsonian Institution* 1923: 453–464.
Oka, R., and C.M. Kusimba. 2008. The archaeology of trading systems, Part 1: Towards a new trade synthesis. *Journal of Archaeological Research* 16: 339–395.
Sahagún, B. de. 1959. *Florentine Codex. General history of the things of New Spain. Book 9 – The merchants.* Trans. C. E. Dibble and A. J. A. Anderson. Salt Lake City, Utah; University of Utah Press.
Sahagún, B. de. 1961. *Florentine Codex. General history of the things of New Spain. Book 10 – The people.* Trans. C. E. Dibble and A. J. A. Anderson. Salt Lake City: University of Utah Press.
Salomón, F. 1978. Pochteca and mindalá: a comparison of long-distance traders in Ecuador and Mesoamerica. *Journal of the Steward Anthropological Society* 9: 231–246.
Schendel, G., J. Amézquita, and M.E. Bustamante. 2014. *Medicine in Mexico: From Aztec herbs to betatrons*. Austin: University of Texas Press.
Vágene, A.J., A. Herbig, M.G. Campana, N.M. Roblers Garcia, C. Warinner, S. Sabin, M.A. Spyrou, A. Andrades Vatueña, D. Huson, N. Tuross, K.I. Bos, and J. Krause. 2018. *Salmonella enterica* genomes from victims of a major sixteenth-century epidemic in Mexico. *Native Ecology Evolution* 2 (3): 520–528.
Villaseñor, J.L. 2016. Checklist of the native vascular plants of Mexico. *Revista Mexicana de Biodiversidad* 87: 559–902.
von Hagen, V.W. 1961. *The ancient sun kingdoms of the Americas*. New York: The World Publishing Company.
Washburn, D.K., W.N. Washburn, P.A. Shipkovaand, and M.A. Pelleymounter. 2014. Chemical analysis of cacao residues in archaeological ceramics from North America: considerations of contamination, sample size and systematic controls. *Journal of Archaeological Science* 50: 191–207.
White, N.M., and R.A. Weinstein. 2008. The Mexican connection and the far west of the U.S. Southeast. *American Antiquity* 73: 227–278.

Chapter 3
Iconography and Phytomorphic Identification

Introduction and Historical Context

The *Voynich Codex,* discovered in 1912 by Wilfred M. Voynich in Italy, had been long assumed to be of pre-Columbian European origin. This conventional dogma was put into question in 1944 by the distinguished botanist Rev. Dr. Hugh O'Neill, who identified in it two New World plants, sunflower (*Helianthus annuus* L.) and capsicum pepper (*Capsicum annuum* L). Despite his credentials as a botanical taxonomist and an expert in Mesoamerican plants, his work was either ignored or ridiculed. In 1991, an obscure note by an Australian, Jacques B.M. Guy, confirmed the sunflower identification and noted not a single European species. He mentioned that a botanical colleague identified passionfruit. Tucker and Talbert, in a seminal paper (2013), extended the identification to 39 plants as either indigenous to the New World or circumboreal, such as *Actaea rubra* (Aiton) Willd. Tucker and Janick (2017) extended the list to 55 species while Tucker and Janick (2018a) added two additional ones. The identification in the *Voynich Codex* of New World plants as well as a mineral and 21 animals (Flaherty et al. 2018) provides hard evidence that the *Voynich Codex* is post-Columbus and not a fifteenth century European work, as had been almost universally assumed by *Voynich Codex* researchers (Zanbergen 2004–2017).

Thus far, all of the species identified are from the New World, with a recent range from Texas to California, south to Nicaragua. Yet, despite these identifications by professional biologists in three refereed scientific publications and a book from a major publisher, some individuals still insist on a pre-Columbian European origin. In the present work, identification of phytomorphs in the *Voynich Codex* is extended to 129 of 131 phytomorphs (98.5%) in the Herbal section plus 41 of 228 phytomorphs (18.0%) in the Pharmaceutical section.

Eurocentric bias may be one explanation for insisting that the provenance of the *Voynich Codex* is from the Old World, not the New. In her comprehensive review of herbals, Agnes Arber (1970) mentioned the *Codex Cruz-Badianus* in two sentences

A. O. Tucker, J. Janick, *Flora of the Voynich Codex*,
https://doi.org/10.1007/978-3-030-19377-5_3

and Hernandez in one sentence; Sahagún's *Florentine Codex* is not mentioned, nor is it included in a chronology of herbals. Wilfrid Blunt (1971) essentially followed suit, though in 1979 Blunt and Sandra Raphael provided an appropriate discussion of the *Codex Cruz-Badianus*. Minta Collins (2000) was even more dismissive of New World materials, not mentioning them at all. Only Frank Anderson (1977) devoted a chapter to Hernandez. Eurocentric bias is further shown by the claim that the first botanic garden in the world was at Padua, Italy, in 1545 (Angarano et al. 2017), completely ignoring the many botanic gardens of the Nahua reported by Hernan Cortés when he entered Mexico in 1519 (Evans 2000, 2005, 2010; Granziera 2001, 2005; Nuttall 1925). Did the Nahua also have nascent herbaria? Arber (1970) claimed that the first herbarium specimen was created by Luca Ghini at Bologna, Italy, prior to 1551, possibly as early as 1532, but the "*maguoey*" as #4 on fol. 100r of the *Voynich Codex* is obviously bent as a dried, pressed specimen (*see* Fig. 3.2).

Aztec palaces were centers of learning, unlike European palaces, whose elite were often uneducated and unbathed. Today, we have only hints of Aztec and Mayan botanical triumphs because their libraries, often on native amate paper, were burnt in a kind of *auto-da-fé* as "heathen" by the Spanish priests (*see* Fig. 2.3). In addition, book burnings by conquerors who rewrote history were routine (Binnqüist 2003; Binnqüist et al. 2012; Marcus 1992; Polastron 2007; von Hagen 1944).

Guidelines for Plant Identification

The appropriate guidelines for identifying phytomorphs were outlined by Tucker and Janick (2017):

1. For any identification of a plant, the strength of evidence depends on the qualifications of the identifier(s).
2. Identification of plants, just as with forensic identification of faces or fingerprints, requires pattern recognition. For botanical subjects, this requires recognition of the appropriate plant family, then to genus, and eventually species. Amateurs of botanical identification often reverse this process and try to identify the species first and ignore the patterns of the plant family, usually leading to false identifications.
3. Any attempt at identification must use currently accepted scientific names and the appropriate vocabulary of terms pertinent to plant taxonomy.
4. Details of habitat or ecological association may provide evidence to support an identification of a phytomorph.
5. Associated names often provide supporting evidence for the identification of a phytomorph.
6. Because of inherent biological variability of plants and differing talents of the artist(s), no identification of a phytomorph can be considered absolute. Rather, degrees of accuracy must be stated, from high to low. Confidence is based upon high percentages of match with a list of features that agree or do not agree with the identification.

7. All previous attempts at identification from books and refereed publications must be presented and discussed as to accuracy. Similar representations of phytomorphs in other original sources should be cited to bolster the identification, along with any ethnobotanical uses or mythology.
8. Identifications of phytomorphs in a transcultural society such as sixteenth century New Spain, pre- and post-Conquest, must consider artistic methods of representation.
9. Pigments (even mineral pigments in European paintings) vary in stability, and the transient nature of vegetable-derived pigments in New World paintings means that the colors seen today may not be the same colors as originally applied.
10. The purpose and background of the portrayal must be considered. Is any mythology associated with it? Are certain diagnostic characteristics emphasized for identification?
11. Any identifications from restored materials should be not be considered as absolute because of past practices with little or no record of the degree of restoration. In the past, some restorations had more in common with forgery, employing identical or similar techniques, along with pigments and pigment binders, used by restorers with questionable ability, who sometimes even distressed the restoration to agree with the rest of the work.

Iconographic Techniques

The *Voynich Codex* utilizes a number of uncommon iconographic techniques, many of which are unique:

1. A limited palette (Fig. 3.1). The small number of pigments available to the artist for the *Voynich Codex* has been previously mentioned (Tucker and Janick 2016). For example, one blue pigment as used for all the tones and shades from blue to purple, such as lilac or lavender.
2. A flat, two-dimensional representation of plants reminiscent of pressed specimens (Fig. 3.2).
3. A foreshortening of large plants, in which young shoots appear "grafted" or inserted upon an older base (Fig. 3.3).
4. Root fantasies in the shape of animals and human faces (Fig. 3.4).
5. Reptiles and amphibians among leaves and roots (Fig. 3.5).
6. Enlarged organs out of proportion to the rest of the plant (Fig. 3.6).
7. A mixture of accurate botanical details versus crude representations (Fig. 3.7).
8. Reconstruction of plants apparently from dried, fragmented samples (Fig. 3.8). The *curanderos/curanderas* today deal mostly with long-lasting dried herbs, so this is not a surprise (notably Torres n.d. a, b; 2006; Torres and Sawyer 2005). As a previous example, fol. 94r matches the repent (creeping) form of *Duranta erecta* L. (*D. repens* L.) (Fig. 3.8). The fruit should be golden yellow in

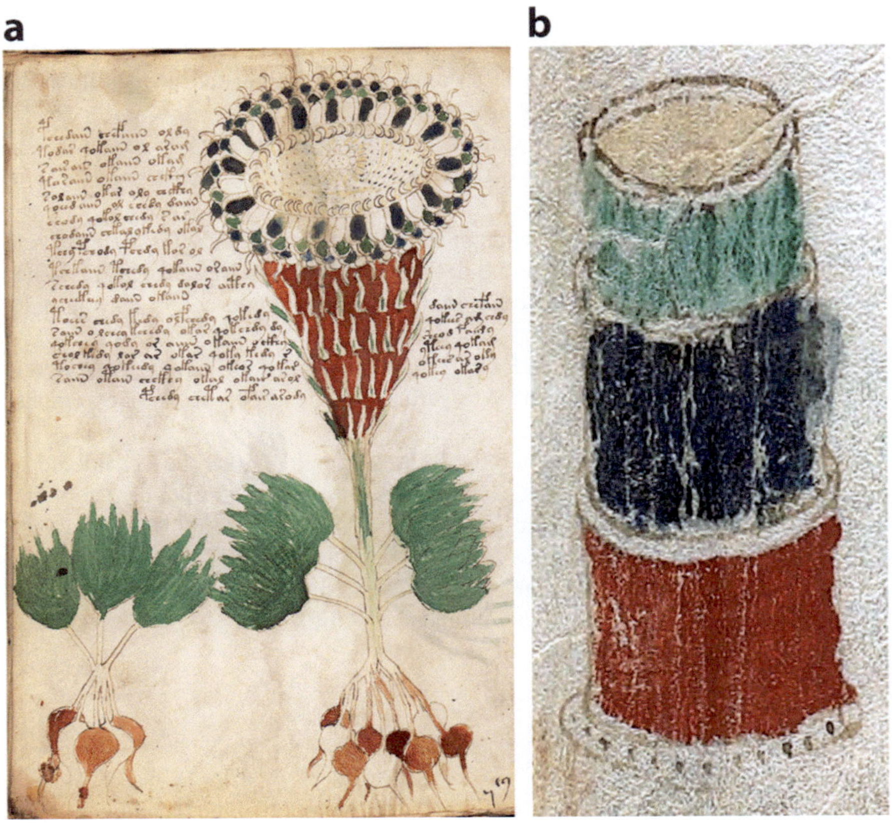

Fig. 3.1 Red-brown, blue, and green pigments in *Voynich Codex* phytomorphs: (**a**) *Smallanthus* sp., folio 40v; (**b**) apothecary jar folio 101r#2

Fig. 3.2 A flat, two-dimensional representation of *Agave atrovirens*, folio 100r #4, which appears as a pressed specimen

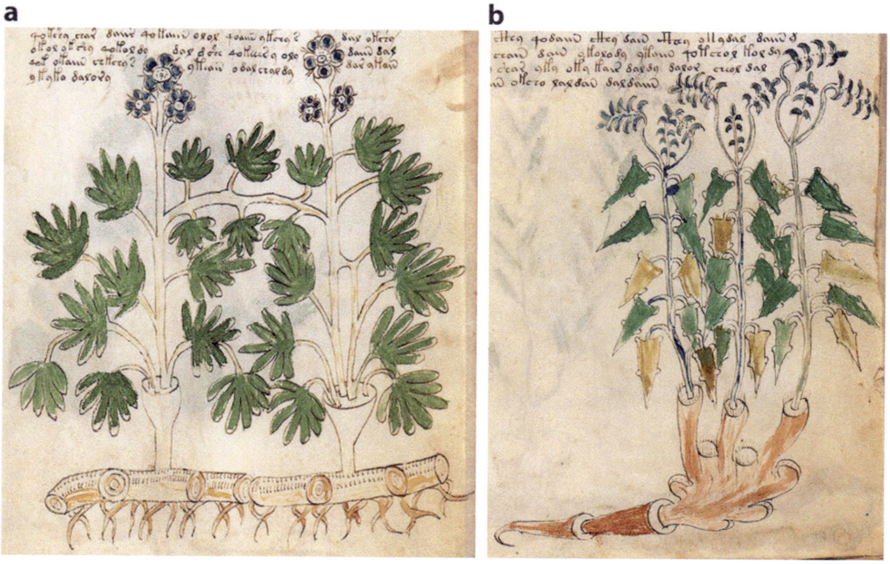

Fig. 3.3 Phytomorphs with apparent insertion of shoots on rootstocks: (**a**) *Ribes malvaceum,* folio 23r; (**b**) *Salvia cacaliifoia,* folio 45r

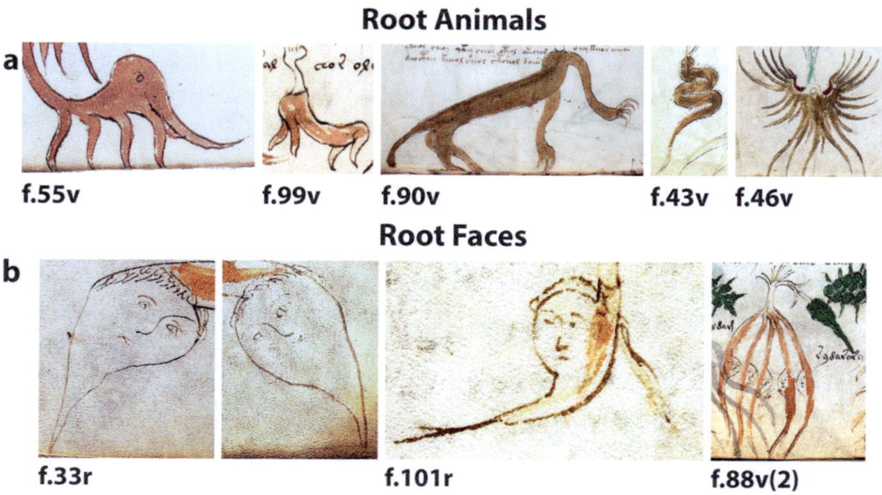

Fig. 3.4 Root fantasies with animal shapes and human faces: (**a**) Animal-like roots. (**b**) Faces in roots

a panicle when fresh, but drying of this species, especially with heat, turns the fruit black, and they often fragment. The painter then re-assembled them into a cyme (Tucker and Janick 2016). Latin American grocers still carry lines of dried medicinal herbs, such as the *gordolobo* (*Pseudognaphalium* spp.) (Fig. 3.9).

Fig. 3.5 Animals associated with phytomorphs: (**a**) Mexican horned lizard probably eating insects from *Dudleya ingens,* folio 25v. (**b**) Caecilians, legless amphibians, intertwined in roots of *Lithophramga affine*, folio 49r

9. Combination of species into a "hybrid" phytomorph (Fig. 3.10). Most of the phytomorphs in the *Voynich Codex* consist of only one species. For example, fol. 9v is *Viola bicolor* in both flowers and foliage. Hybrid phytomorphs represent at least two different species united into one, similar to Fol.33v, which may be *Psacalium* sp.? with *Pippenalia* sp.? We propose that the artist drew these hybrid phytomorphs when confronted with a mixture of dried specimens, and created artificial devices (like the cupule in Fol. 15v, a morphological component that has not existed since the pteridosperms). Those who need convincing of the concept of hybrid phytomorphs need only to examine fol. 90r, where a flower has been grafted onto a fern. These hybrid phytomorphs then consist of two species that are used in combination rather than alone. The concept of mixtures of herbs to cure a particular problem is ingrained in the *curanderos/curanderas*, as exemplified by the herbal remedies from present-day Mexican markets seen in Fig. 3.9b.

The concept of hybrid phytomorphs in the *Voynich Codex* was first elucidated by Elizebeth Smith Friedman in 1962. In a newspaper article, she wrote: "The root system belongs to one plant, the stem system to another, the leaves and flowers to still others." We cannot document three species in one hybrid phytomorph, but we

Fig. 3.6 Enlarged organs in phytomorphs *of Cirsium* sp., folio 50r. See also Fig. 3.1

do document two species. Recognition of hybrid phytomorphs is of particular importance and is unique to the *Voynich Codex*. Perhaps this was a way to indicate the powerful effect of combinations of two herbs versus single herbs, or combinations of many herbs in the Pharmaceutical section.

Geographic Range

Previous research (Tucker and Talbert 2013; Tucker and Janick 2018; Janick and Tucker 2018) has shown a range of potential plants from Texas to California, south to Nicaragua. Thus, matches to the phytomorphs were sought from species in this geographical area. Some of the potential identifications are relatively accurate because the phytomorphs include diagnostic characters, such as *Brasenia schreberi* on Fol. 49v. Others, because the phytomorphs are crude or seem to have been "re-created," are guesses. The latter are indicated by "cf." ("compares favorably") or "?." This publication includes only some of the abbreviated phytomorphs of the roots and leaves in the Pharmaceutical section, which are restricted in many

Fig. 3.7 Accurate drawing of phytomorph: (**a**) *Viola bicolor*, folio 9v, versus stylized drawing (**b**) *Eryngium* sp., folio 16v

cases to only roots, leaves, or shoots. Many of these phytomorphs will probably remain unidentified forever unless somebody can make lucky guesses from the names, because many of the phytomorphs in this section are labeled with Voynichese symbols (Fig. 1.2).

Plant Uses

Identification of phytomorphs in the *Voynich Codex* shows mostly medicinals, but there are also food, ornamental, and industrial plants. One would have expected maize to be included, since it was so important to the culture of the indigenous

Fig. 3.8 Phytomorph of *Duranta erecta*, folio 94r, probably from a dried sample, based on black fruit color

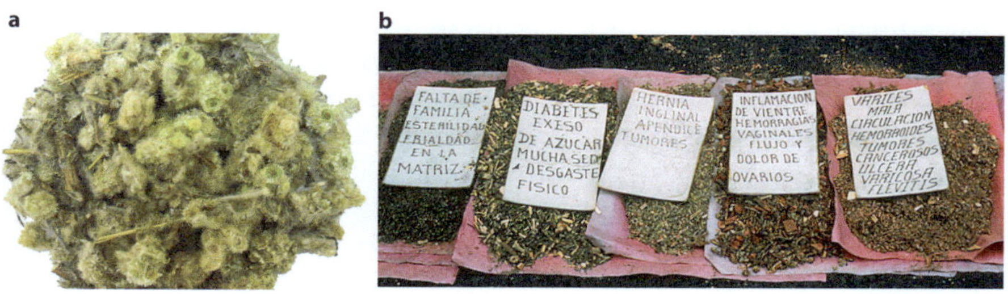

Fig. 3.9 Dried mixtures of herbs in Mexico: (**a**) *Gordolobo* is a dried mixture of *Pseudognaphalium* spp. that have been used for centuries in Mexico as expectorants. (Source: http://www.papajimsbotanica.com/product_info.php/products_id/2548); (**b**) Mixtures of herbs found in present-day Mexican markets. (Source: http://www.mexicolore.co.uk/images-3/396_17_2.jpg)

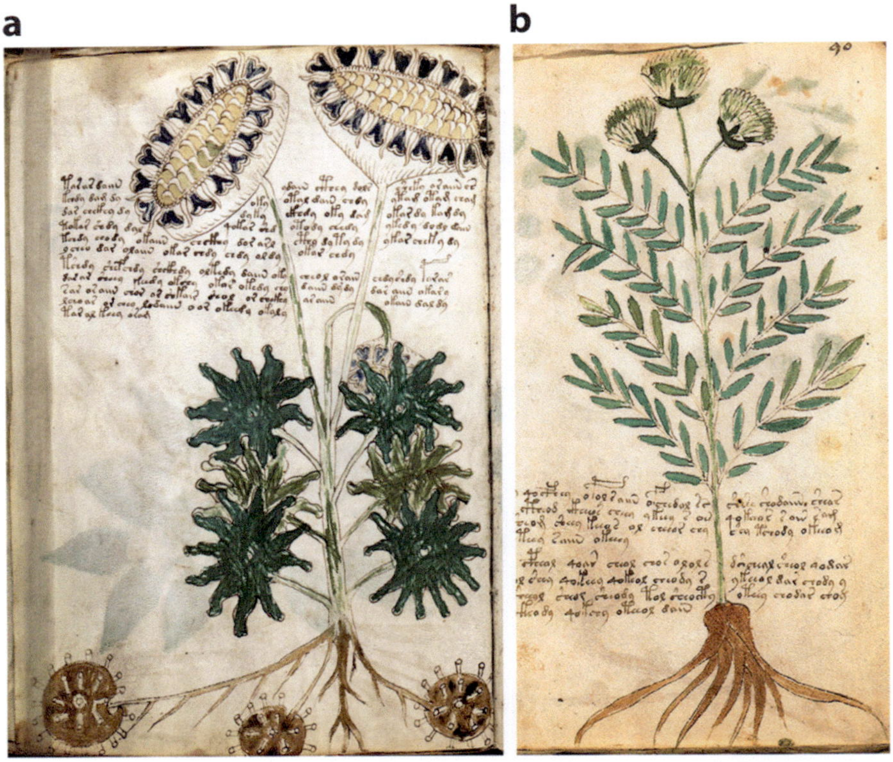

Fig. 3.10 Hybrid phytomorphs: (**a**) folio 33v, *Psacalium* sp. + *Pippenalia* sp.? (see Fig. 4.150); (**b**) folio 90r, *Osmunda regalis* var. *spectabilis* + *Vernonia* sp. (see Fig. 4.161)

population, but its lack may be explained by the 14 missing folios (28 pages) of the Herbal section. However, the Pharmaceutical section includes an image of a *Zea* sp., probably teosinte, considered a parent of maize (Fig. 4.124).

References to medicinal plants of Mexico seem obvious (e.g., Martinez 1969) because of the widespread use of these plants in the ancient Aztec empire, but we included references to the medicinal plants of the West Indies (e.g., Asprey and Thornton 1953, 1954; Ayensu 1981). The putative author, Gaspar de Torres, was born in Santo Domingo, and some Taino names have been noted in the text (Tucker and Janick 2018a, Chap. 4). Undoubtedly Torres searched for "botanical cognates" in Mexico from his previous experience in Santo Domingo, much as the Spanish substituted local species for oregano, garlic, bay laurel, cinnamon, and other herbs and spices in the New World. Because the range of these plants includes what is now known as the United States of America, Moerman's 1998 work on Native American medicinal plants is also cited.

Sources and Techniques

The present publication extends the 55 plant species to ca. 168, and the 34 families to 64 in the following chapter. All of the plants, except for the introduced weedy salsify, *Tragopogon porrifolius*, and a montane species, *Oclemena acuminata*, are native from the southeastern United States to California, and south to Nicaragua. Herbs from outside central Mexico could have been at one of the Nahua botanic gardens (which were established prior to European botanic gardens) or, more likely, were traded as dried samples in the vast trading networks. This publication also identifies a major psychoactive plant of the New World, *Nymphaea ampla.* A number of these species also occur in the West Indies with medicinal uses.

The species overlap with other herbals of this period, such as the *Codex Cruz-Badian*us, Hernández' botanical texts, Sahagún's *Florentine Codex,* and Gerard's *Herball*, will be discussed in Chap. 5. Common themes among the medicinal uses quoted are reproductive problems (especially female problems, such as menstruation, breastfeeding, contraception, delivery), colds, diarrhea, and skin problems. The Nahua marketplace was dominated by women (Truitt 2018); this implies that not only were the herbs sold by women but the knowledge of their use was also transmitted by women. Women's problems were solely handled by women (Farfán and Elferink 2010), and we thus have almost no record of the herbs used for these medical conditions, because chroniclers were exclusively men. The idea that this could be a women's herbal, at least in part, was first promulgated by L. C. Strong and E. L. McCawley in 1947. It is intriguing that the very first word in the *Voynich Codex* is , which is transliterated as *huomin*, pronounced "women."

The Mexican and European influences on Gaspar de Torres are obvious, but what about his formative years in Santo Domingo? In the mid-1500s, Bartolomé de las Casas, a Dominican friar, reported three mutually unintelligible Arawak-derived languages in Hispaniola (a common language, Macorix, and Ciguayo); he also indicated what we call "Taino" was a *lingua franca* understood throughout Hispaniola (Granberry and Vesceloius 2004). Conventional historians tell us that the Taíno are extinct, but modern DNA evidence shows us that some populations survived in remote areas not affected by disease or Spanish enslavement. And not only did the Taíno survive, but so did their cultural heritage. The medicinal practice called *Agua Dulce* emphasizes women, bathing, four directions, and herbs (Baracutei Estevez 2008, 2016, 2018; González 2018; Schroeder et al. 2018; Wade 2018; Woodaman 2017), all in common with the *Voynich Codex.*

Nahuatl Plant Names

The names of the plants in the Herbal section of the *Voynich Codex* have not been located in the text, but 188 of the 228 plants in the Pharmaceutical section are labeled and transliterated (Tucker and Janick 2018b, Chap. 5). Nahuatl names for plants are typically of four types (Clayton et al. 2009):

1. Primary "folk-generic" names that cannot at present be analyzed but which are likely to have been known widely and to be present as cognates in the modern Nahua languages. One example from the *Voynich Codex* would be *nochtli,* or the phonetic orthographic variant *nāshti.*
2. Compound "folk generic" names
3. "Folk-specific" names composed of a generic term plus a qualifying epithet (which may be compounded into the name), a class less likely to be widespread
4. "Descriptive phases" that may have been coined by Gaspar de Torres himself and "which are least likely to have been shared widely and to have been reserved in contemporary languages." An example from the *Voynich Codex* would be *māncuānaca* or its phonetic orthographic variant *māncuonoca,* perhaps derived from *mano* (*hand* in Spanish) plus *cuānaca* (*chicken foot* in Nahuatl). Indeed, the phytomorph root in folio 100v#13 does resemble a chicken foot.

The names of the phytomorphs in the *Voynich Codex* are also in other nearby languages, such as the Taino *maguey,* or its phonetic orthographic variant māgueoy (fol. 100r#1). This is in addition to liberal use of Spanish, such as *ahumi,* or *I smoke* (fol. 100v(2)#6).

The names attached to the phytomorphs in *the Voynich Codex* in symbols typical to shorthand of sixteenth century Spanish correspondence (Janick and Tucker 2018), represent a private, personal phonetic spelling of the languages heard by Gasper de Torres in a Spanish alphabet (which notably lacks a "w"). These are not the same spellings written by Molina and other Spanish priests in their compilations of Nahuatl dictionaries. To characterize Gaspar de Torres as a poor linguist or poor speller would not be accurate, since his codex was obviously not written to be widely shared. The plant names add further support to our leading hypothesis that Voynichese is derived from a Nahuatl *lingua franca,* which was a language of commerce that became extinct when it was replaced by Spanish (Dakin 1981).

Conclusions

The identification of plants in the *Voynich Codex* represents a unique challenge. Amazingly, few professional botanists, except for Rev. H. O'Neill in 1944, have tackled the phytomorphs in the *Voynich Codex* since its reappearance in 1912, even though numerous botanists have attempted to identify the equally nebulous phytomorphs of the *Codex Cruz-Badianus* (Emmart 1940), Sahagún's *Florentine Codex* (1951–1982), and Hernandez et al. (1651). The difference is that these documents usually have descriptions that allow some degree of identification, so it is not necessary to rely solely upon stylized sixteenth century botanical drawings. We also have a recorded provenance for these works, which are not mysterious, like the *Voynich Codex*. The identifications of phytomorphs of the *Voynich Codex* in the next chapter are the fruition of efforts by an academic botanist and horticulturist, and we fully expect that they will be refined in the future.

Literature Cited

Anderson, F.J. 1977. *An illustrated history of the herbals*. New York: Columbia University Press.

Angarano, A., B. Baldan, R. Marcucci, T. Pievani, F. Piovan, M. Villani, and S. Zaggaia. 2017. *Hortus Mirabilis: Discovering the world's oldest botanical garden*. New York: Rizzoli.

Arber, A. 1970. *Herbals: Their origin and evolution, a chapter in the history of botany 1470–1670*. Darien: Hafner Publishing Co.

Asprey, G. F., and P. Thornton. 1953, 1954. Medicinal plants of Jamaica. Parts I and II. *West Indian Medical Journal* 2(4):233–252, 3(1):17–41.

Ayensu, E.S. 1981. *Medicinal plants of the West Indies*. Algonac: Reference Publications.

Baracutei Estevez, J. 2008. Amerindian mtDNA in Puerto Rico: When does DNA matter? *Centro Journal* 20: 218–228.

Baracutei Estevez, J. 2016. Buyío Taimani—Agua Dulce: Continuing Taino Traditions in the Dominican Republic–Part 1 of 2. https://www.researchgate.net/publication/311946428. Accessed 12 Sept 2018.

Baracutei Estevez, J. 2018. Agua dulce in Kiskeya: Tobacco, casabe, water and stone. *American Indian* 19 (3): 24–27.

Binnqüist, C. L. 2003. The endurance of Mexican amate paper: Exploring additional dimensions to the sustainable development concept. D. Thesis. University of Twente, Enschede, Netherlands.

Binnqüist, C.L., A. Quintanar-Islas, and M.V. Meeren. 2012. Mexican bark paper: Evidence of history of tree species used and their fiber characteristics. *Economic Botany* 66: 138–148.

Blunt, W. 1971. *The art of botanical illustration*. London: Collins.

Blunt, W., and S. Raphael. 1979. *The illustrated herbal*. London: Thames and Hudson.

Clayton, M., L. Guerrini, and A. de Avila. 2009. *Flora: The Aztec herbal*. London: Royal College Enterprises.

Collins, M. 2000. *Medieval herbals: The illustrative traditions*. London: The British Library.

Dakin, K. 1981. The characteristics of a Nahuatl *lingua franca*. In *Nahuatl studies in memory of Fernando Horcasitas*, ed. Frances Karttunen. Texas Linguistic Forum 18:5–67.

Emmart, E.W. 1940. *The Badianus manuscript (Codex Barberini, Latin 141) Vatican Library. An Aztec Herbal of 1552*. Baltimore: Johns Hopkins Press.

Evans, S.T. 2000. Aztec royal pleasure parks: Conspicuous consumption and elite status rivalry. *Studies in the History of Gardens and Designed Landscapes* 20: 206–228.

Evans, S.T. 2005. The Aztec palace under Spanish rule. Disk motifs in the Mapa de México de 1550 (Uppsala Map or Mappa de Santa Cruz). In *The Postclassic to Spanish-era transition in Mesoamerica*, ed. S. Kepecs and R.T. Alexander, 13–33. Albuquerque: University of New Mexico Press.

Evans, S.T. 2010. The garden of the Aztec philosopher-king. In *Gardening: Philosophy for everyone*, ed. D. O'Brien, 207–219. London: Blackwell-Wiley.

Farfán, J.A.F., and J.G.R. Elferink. 2010. *Ethnobotany and Aztec sexuality*. Muenchen: Lincom Europa.

Flaherty, E., A.O. Tucker, and J. Janick. 2018. Zoomorph identification. Chapter 6. In *Unraveling the Voynich Codex: a 16th century Mexican manuscript*, ed. J. Janick and A.O. Tucker. Cham: Springer Nature.

González, C.M. 2018. Abuelas, ancestors and atabey: The spirit of Taíno resurgence. *American Indian* 19 (3): 18–23.

Granberry, J., and G.S. Vescelius. 2004. *Languages of the pre-Columbian Antilles*. Tuscaloosa: University of Alabama Press.

Granziera, P. 2001. Concept of the garden in pre-Hispanic Mexico. *Garden History* 29: 185–213.

Granziera, P. 2005. Huaxtepec: The sacred garden of an Aztec emperor. *Landscape Research* 30 (1): 81–107.

Guy, J.B.M. 1991. Voynich revisited. *Cryptologia* 15: 161–166.

Hernández, F., F. Celsi, F. Colonna, B. Deversini, J. Faber, J. Greuter, V. Mascardi, N.A. Recchi, and J. Terentius. 1651. *Rerum medicarum Novae Hispaniae Thesarus, seu, Plantarum animalium mioneralium Mexicanorum historia*. Romae: Vitalis Mascardi.

Janick, J., and A.O. Tucker, eds. 2018. *Unraveling the Voynich Codex: Mexican*. Cham: Springer Nature.

Marcus, J. 1992. *Mesoamerican writing systems: Propaganda, myth, and history in four ancient civilizations*. Princeton: Princeton University Press.

Martinez, M. 1969. *Las plantas medicinales de Mexico*. 5th ed. Mexico: Ediciones Botas.

Moerman, D.E. 1998. *Native American ethnobotany*. Portland: Timber Press.

Nuttall, Z. 1925. The gardens of ancient Mexico. *Annual Report of the Smithsonian Institution* 1923: 453–464.

O'Neill, H. 1944. Botanical observations on the Voynich MS. *Speculum* 19: 126.

Polastron, L.X. 2007. *Books on fire: The destruction of libraries throughout history*. Rochester: Inner Traditions.

Sahagún, B. de. 1951–1982. *Florentine Codex. General history of the things of New Spain.* [1540–1585] 12 vol. Transl. A. J. O. Anderson and C.E. Dibble. Salt Lake City: University of Utah Press.

Schroeder, H., M. Sikora, S. Gopalakrishnan, L.M. Cassidy, P.M. Delser, M.S. Velasco, J.G. Schraiber, S. Rasmussen, J.R. Homburger, M.C. Ávila-Arcos, M.E. Allentoft, J.V. Moreno-Mayar, G. Renaud, A. Gómez-Carballa, J.E. Laffoon, R.J.A. Hopkins, T.F.G. Higham, R.S. Carr, W.C. Schaffer, J.S. Day, M. Hoogland, A. Salas, C.D. Bustamant, R. Nielsena, D.G. Bradley, C.L. Hofman, and E. Willerslev. 2018. Origins and genetic legacies of the Caribbean Taino. *Proceedings of the National Academy of Sciences* 115: 2341–2346.

Strong, L.C., and E.L. McCawley. 1947. A verification of a hitherto unknown prescription of the 16th century. *Bulletin of the History of Medicine* 21: 898–904.

Torres, E.C. n.d.-a. *Green medicine: Traditional Mexican-American herbal remedies*. Kingsville: Nieves Press.

Torres, E.C. n.d.-b. *The folk healer: The Mexican-American tradition of the curanderismo*. Albuquerque: Nieves Press.

Torres, E.C. 2006. In *Healing with herbs and rituals: A Mexican tradition*, ed. T.L. Sawyer Jr. Albuquerque: University of New Mexico Press.

Torres, E.C., and T.L. Sawyer Jr. 2005. *Curandero: A life in Mexican folk healing*. Albuquerque: University of New Mexico Press.

Truitt, J. 2018. *Sustaining the divine in Mexico Tenochtitlan: Nahuas and Catholicism 1523–1700*. Oceanside: Academy of American Franciscan History.

Tucker, A.O., and J. Janick. 2016. Identification of phytomorphs in the Voynich Codex. *Horticultural Reviews* 44: 1–64.

Tucker, A.O., and J. Janick. 2017. Identification of plants in the 1584 murals of the Casa del Deán, Puebla, Mexico. *Notulae Botanicae Horti Agrobotanici Cluj-Napoca* 45 (1): 1–8. http://www.notulaebotanicae.ro/index.php/nbha/article/view/10692/7966.

Tucker, A.O., and J. Janick. 2018a. Phytomorph and geomorph identification, chapter 4. In *Unraveling the Voynich Codex*. Cham: Springer Nature.

Tucker, A.O., and J. Janick. 2018b. Phytomorphs in the pharmaceutical section: The Rosetta stone of the Voynich codex. Chapter 5. In *Unraveling the Voynich Codex*. Cham: Springer Nature.

Tucker, A.O., and R.H. Talbert. 2013. A preliminary analysis of the botany, zoology, and mineralogy of the Voynich manuscript. *HerbalGram* 100: 70–85.

von Hagen, V.W. 1944. *The Aztec and Maya papermakers*. New York: J. J. Augustin Publisher.

Wade, L. 2018. 'Extinct' Caribbeans have living descendants: Ancient DNA from Taino woman shows kinship to modern genomes. *Science* 359 (6378): 858.

Woodaman, R. 2017. Taíno survival: Back into history. *National Museum of the American Indian Winter*: 14–21.

Zandbergen, R. 2004–2017. The Voynich manuscript. www.voynich.nu/. Accessed 6 June 2017.

Chapter 4
Phytomorph Identification in the *Voynich Codex*

The present chapter more than triples the number of plants identified in the *Voynich Codex*, from 4 species identified in 1944 by Rev. Hugh O'Neill, to 37 by Tucker and Talbert in 2013, to 55 species by Tucker and Janick in 2016, and to 57 species in Tucker and Janick (2018, Chap. 4). The phytomorphs in the *Voynich Codex* are listed alphabetically by family and species, and the folio designation is listed. Each identification is divided into four subsections:

Description: a botanical name including formal botanical identification
Distribution: ecology and range
Names: English and Spanish names, and names in Mesoamerican when known
Uses: medicinal, culinary, and other uses of these and related species

Each identification is compared with contemporary photographic images, including herbarium sheets or botanical drawings, or sixteenth century images, when appropriate. The names transliterated from Voynichese follow Janick and Tucker (2018).

The reliability and accuracy of identification will be explained in the text. Those phytomorphs in which the species cannot be accurately determined will be identified as "generic name sp." indicating the specific epithet is unknown. Species in doubt will be followed by a question mark (?). Where an identification is suggested or likely, the species will be followed by the abbreviation "cf." meaning "compares favorably," and a suggested species will be listed.

Nomenclature follows our interpretation of the cited revisions, and/or the Germplasm Resources Information Network (GRIN) (USDA, ARS 2015), and/or the collaboration of the Royal Botanic Gardens, Kew, and the Missouri Botanical Garden (Plant List 2013).

A. O. Tucker, J. Janick, *Flora of the Voynich Codex*,
https://doi.org/10.1007/978-3-030-19377-5_4

Lycophytes and Ferns

Huperziaceae/Lycopodiaceae

Fol. 100v(2)#6, Fol. 102r#11. *Huperzia beiteliana* (Fig. 4.1)

Description These phytomorphs have tiny, linear leaves on a branched stem clearly of the Huperziaceae/Lycopodiaceae. The best fit is the terrestrial club-moss *Huperzia beiteliana* Mickel (Fig. 4.1a–c).

Distribution Pine-oak forests of Mexico (Mickel and Smith 2004).

Names A club-moss (English). The word accompanying the phytomorph on fol. 102r#11, , is transliterated as *āhu/gumi,* probably derived from the Spanish, *ahumi,* I smoke, perhaps referring to the release of spores, which resemble smoke (Fig. 4.1d).

Uses *Huperzia* spp. continue to be used medicinally (Ma et al. 2007).

Fig. 4.1 *Huperzia beiteliana*: (**a**) Fol. 100v(2)#6; (**b**) fol. 102r#11; (**c**) *Huperzia beiteliana*. (Source: http://www.fernsoftheworld.com/2014/12/07/huperzia-beiteliana); (**d**) smoky discharge of spores from a related species, *Lycopodium lagopus* (Laestadius ex C. Hartman); G. Zinserling ex Kuzeneva-Prochorova. (Source: Jordan Metzgar)

Fol. 100v(1)#13. *Huperzia dichotoma* (Fig. 4.2)

Description This phytomorph has tiny, linear leaves clearly of the Huperziaceae/ Lycopodiaceae. This phytomorph (Fig. 4.2a) is the pendant epiphyte *H. dichotoma* (Jacq.) Trevis. (Fig. 4.2b).

Distribution Native of wet montane forests from Florida to Paraguay and the Caribbean (Mickel and Smith 2004).

Names A club-moss (English). The word accompanying this phytomorph,  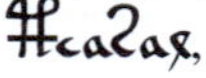, transliterates to *māncuonoca*, which may be derived from *mano* (Spanish) + *cuānaca* (Nahuatl) for chicken hand/foot (Fig. 4.2c) (Herrera 2004; Karttunen 1983; Siméon 2010), alluding to the epiphytic attachment of this species. This may be the same as *quamiauatl/quamiahoatl* in Hernández et al. (1651:258) and Hernández (1942:1034) (Batres et al. 2012) or *tlamacazqui ipapa* in Sahagún's *Florentine Codex* (Díaz 1976; Estrada Lugo 1989).

Uses *Huperzia* spp. continue to be used medicinally (Ma et al. 2007). Recently, this species has been found to have anticholinesterase activity because of the huperzine A content (Vázquez García et al. 2017).

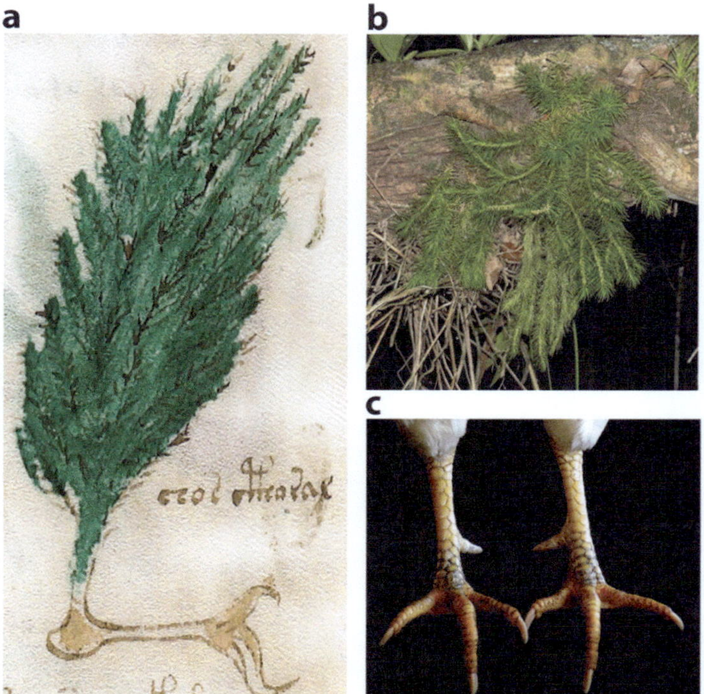

Fig. 4.2 *Huperzia dichotoma*: (**a**) fol. 100v(1)#13; (**b**) *Huperzia dichotoma*. (Source: permission of Bobby Hattaway, PhD); (**c**) chicken foot. (Source: http://www.paulmidler.com/wheres-my-chicken-foot)

Fol. 100v(1)#12, 101r#9. *Huperzia reflexa* (Fig. 4.3)

Description These phytomorphs (Fig. 4.3a, b) have tiny, linear leaves clearly of the Huperziaceae/Lycopodiaceae. The best fit is *Huperzia reflexa* (Lam.) Trevis. (Fig. 4.3c).

Distribution Native of wet montane forests from Mexico to Paraguay and the Caribbean; terrestrial, rarely epiphytic (Mickel and Smith 2004).

Names A club-moss (English). The word accompanying the phytomorph, 8ɑlf-ctaɣ, on fol. 100v(1)#12 transliterates to *chitlmoca,* perhaps derived from *mocacoquitl,* Nahuatl for "full of mud" (Herrera 2004; Karttunen 1983; Siméon 2010), perhaps indicating a terrestrial habitat or visible fungi at the base. The Lycopodiaceae have associated mycorrhizal associations that can persist in the photosynthetic sporophytes (Winther and Friedman 2008); note the fungi at the base of Fig. 4.3c.

Uses *Huperzia* spp. continue to be used medicinally (Ma et al. 2007) and have recently been shown to have alkaloids that inhibit acetylcholinesterase activity in the brain (Konrath et al. 2012).

Fig. 4.3 *Huperzia reflexa*: (**a**) fol. 100v(1)#12; (**b**) fol. 101r#9; (**c**) *H. reflexa*. (*Source:* http://www.plantsystematics.org/imgs/robbin/r/Lycopodiaceae_Huperzia_reflexa_39169.html)

Hymenophyllaceae

Fol. 101v(1)#8. *Trichomanes ekmanii* (Fig. 4.4)

Description This phytomorph (Fig. 4.4a) is of two fronds on a rhizome; this matches *Trichomanes ekmanii* Wess. Boer (Fig. 4.4b), which is mostly one cell thick and typically found in very wet areas. This species has recently been transferred to *Dimoglossum ekmanii* (Wess. Boer) Ebihara & Dubuisson as a result of chlorplast sequences (Ebihara et al. 2006).

Distribution Native from Mexico to Brazil and the Caribbean (Mickel and Smith 2004). Older leaves are typically rotund, as in the phytomorph.

Names Filmy fern (English). The word accompaning this, 8oɣa?9, *chācaoei*, might be derived from the Nahuatl *chica,* to spit (Herrera 2004; Karttunen 1983; Siméon 2010), perhaps describing the moist air of the habitat where this species is found.

Uses Unknown.

a b

Fig. 4.4 *Trichomanes ekmanii*: (**a**) fol. 101v(1)#8; (**b**) *T. ekmanii*. (Source: Emily Lisborg)

Ophioglossaceae

Fol. 100v#5. *Ophioglossum palmatum* (Fig. 4.5)

Description O'Neill (1944) identified the eusporangiate fern *Botrychium lunaria* (L.) Sw., moonwort, in the *Voynich Codex* on fol. 100v. O'Neill did not designate the phytomorph number, only the folio number (100v), but only #5 would match with a member of the Ophioglossaceae. The fronds with both a fertile and a sterile portion, the fertile portion bladelike and palmately lobed into multiple segments (Fig. 4.1b) suggest instead a specimen of the eusporangiate fern *Ophioglossum palmatum* L. A photograph of this species (Fig. 4.1b) confirms the identification.

Distribution Epiphytic in dense, wet forests at low to middle elevations from Florida to Brazil (Mickel and Smith 2004).

Names Hand fern (English). The word accompanying this phytomorph, 2oϗccc-2oc, *nācâăăanăă*, has an unlnown meaning.

Uses Unknown.

Fig. 4.5 *Ophioglossum palmatum*: (**a**) fol. 100v#5; (**b**) *O. palmatum*. (Source: Robbin Moran, New York Botanical Garden)

Osmundaceae

Fol. 100v(1)#6. *Osmunda regalis* var. *spectabilis* (Fig. 4.6)

Description This phytomorph (Fig. 4.6a) is a doubly pinnately compound leaf with oblong leaflets and attached rhizome. This is similar to fol. 90r, see Fig. 4.161, *Osmunda regalis* L. var. *spectabilis* (Willd.) A. Gray (Fig. 4.6b).

Distribution Native to North and South America. In Mexico, this species is especially abundant in Chiapas (Mickel and Smith 2004), where Mam, a Mayan language, is spoken.

Names Royal fern or flowering fern (English). The word accompanying this phytomorph, *itlmamchi,* might reflect the language of Chiapas. The *itl*-prefix may be derived from Nahuatl *itla,* thing, while the *-chi* suffix is probably derived from *chi*, make (Herrera 2004; Karttunen 1983; Siméon 2010).

Uses The medicinal value of *O. regalis* by the Iroquois is discussed by Moerman (1998). Although found worldwide, the medicinal value varies, primarily centered upon the astringent qualities of the rhizome.

Fig. 4.6 *Osmunda regalis* var. *spectabilis*: (**a**) fol. 100v(1)#6; (**b**) *O. regalis* var. *spectabilis*. (Source: http://wisflora.herbarium.wisc.edu/imagelib/imgdetails.php?imgid=15457)

Pteridaceae

Fol. 100v(1)#10. *Ceratopteris pteridoides*? (Fig. 4.7)

Description This phytomorph (Fig. 4.7a) consists of three fern fronds, but it is unusual in not being deeply divided. This might match the aerial leaves of *Ceratopteris pteridoides* (Hook.) Hieron. (Fig. 4.7b), a water fern of subfamily Ceratopteridoideae of the Pteridaceae. Aerial leaves are only lobed, but submerged leaves are 2–3× pinnatifid.

Distribution Pantropical, native from southeast United States to Brazil and the Caribbean (Mickel and Smith 2004).

Names Water fern (English). The accompanying word, *ātlmaāe,* includes the Nahuatl *atl,* water, and perhaps *mayaui,* to remove (Herrera 2004; Karttunen 1983; Siméon 2010), suggesting its use as a diuretic.

Uses This species has been long used in Colombia as a diuretic, whose use has recently been reaffirmed (Alviz et al. 2013).

Fig. 4.7 *Ceratopteris pteridoides*?: (**a**) fol. 100v(1)#10; (**b**) *C. pteridoides*. (Source: https://rybicky.net/atlasrostlin/ceratopteris_pteridoides)

Schizaeaceae

Fol. 101v(1)#12. (Fig. 4.8). *Schizaea elegans* (Fig. 4.8)

Description This phytomorph (Fig. 4.8a) shows a dichotomously branched frond of a fern. *Schizaea elegans* (Vahl) Sw. fits rather well (Fig. 4.8b).

Distribution Native from Mexico to Brazil (Mickel and Smith 2004).

Names Comb fern (English).

Uses Unknown.

Fig. 4.8 *Schizaea elegans*: (**a**) folio 101v(1)#12; (**b**) *Schizaea elegans*. (Source: https://alchetron.com/Schizaea-4418061-W)

Gymnosperms

Pinaceae

Fol. 100r#9. *Picea* sp.? (Fig. 4.9)

Description The phytomorph (Fig.4.9a) appears to be to be an excurrent tree (Fig. 4.9b) with pendulous branches. Typically, excurrent trees are conifers of genera such as *Abies, Picea, Pinus,* or *Pseudotsuga.* None of these species exhibit pendulous growth when young, but older specimens may be somewhat pendulous; an older stand of *Picea chihuahuana* Martinez (Fig. 4.9b) might be an example.

Distribution *Picea chihuanhuana* is native to northwest Mexico.

Names This phytomorph is labeled , *nanoya,* Spanish for grandmother, a sobriquet sometimes applied to the oldest member of the forest.

Uses Moerman (2009) records medicinal uses for *Picea* spp.

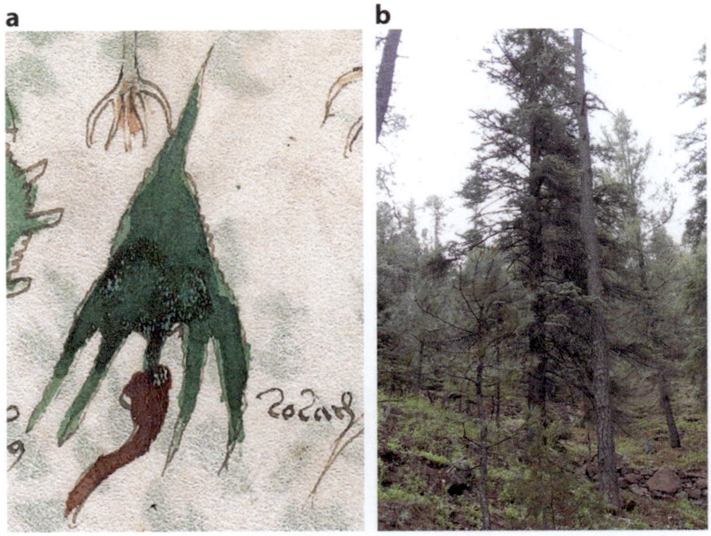

Fig. 4.9 *Picea* sp.?: (**a**) fol. 100r#9; (**b**) *Picea chihuahuana*. (Source: http://www.inforural.com.mx/peligro-extincion-arbol-picea-la-tala-ilegal)

Taxodiaceae

Fol. 100r#15. *Taxodium* sp., cf. *Taxodium mucronatum* (*T. huegelii*, *T. mexicanum*)? (Fig. 4.10)

Description This phytomorph (Fig. 4.10a) is very crude but appears to be either the cones or whole plant outlines of the Mexican cypress, *T. mucronatum* Ten. (*T. huegelii* hort. ex P. Lawson & C. Lawson; *T. mexicanum* Carrière). The cones and forked tree trunk of *T. mucronatum* are shown in Fig. 4.10b, c. This species is often multi-trunked in older specimens such as The Tule tree, or *el rbol del Tule* on the grounds of a church in Santa María del Tule in the Mexican state of Oaxaca, ca. 2000 years old.

Distribution Southwestern United States, Mexico, and Guatemala.

Names Montezuma bald cypress, Montezuma cypress (English); *sabino* (Spanish); *ahoehoetl/aueuetl/ahuehuetl, ahuehuecuahuitl,* or *ahuehuete/ahuehuetl* (Nahuatl) (Hernández et al. 1651; Hernández 1942; Dressler 1953; Díaz 1976; Farfán and Elferink 2010). The name accompanying this phytomorph, [illegible], transliterates to *ātlācân,* perhaps derived from *atlacaneci*, Nahuatl for bestial man, from the resemblance of the phytomorph to testicles.

Uses Medicinal uses in Mexico are recorded by Martinez (1969). Ornamental use by Aztecs in garden paths.

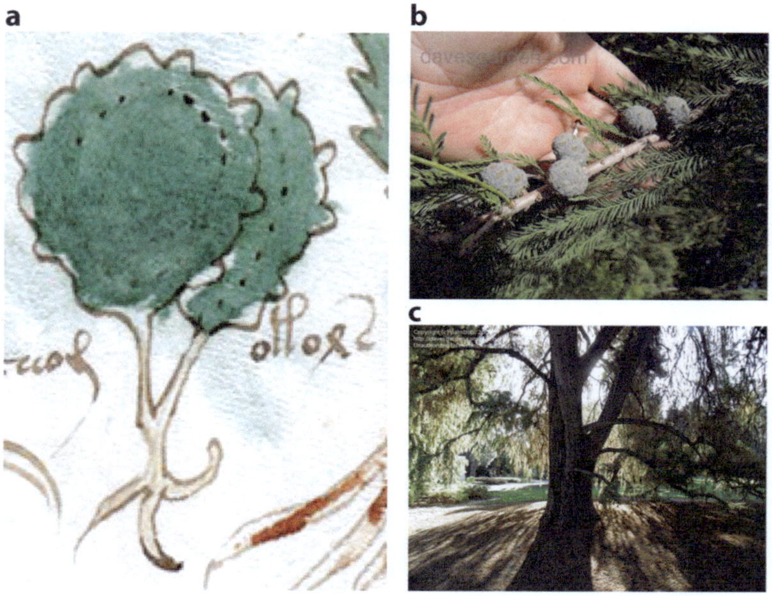

Fig. 4.10 *Taxodium* sp.: (**a**) fol. 100r#15); (**b**, **c**) grouped strobili (cones) and forked tree trunk of *T. mucronatum*, respectively. (Source: Geoff Stein)

Angiosperms

Acanthaceae

Fol. 11v. *Ruellia tuberosa* (Fig. 4.11)

Description This phytomorph (Fig. 4.11a) has an actinomorphic bluish corolla with cuneate corolla lobes. The leaves are oval to oblong with cuneate edges, varying from green to coppery on a rhizomatous base with tubers. The best match fits the variability of *R. tuberosa* L. (Fig. 4.11b). Though normally green, the leaves can be coppery in dry situations or when dried improperly. In the phytomorph (Fig. 4.11a) they appear as if re-arranged artificially after falling from dried specimens.

Distribution Native range, Central America, but naturalized from the Caribbean to northern and western South America.

Names Minnieroot, fever root, duppy gun, snapdragon root, sheep potato (English).

Uses Ayensu (1981) records "Roots in tea for oliguria, heat, flu; in tisanes and infusions for venereal diseases and constipation…Roots: purgative, emetic, febrifuge. Leaves: febrifuge, diaphoretic. Roots and leaves used for intestinal inflammation, colds, coughs, cooling drink. Tubers and leaves used with *Cordia obliqua* and *Annona muricata* for blood disorders. Was used for cystitis and enteritis." Naturalized in the tropics around the world, it has become a major medicinal plant and is used for anticancer activity; antidiabetic, antihyperlipidemic, and hepatoprotective activity; antimicrobial activity; antinociceptive and anti-inflammatory activity; antioxidant activity; antispermatogenic activity; anti-ulcer activity; cardiovascular and hypertensive activity; and purgative and cholinergic activity (Afzal et al. 2015; Khachitpongpanit et al. 2016; Pardeshi et al. 2017).

Fig. 4.11 *Ruellia tuberosa*: (**a**) fol. 11v; (**b**) *R. tuberosa* inflorescence. (Source: J.M. Garg CC BY 3.0); (**c**) tuberous roots. (Source: https://resepamelia.blogspot.com/2015/04/obat-diabetes-tanaman-pletekan-atau.html)

Amaranthaceae

Fol. 3r. *Amaranthus* sp., cf. *A. cruentus* (Fig. 4.12)

Description This phytomorph (Fig. 4.12a) is foliar only, with leaves of green, white, and red; it matches a cultivar of *Amaranthus* cf. *cruentus* L., (Fig. 4.12b) with red and green leaves.

Distribution Native to Central America but now naturalized worldwide.

Names Blood, red, or purple amaranth (English); *huautli* (Nahuatl). Probably illustrated as *uauhquilitl/uauhtli* of pl. 462 of the *Florentine Codex* (Sahagún 1963): "It has no stems. It is slender. It has spikelets; [these] are like amaranth. For this reason it is called *uauhçacat*; it has many seeds, much like amaranth. And its special attribute is to cover the plots seeded with amaranth."

Uses In addition to being a traditional food and medicinal, *Amaranthus cruentus* is an antioxidant (Nana et al. 2012).

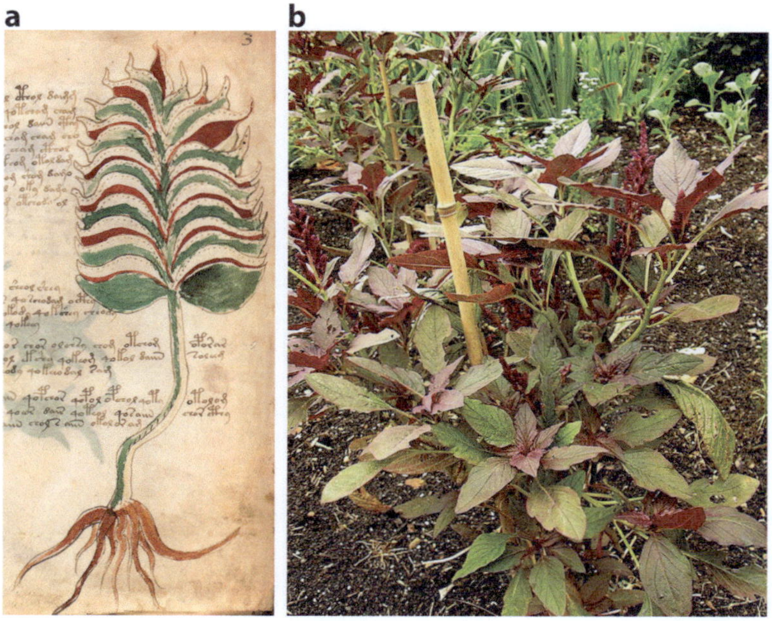

Fig. 4.12 *Amaranthus sp.*: (**a**) fol. 16v; (**b**) *A. cruentus*. (Source: http://www.wikiwand.com/en/Amaranthus_cruentus)

Apiaceae

Fol. 87v, Right Side. *Cymopterus* sp.? cf. *C. multinervatus* (Fig. 4.13)

Description This phytomorph (Fig. 4.13a) has a cyme-like inflorescence that might be Apiaceae. The foliage is once-pinnatifid, the roots ample. This might be a *Cymopterus* sp., *Cymopterus multinervatus* (J.M. Coult. & Rose) Tidestrom (Fig. 4.13b).

Distribution Native from California to Mexico.

Names Purple-nerve spring parsley (English)

Uses The culinary and medicinal value of *Cymopterus* species is discussed by Moerman (1998) and Moore (1990).

Fig. 4.13 *Cymopterus* sp.?: (**a**) fol. 87v; (**b**) *C. multinervatus*. (Source: http://www.fireflyforest.com/flowers/2742/cymopterus-multinervatus-purplenerve-springparsley)

Fol. 16v. *Eryngium* sp., cf. *E. heterophyllum* (Fig. 4.14)

Description Probably the most phantasmagoric phytomorph in the *Voynich Codex* is the *Eryngium* sp. portrayed on fol. 16v (Fig. 4.14a). The inflorescence is colored blue, the leaves red, and the rhizome ochre, but the features verge on a stylized appearance rather than the botanical accuracy of the *Viola bicolor* of fol. 9v. This lack of technical attention makes identification beyond genus difficult. However, a conjecture might be *Eryngium heterophyllum* Engelm (Mathias and Constance 1941). This species has similar blue inflorescences (Fig. 4.14b), involucral bracts (Fig. 4.14c), and stout roots, and it also develops rosy coloring on the stems and basal leaves (Fig. 4.14c). This lack of concern about the shape of the leaves also plagues identifications in the *Codex Cruz-Badianus* (Clayton et al. 2009). However, *E. heterophyllum* has pinnately compound leaves, not peltate, but leaves subtend the inflorescence and cover the stem, suggesting that the phytomorph was drawn from a dried, fragmented specimen.

Distribution Native to Mexico and to Arizona, New Mexico, and Texas in the United States.

Names Mexican thistle, Wright's eryngo, Mexican eryngo (English).

Uses Gallstone treatment in Mexico; experimental evidence for a hypocholesteremic effect (Navarrete et al. 1990).

Fig. 4.14 *Eryngium* sp.: (**a**) fol. 16v; (**b**) inflorescence of *E. heterophyllum.* (Courtesy of Blooms of Bessihngham®); (**c**) leaves of *E. comosum* R. Delaroche illustrating anthocyanin accumulation. (Courtesy of plantae mexicanae tenorianae, Pedro Tenorio-Lezama)

Folio 41v. *Lomatium* sp., cf. *L. dissectum* (Fig. 4.15)

Description This phytomorph (Fig. 4.15a) shows a plant that bears a dense fruiting umbel with schizocarps that are flat, which is characteristic of the Apiaceae. These schizocarps have a dark center and white wings. The growth habit is a rosette bearing coppery tuberous roots. The leaves are deeply dissected and palmate. The best match would be biscuitroot, a *Lomatium* sp., most probably *dissectum* (Nutt.) Mathias & Constance (Fig. 4.15b–d). The fruits (Fig. 4.15b) are flat schizocarps with papery wings born in a full umbel (Fig. 4.15c). The growth habit is a rosette and it bears tuberous roots (Fig. 4.15d) that are coppery when cleaned. The leaves vary

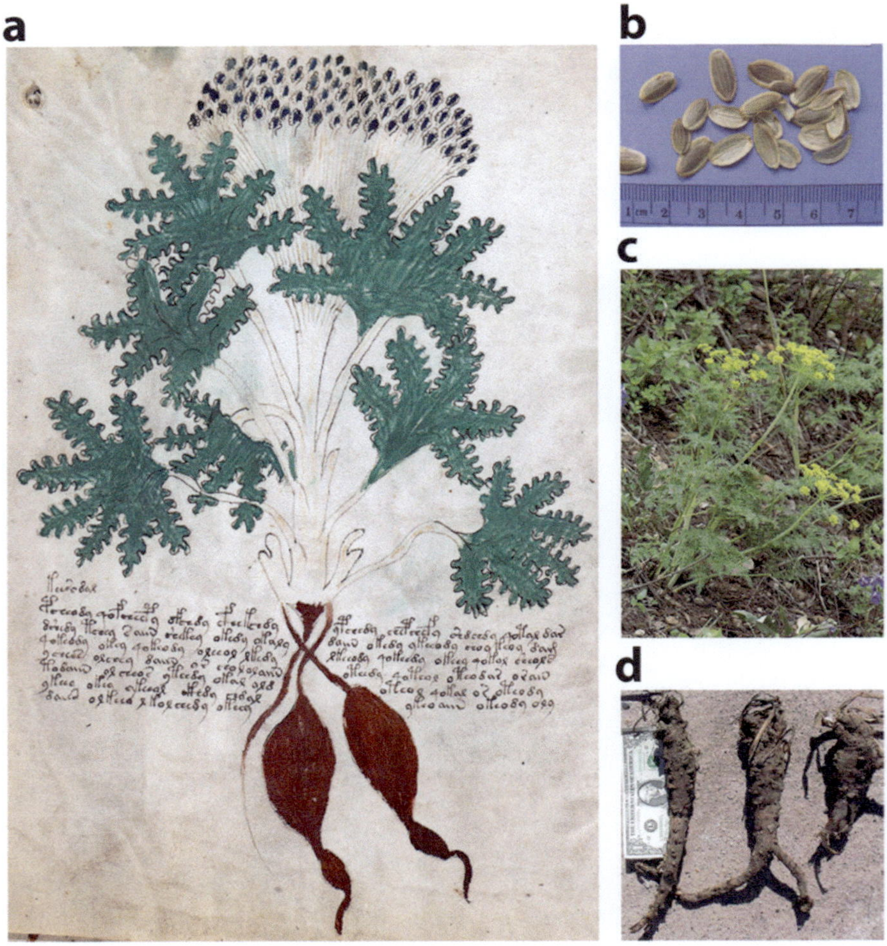

Fig. 4.15 *Lomatium dissectum*: (**a**) fol. 41v; (**b**) fruits; (**c**) foliage and flowers; (**d**) tuberous roots (Tilley et al. 2010)

from coarsely to deeply dissected, but are best described as pinnatifid, not palmate. This confusion of palmate versus pinnate is common in both the *Voynich Codex* and the *Codex Cruz-Badianus*.

Distribution Native from British Columbia to Mexico

Names Fernleaf biscuitroot (English).

Uses While sometimes taken for influenza, it can induce a rash (Marshall and Thornton 2018).

Apocynaceae

Fol. 100r#14. *Gonolobus chloranthus* (Fig. 4.16)

Description This phytomorph (Fig. 4.16a) appears to be the ridged fruit of an asclepiad, possibly *Gonolobus chloranthus* Schltdl. (Fig. 4.16b).

Distribution *Gonolobus* is native to South and Central America and Mexico; *G. chloranthus* is a Mexican species.

Names The *tlalayotli* in the *Florentine Codex* (Sahagún 1963: pl. 488a) with a similar illustration of the fruit (but with smooth ribs) is nominally accepted as the related species *G. erianthus* Decne., or *calabaza silvestre*. The name accompanying this phytomorph, oɣcɛccoɣ, is transliterated as *ācâmăāh*, perhaps derived from *acamaya*, Nahuatl for crayfish or crab, alluding to the knobs on the ribs.

Uses The roots of *G. niger* (Cav.) Schult. are used today in Mexico to treat gonorrhea (González Stuart 2004).

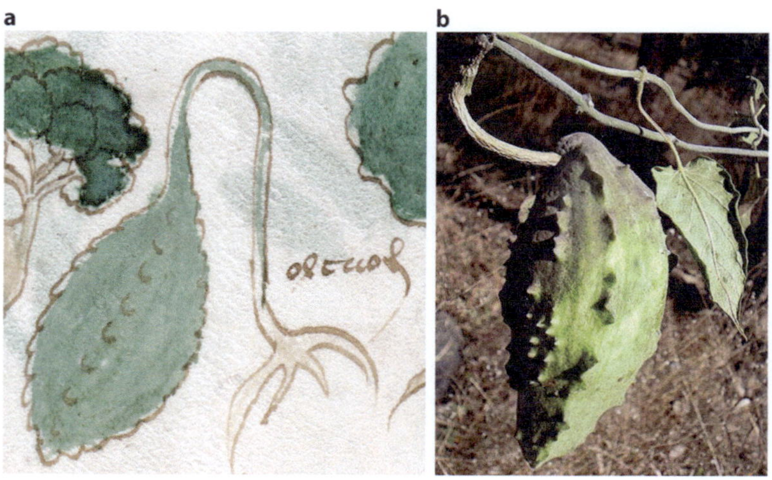

Fig. 4.16 *Gonolobus chloranthus*: (**a**) fol. 100r#14; (**b**) *G. chloranthus* fruits. (Courtesy of Guadalupe Cornejo-Tenorio and Guillermo Ibarra-Manríquez, The Field Museum)

Araceae

Fol. 100r#7. *Philodendron* sp. cf. *P. goeldii* (Fig. 4.17)

Description This phytomorph (Fig. 4.17a) appears to be the leaf or stem of an aroid, most probably a species of *Philodendron*, but the crudeness of the drawing belies whether this is a stem or compound leaf. If the latter, it may be a crude representation of the pedately compound leaf of *Philodendron goeldii* G.M. Barroso (Fig. 4.17b).

Distribution *Philodendron goeldii* is native to northern South America.

Names The name accompanying this phytomorph, , transliterates as *macanol*, perhaps derived from *macana*, the Taino wooden sword studded wth obsidian (*macuahuitl* in Nahuatl (Fig. 4.17c).

Uses Unknown.

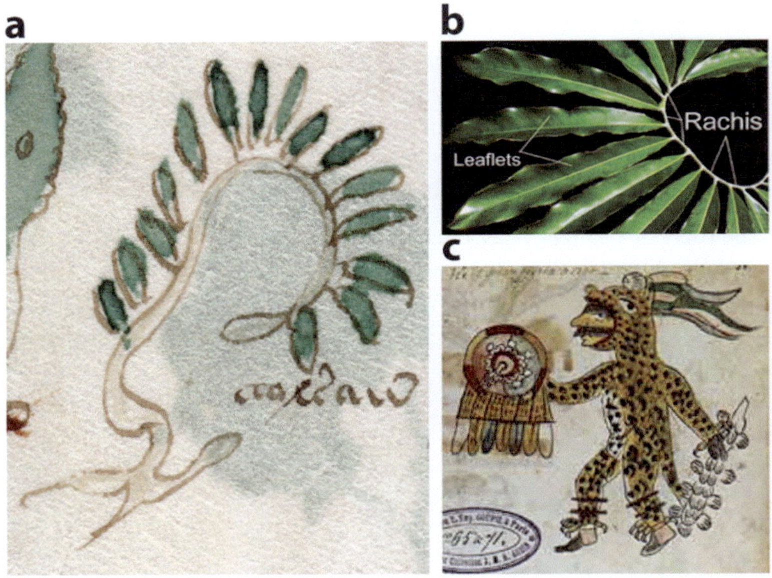

Fig. 4.17 *Philodendron* sp.: (**a**) fol. 100r#7; (**b**) *P. goeldii*. (Courtesy of Steve and Janice Lucas); (**c**) *macuahuitl* of the Aztecs from the *Codex Ixtlilxochitl*. (Source: http://www.mexico-france.com/codices_esp.html)

Fol. 100r#2. *Philodendron mexicanum* (Fig. 4.18)

Description This phytomorph (Fig. 4.18a), appears to be a vining aroid with hastate leaves, ripped from a tree, most probably *Philodendron mexicanum* Engl. (Fig. 4.18b, c).

Distribution Ranges from Mexico south through Central America and into Colombia and Ecuador.

Names Known as *huacalazochitl* in Nahuatl (Zepeda and White 2008); known as *huacalxochitl/huacalxōchitl* (huacal flower) in fol. 10v, 18v, 27v, 39v, and 56v of the *Codex Cruz-Badianus* (Emmart 1940; Cruz and Badiano 1991; Alcántara Rojas 2008; Gates 2000; Clayton et al. 2009) or *huacalazochitl* (Zepeda and White 2008). The name accompanying this phytomorph, 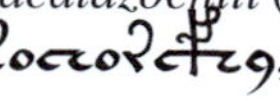, *nāmāepi*, may be Mixtec for soap plant.

Uses Bown (1988) writes of the Araceae in general: "Most of the species of Araceae which are used internally for bronchial problems contain saponins, soap-like glycosides which increase the permeability of membranes to assist in the absorption of minerals but also irritate the mucous membranes and make it more effective to cough up phlegm and other unwanted substances in the lungs and bronchial passages."

Fig. 4.18 *Philodendron mexicanum*: (**a**) fol. 100r#2; (**b**, **c**) leaves of *P. mexicanum*. (Source: Dave's Garden and Steve and Janice Lucas, respectively)

Fol. 100v(2)#3. *Philodendron* sp.? cf. *P. subincisum* (Fig. 4.19)

Description This phytomorph (Fig. 4.19a) is entirely vegetative, bent to one side, i.e., a liana or vine. Leaves are folded in half but possibly sagittate with lobed margins and acute tips. This is colored blue, which could mean green suffused with pink, purple, or blue. This matches a *Philodendron subincisum* Schott (Fig. 4.19b). Color varies from green to maroon, and maroon variants are quite common in the genus.

Distribution *Philodendron subincisum* is native to Mexico.

Names Giant aroid (English).

Uses Unknown.

Fig. 4.19 *Philodendron* sp.: (**a**) fol. 100v(2)#3; (**b**) *P. subincisum*. (Source: Peter Richardson)

Fol. 89v(1)#5. *Spathiphyllum cannifolium*? (Fig. 4.20)

Description This phytomorph (Fig. 4.20a) appears to be an Araceae with a white spathe, pale greenish spadix, and ample roots. One match might be a *Spathiphyllum* sp., such as *Spathiphyllum cannifolium* (Dryand. ex Sims) Schott (Fig. 4.20b).

Distribution Native from Colombia to Venezuela.

Names Peace lily (English). The name accompanying this phytomorph, , trasliterates to *āgu/huācaācaoll.* This might be derived from the Spanish *agua* plus Nahuatl, *caccolli*—watery hardness (Herrera 2004; Karttunen 1983; Siméon 2010); this may allude to its aid in kidney stones.

Uses Widely planted in the tropics as an ornamental. This species is antibacterial and anti-inflammatory (Abdullah et al. 2011; Ahmad Rausa et al. 2015). The fleshy inflorescences of species of *Spathiphyllum* are eaten as a vegetable in Guatemala and El Salvador (Chizmar Fernández 2009).

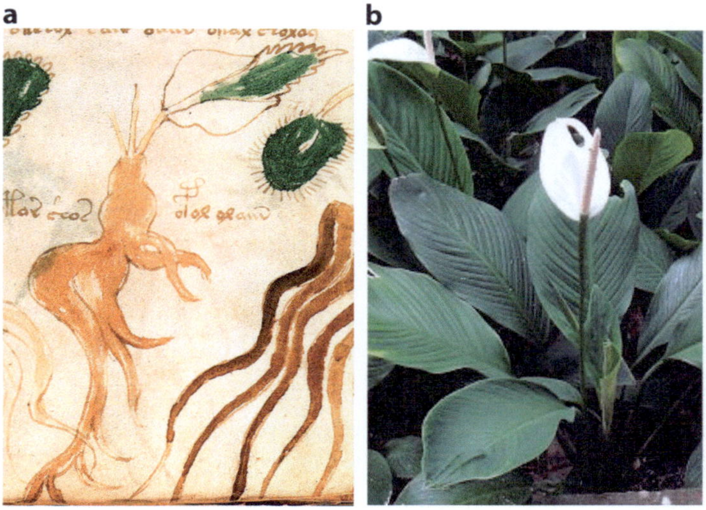

Fig. 4.20 *Spathiphyllum cannifolium*?: (**a**) Fol. 89v(1)#5; (**b**) *S. cannifolium*. (Source: Ong Kwan Han @ natureloveyou.sg)

Fol. 42r (Left Side). *Syngonium podophyllum* (Fig. 4.21)

Description This phytomorph (Fig. 4.21a) is entirely vegetative and consists of a sagittate leaf that is green with white veins. This is most probably a leaf of the Araceae, and *Syngonium podophylllum* Schott matches rather well (Fig. 4.21b).

Distribution Native from Mexico to central South America.

Names Syngonium, nephthytis (English).

Uses Medicinal among the Huastecs (Alcorn 1984). Today, it has been documented as antibacterial (Camporese et al. 2003).

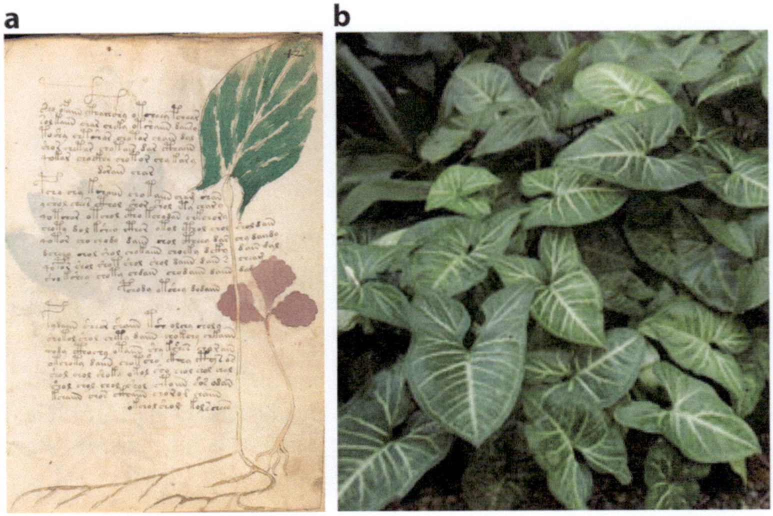

Fig. 4.21 *Syngonium podophyllum*: (**a**) fol. 42r, left side; (**b**) *S. podophyllum* foliage. (Source: Arseny Peisakhov)

Araliaceae

Fol. 35v. *Polyscias guilfoylei* (Fig. 4.22)

Description This phytomorph (Fig. 4.22a) has pale green to brown fruits of the Araliaceae with dark green, pinnately lobed leaves and ample roots. Only the extremely variable *P. guilfoylei* (W. Bull) L.H. Bailey, is similar (Fig. 4.22b).

Distribution This species is found in both the paleotropics and neotropics, from Central America to South America and the Caribbean.

Names Geranium aralia (English).

Uses Ayensu (1981) records for *P. guilfoylei*: "Tea of leaves for colds; leaves for head poultice." Asprey and Thornton (1953, 1954) record for this species: "The leaves are used in Jamaica to prepare tea for colds."

Fig. 4.22 *Polyscias guilfoylei*: (**a**) fol. 35v; (**b**) leaves of *P. guilfoylei*. (Source: https://www.pinterest.com/pin/335940453433264062)

Asparagaceae/Agavaceae

Fol. 100r#1. *Agave* sp. cf. *A. ellemeetiana*; Fol. 100r#4. *Agave* sp. cf. *A. atrovirens* (Fig. 4.23)

Description These two phytomorphs (Fig. 4.23a, c) both resemble the genus *Agave*. The one with smooth leaf margins, suggestive of a closed hand, is probably *A. ellemeetiana* Jacobi (Fig. 4.23b), while the other appears to be a pressed specimen with serrated or toothed leaf margins, quite possibly *A. atrovirens* Karw. ex Salm-Dyck (Fig. 4.23d).

Names Agave, century plant (English), *maguey* (Spanish); Nahuatl names are acametl, amole/amolli, macoztic metl, metl, metl coztli/mecoztli, mexcalmetl, nequametl, quametl, quetzalichtli, teohamatl/teoamatl, teometl, theometl, tlacametl, tlacoamatl/tlacametl, and xolometl (Hernández et al. 1651; Hernández 1942; Reko 1947; Dressler 1953; Sahagún 1963; Diaz 1976; Ocaranza 2011). Phytomorph 101r#1 is labeled, mānoeātzācâ, which might be derived from *mano* (Spanish), hand, and *tzacua* (Nahuatl), to close or enclose (Herrera 2004; Karttunen 1983; Siméon 2010), whereas fol. 100r#4 is labeled, māguoey, suggestive of the Spanish loan word *maguey* from the Taino word for agave.

Distribution Agave is native to Mexico.

Uses The juice of agave, when fermented, is the source of pulque, mesal, and tequila, a fermented alcoholic beverage considered sacred by the Aztecs and widely consumed in sixteenth century colonial New Spain (Hough 1908; Dressler 1953).

Fig. 4.23 *Agave* sp.: (**a**) fol. 100r#1; (**b**) *A. ellemeetiana*. (Source: http://www.agaveville.org/viewtopic.php?f=19&t=343); (**c**) fol. 100r#4; (**d**) *A. atrovirens*. (Source: Jeno Kapitany)

Asteraceae

Fol. 48v. *Acourtia runcinata*? (Fig. 4.24)

Description This phytomorph (Fig. 4.24a) has four terminal asterid heads, two per stem, possibly subfamily *Mutisioideae,* with bluish and yellow heads and whitish involucres. Leaves are pinnately lobed. Roots are ample. A match might be *Acourtia runcinata* (Lag. ex D. Don) B.L. Turner (Fig. 4.24b).

Distribution Native from Texas to Mexico.

Names Featherleaf, desert peonia (English).

Uses Unknown.

Fig. 4.24 *Acourtia runcinata*?: (**a**) fol. 48v; (**b**) *A. runcinata*. (Source: http://museum2.utep.edu/chih/gardens/plants/acourtiaruncinata.htm)

Fol 8v. *Acourtia thurberi* (Fig. 4.25)

Description This phytomorph (Fig. 4.25a) has four terminal asterid heads, possibly subfamily *Mutisioideae,* with light reddish heads and bluer involucres. Leaves are very slightly serrate. A good match is *Acourtia thurberi* (A. Gray) Reveal & King (Fig. 4.25b).

Distribution Native from Arizona and New Mexico to Chihuahua, Durango, and Sonora.

Names Thurber's desert peony (English).

Uses A decoction of this species is antidiabetic and antihyperalgesic (Martínez et al. 2017).

Fig. 4.25 *Acourtia thurberi*: (**a**) fol. 8v; (**b**) *A. thurberi*. (Source: T.R. Van Devender CC BY-NC 3.0)

Fol. 41r. *Ageratina adenophora*? (Fig. 4.26)

Description This phytomorph of subfamily Asterideae tribe Eupatorieae (Fig. 4.26a) has a whitish inflorescence without foliage. Roots are ample but not tuberous. This could be any genus within this tribe, but *Ageratina adenophora* (Spreng.) R.M. King & H. Rob., crofton weed, has many similarities (Fig. 4.26b).

Distribution Native to Mexico and possibly Central America.

Names Crofton seed, sticky snakeroot (English).

Uses The essential oil and aqueous extract of the leaves are antibacterial (Rajalakshmi et al. 2016; Subba and Kandel 2012).

Fig. 4.26 *Ageratina adenophora*?: (**a**) fol. 41r; (**b**) *A. adenophora*. (Source: http://keys.lucidcentral.org/keys/v3/eafrinet/weeds/key/weeds/Media/Html/Ageratina_adenophora_(Crofton_Weed).htm)

Fol. 37r. *Ageratina ligustrina* (Fig. 4.27)

Description This Asteraceae phytomorph (Fig. 4.27a) of subfamily Asterideae tribe Eupatorieae has a whitish inflorescence and elliptic leaves with an entire margin. This matches the variability of an *Ageratina* sp., most probably *A. ligustrina* (DC.) R.M. King & H.Rob. (Fig. 4.27b).

Distribution Native from Mexico to Central America.

Names Privet-leaved ageratina, privet-leaved snakeroot (English).

Uses It is used for gastrointestinal diseases by the Highland Maya of Chiapas, Mexico (Berlin and Berlin 1996).

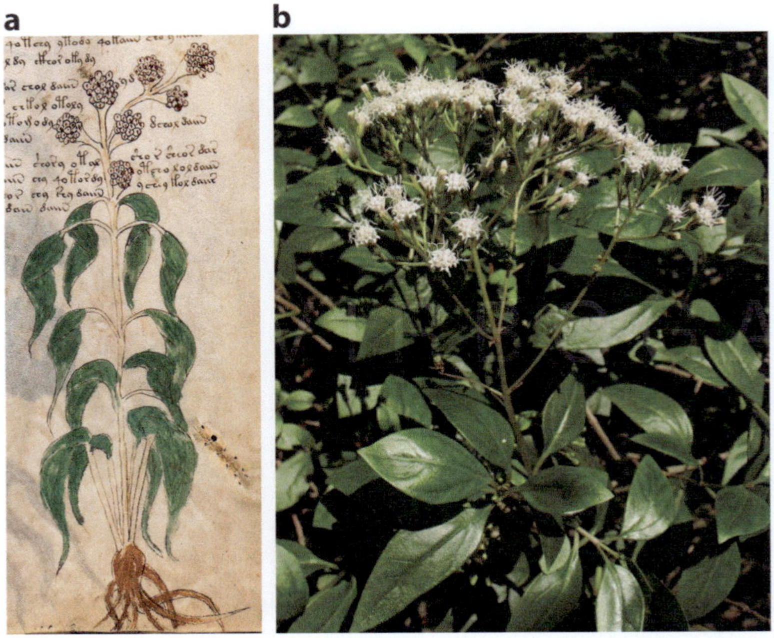

Fig. 4.27 *Ageratina ligustrina*: (**a**) fol. 37r; (**b**) *A. ligustrina*. (Source: http://www.virboga.de/Ageratina_ligustri; (b_a.htm)

Fol. 53r. *Ambrosia* sp., cf. *A. ambrosioides* (Fig. 4.28)

Description This phytomorph (Fig. 4.28a) has composite heads borne on a raceme. Leaves are brown to green and sharply serrate. Roots are brown and segmented. This matches the variability of *Ambrosia ambrosioides* (Delpino) W.W. Payne (Fig. 4.28b, c).

Distribution Native from Arizona and California to Mexico.

Names Canyon ragweed (English); *chiura* (Spanish).

Uses Leaves are anticarcinogenic from a cytotoxic component, damsin (Doskotch and Hufford 1969). Moerman (2009) records that it is used among Native Americans as an analgesic, antihemorrhagic, and cough medicine. He writes: "Gynecological Aid Decoction of crushed roots taken by women for pains and menstrual hemorrhage."

Fig. 4.28 *Ambrosia* sp.: (**a**) fol. 53r; (**b**, **c**) flowers, inflorescences, and leaves of *A. ambrosioides*. (Source: T. Beth Kinsey and This work by Cabeza Prieta Natural History Association (http://cabezaprieta.org) is licensed under a Creative Commons Attribution-NonCommercial 4.0 International License, respectively)

Fol. 46v. *Bartlettina* sp. cf. *B. sordida* (Fig. 4.29)

Description This phytomorph (Fig. 4.29a) belongs to Asteraceae, subfamily Asterideae, tribe Eupatorieae. Inflorescence has serrate, somewhat hastate leaves. Phyllaries are painted bluish. Roots are ample but not tuberous. This bears an uncanny resemblance to the shrubby *Bartlettina sordida* (Less.) R.M. King & H. Rob. (Fig. 4.29b). This species may be *quauhalahuac/cuauh-alāhua-c* of fol. 57v and 59r of the *Codex Cruz-Badianus* (Guerra 1952; Reko 1947). However, Mexico has 24 species of this genus (Villaseñor 2016), so it is difficult to pin down.

Distribution Native from Mexico to South America.

Names Purple torch, blue mist flower (English).

Uses Ornamental.

Fig. 4.29 *Bartlettina* sp.: (**a**) fol. 46v; (**b**) *B. sordida*. (Source: Raulbot CC BY-SA 3.0)

Fol. 46r. *Blumea viscosa*? (Fig. 4.30)

Description This phytomorph (Fig. 4.30a) shows two palmate leaflets/leaf, bent to one side, along with white ray florets and red disk florets unlike any plant with an Asteraceae inflorescence subfamily Asterideae tribe Inuleae. The rhizome is substantial. This may have been assembled from dried, fragmented leaves. Leaves aside, the closest match may be *Blumea viscosa* (Mill.) V.M. Badillo (Fig. 4.30a), which is perennial and smells of turpentine. This may be the same as *tezompotonic*/"*tzonpotonic*" of Hernández et al. (1651:1651) and Hernández (1942:135). An additional Nahuatl name might be *tlacochichic* (Díaz 1976).

Distribution Weedy and of uncertain nativity, probably South America, the type locality is Mexico (Nesom 1983).

Names Clammy false oxtongue (English).

Uses Naturalized in Benin, it is used as a diuretic, principally by pregnant women (Achigan-Dako et al. 2010).

Fig. 4.30 *Blumea viscosa*?: (**a**) fol. 46r; (**b**) *B. viscosa*. (Source: Surajit Koley); (**c**) foliage. (Source: Copyright © 2009 LasEcomujeres.org)

Fol. 48r. *Chaenactis* sp.? (Fig. 4.31)

Description This phytomorph (Fig. 4.31a), with an Asteraceae subfamily Asterideae tribe Helianthae inflorescence, has leaves that are palmately lobed at the apex, a condition not found in nature. Roots are ample but not tuberous. This might be a *Chaenactis* sp., but eight species of this genus are native to Mexico (Villaseñor 2016), so it is difficult to give an exact species with such a crude drawing. It may be *Chaenactis stevioides* Hook. & Arn. (Fig. 4.31b).

Distribution *Chaenactis stevioides* is native from California to Mexico.

Names Desert pincushion (English).

Uses The medicinal value of this genus is discussed by Moerman (1998).

a

b

Fig. 4.31 *Chaenactis* sp.?: (**a**) fol. 48r; (**b**) *C. stevioides*. (Source: http://snowbirdpix.com/sonoran_desert_plant_page.php?id=1073)

Fol. 87v (Left Side). *Chaptalia* sp.? cf. *Chaptalia tomentosa* (Fig. 4.32)

Description This phytomorph (Fig. 4.32a) of Asteraceae subfamily Mutisioideae has a terminal whitish inflorescence with entire spatulate leaves from a rosette on ample roots. The overall resemblance is to a *Chaptalia* sp., e.g., *Chaptalia tomentosa* Ventenat (Fig. 4.32b).

Distribution Native from North Carolina to Texas.

Names Pineland daisy (English).

Uses Ayensu (1981) writes of *C. nutans* (L.) Polák: "Tea for tension and laryngitis. Poultices of leaves on sores; leaf tea for amenorrhea. Diuretic. Decoction for colic, colds, convulsions, drink for childbirth and to provoke menses. Leaf, boiled or fresh, ground up with oil or fat, as a dressing for boils. Wounds." Asprey and Thornton (1953, 1954) also record for this species: "Browne 'It is reckoned an excellent diuretic and is used as such by many people.' Sloane believed the decoction to be of value for colic and wind, colds, convulsions; as a drink for women at childbirth and to provoke the menses. More recently, Beckwith reports the use of the leaf, either boiled or fresh, ground up with oil or fat as a dressing for boils." The medicinal value of this species is also discussed by Moerman (1998).

Fig. 4.32 *Chaptalia* sp.?: (**a**) fol. 87v, left side); (**b**) *C. tomentosa*. (Used by Permission of the North Carolina Native Plant Society)

Fol. 50r. *Cirsium* sp.? (Fig. 4.33)

Description Asteraceae inflorescences of subfamily Carduoideae typically blast apart from the center when dried, leaving behind an outer rim. This apparently happened with this phytomorph (Fig. 4.32a), and the artist merely painted what he saw. Fortunately, the cypselae are displayed. The foliage, however, is not as accurate, painted blue in the center with a crenulated margin that defies current botanical descriptors. It might match a *Cirsium* sp., such as *C. acantholepis* (Hemsl.) Petr. (Fig. 4.33b).

Distribution Native to Mexico.

Names *Cirsium acantholepis* is known as spiny or scaly thistle (English).

Uses The medicinal value of this genus is also discussed by Moerman (1998).

Fig. 4.33 *Cirsium* sp.?: (**a**) fol. 50r; (**b**) *C. acantholepis*. (Source: Anne-Marie Veith)

Fol. 34v. *Corethrogyne filaginifolia*? (Fig. 4.34)

Description This phytomorph (Fig. 4.34a) has five Asteraceae inflorescences of subfamily Asterideae tribe Astereae. Phyllaries and ray florets are whitish; disk florets are yellowish. Leaves are rotund to orbicular, green to brown, hairy, on side branches. Roots are ample but not tuberous. No Asteraceae exactly matches this description, but if created from dried, fragmented specimens, this falls within the variability of *Corethrogyne filaginifolia* (Hook. & Arn.) Nutt. [*Lessingia filaginifolia* (Hook. & Arn.) M.A. Lane]. Ray florets range from white to purple. Leaves vary from sessile to winged-petioled, linear to oblanceolate, spoon-shaped, or ovate, entire or toothed, hairy (Fig. 4.34b, cultivar 'Silver Carpet').

Distribution California aster, native from Oregon to Mexico.

Names California aster (English).

Uses Moerman (1998) says of this species: "Kawaiisu *Cold Remedy, Diaphoretic,* and *Herbal Steam* Infusion of twigs and leaves used as a steam bath to induce sweating for colds."

Fig. 4.34 *Corethrogyne filaginifolia*?: (**a**) fol. 34v; (**b**) California aster. (Source: http://davesgarden.com/guides/pf/go/106755)

Fol. 35r. *Cosmos sp.* cf. C. *bipinnatus* (Fig. 4.35)

Description This phytomorph has an Asteraceae inflorescence, probably subfamily Asterideae tribe Astereae, with eight reflexed, slightly tubular bluish ray florets and numerous reddish disk florets (Fig. 4.35a). Foliage is linear and arranged in a basket-like formation. Roots are ample and slightly tuberous. These characters can only match a *Cosmos* sp., probably allied with the tubular or quilled forms of *C. bipinnatus* Cav. (Fig. 4.35b). This species has high morphological variability (Paniagua-Ibáñez et al. 2015).

Distribution *C. bipinnatus* is native to Mexico.

Names *Grasol morado* (Spanish), Mexican aster (English).

Uses The medicinal value of this genus is discussed by Moerman (1998). The essential oil of *C. bipinnatus* is antibacterial (Olajuyigbe and Ashafa 2014).

Fig. 4.35 *Cosmos* sp.: (**a**) fol. 35r; (**b**) *C. bipinnatus*. (Source: http://b-and-t-world-seeds.com/cartall.asp?species=Cosmos%20bipinnatus%20Sea%20Shells%20-%20Stars%20Mix&sref=29083)

Fol. 99v#18. *Dahlia* sp.? (Fig. 4.36)

Description This root phytomorph (Fig. 4.36a) is labelled [illegible], transliterated as *ăcatli*, or reed, and might refer to the roots of a *Phragmites* sp. or another member of the Poaceae. However, this is a conundrum, as tuberous roots of the Poaceae would be expected to be segmented. Though de Ávila Blomberg (2012), Díaz (1976), and Reko (1947) have stated that *acatl* refers to *Phragmites australis* (Cav.) Trin. ex Steud., Díaz (1976) also suggests *Quercus laurina* Bonpl., *Q. polymorpha* Schldtl. & Cham., *Q. rugosa* Née, or *Q. peduncularis* Née. None of these species have tuberous roots that match the phytomorph illustrated as *acatli*. A hypothesis is that these are the tubers of a *Dahlia* sp. (Fig. 4.36b), as one name for this genus in Nahuatl is *acocotli* (Díaz 1976; Král 2014).

Distribution Native to Mexico.

Names Dahlia (English), *dalia* (Spanish).

Uses Still used in gastrointestinal diseases by the Highland Maya of Chiapas, Mexico (Berlin and Berlin 1996) and was also used for many medicinal and nutritive properties by the Aztecs (Whitley 1985).

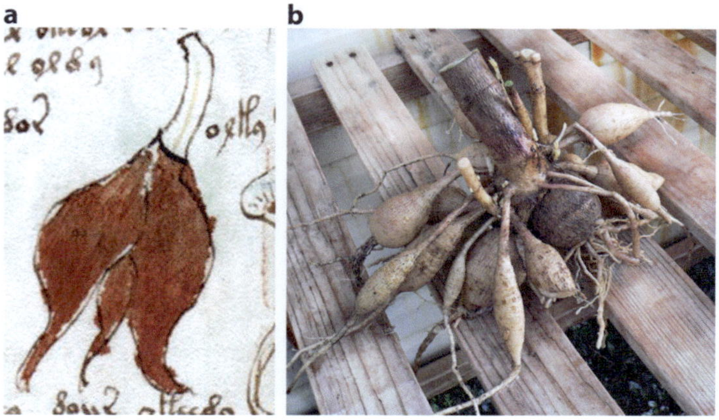

Fig. 4.36 *Dahlia* sp.: (**a**) fol. 99v#18; (**b**) roots. (Source: F.D. Richards CC BY-SA)

Fol. 3v. *Emilia fosbergii*? (Fig. 4.37)

Description This phytomorph (Fig. 4.37a) has two blue Asteraceae inflorescences of subfamily Senecioneae with deeply pinnatifid green leaves. The tapering tap root or rhizome is slightly pubescent. The bent inflorescence might indicate a bird-pollinated species (ornithophilous), but no member of the Asteraceae in Mesoamerica matches this description (Vogel 2015). One possibility is the highly variable *Emilia fosbergii* Nicolson (Fig. 4.37b–d). Floral colors range from magenta to orange.

Distribution The native range is unknown, but it is found from Florida to South America.

Names Florida tasselflower (English).

Uses Ornamental.

Fig. 4.37 *Emilia fosbergii*?: (**a**) fol 3v; (**b**) *E. fosbergii* flowers. (Source: Richard A. Howard, USDA Natural Resources Conservation Service); (**c**). *E. fosbergii* root. (Source: southeasternflora.com); (**d**) *E. fosbergii* herbarium sheet. (Source: http://www.brevardfl.gov/NaturalResources/EnvironmentalResources/Plants/page/12)

Fol. 50v. *Erigeron* sp.? (Fig. 4.38)

Description This phytomorph (Fig. 4.38a) of Asteraceae subfamily Asterideae tribe Astereae inflorescence is painted with blue pigment so that the three buds are merged into the principal unopened flower. Leaves are elliptic and slightly crenate. The roots seem to be infected with root knot nematode. This matches the variability of an *Erigeron* sp.; Fig. 4.38b, c show the flower and leaves of *E. karvinskianus* DC.

Distribution Genus is cosmopolitan but has high diversity in North America.

Names *Erigeron karvinskianus* is called Mexican fleabane (English).

Uses The medicinal value of this genus is discussed by Moerman (1998). The medicinal value of *E. karvinskianus* is discussed by Chávez (2006) and González Stuart (2004).

Fig. 4.38 *Erigeron* sp.: (**a**) fol. 50v; (**b**) *E. karvinskianus* flowers. (Source: Hectonichus CC BY-SA 3.0 & www.iewf.org) and (**c**) leaves. (Source: www.iewf.org)

Fol. 93r. *Helianthus annuus* (Fig. 4.39)

Description The phytomorph has a large, single yellow asterid head borne on a stout, thick stem (Fig. 4.39a). Leaves are green, alternate, ovate-lanceolate, acute, and entire. Petioles are short, with lines drawn down along the stem, possibly indicating a clasping base of the petiole or ridges along the stem. Roots are multiple, primary, and unbranched. The best match is sunflower, *Helianthus annuus* L. (Fig. 4.39b–d). This identification was first made by Hugh O'Neill (1944), who was a plant taxonomist and Curator of the Catholic University herbarium (LCU), who confirmed the determination with six botanists. Lincoln Taiz (Taiz and Taiz 2011), emeritus plant physiologist at the University of California, Santa Cruz, confirms the resemblance, and *Helianthus* authorities Robert Bye (personal communication 2014), distinguished ethnobotanist at Universidad Nacional Autónoma de México (UNAM) in Mexico City, and Billie Turner (personal communication 2014), one of the world's leading experts on Mexican Asteraceae and former Curator of the Herbarium (TEX) of the University of Texas at Austin, also confirm this identification. Sunflower researcher Jessica Barb (personal communication 2015) of Iowa State University notes that inbred lines of sunflower have very short petioles (see Fig. 4.39c), and that leaf variation is quite high. An illustration from the *Florentine Codex* is shown in Fig. 4.39d.

Distribution The preponderance of evidence points to Mexico as the center of domestication for sunflower (Harter et al. 2004; Heiser 2008; Lentz et al. 2008a, b; Rieseberg and Burke 2008; Bye et al. 2009; Blackman et al. 2011; Moody and Rieseberg 2012). It is widely naturalized in the United States.

Names Sunflower (English), *girasol* (Spanish). In Mexico, Nahuatl names in period literature are *chilamacatl* (Sahagún 1963), *chimalacatl* or *chimalacaxochitl* (Hernández 1942; Sahagún 1963), and *chimalatl peruina* (Hernández et al. 1651). *Chimal* means shield in Nahuatl, and sunflower is often called shield flower. Additional Nahuatl names are *acahualli* (Ramírez and Alcocer 1902; Dressler 1953) and *chimalxochitl* (Zepeda and White 2008).

Uses Medicinal uses in Mexico are recorded by Martinez (1969). Food source (oil, pseudocereal, starch, seed).

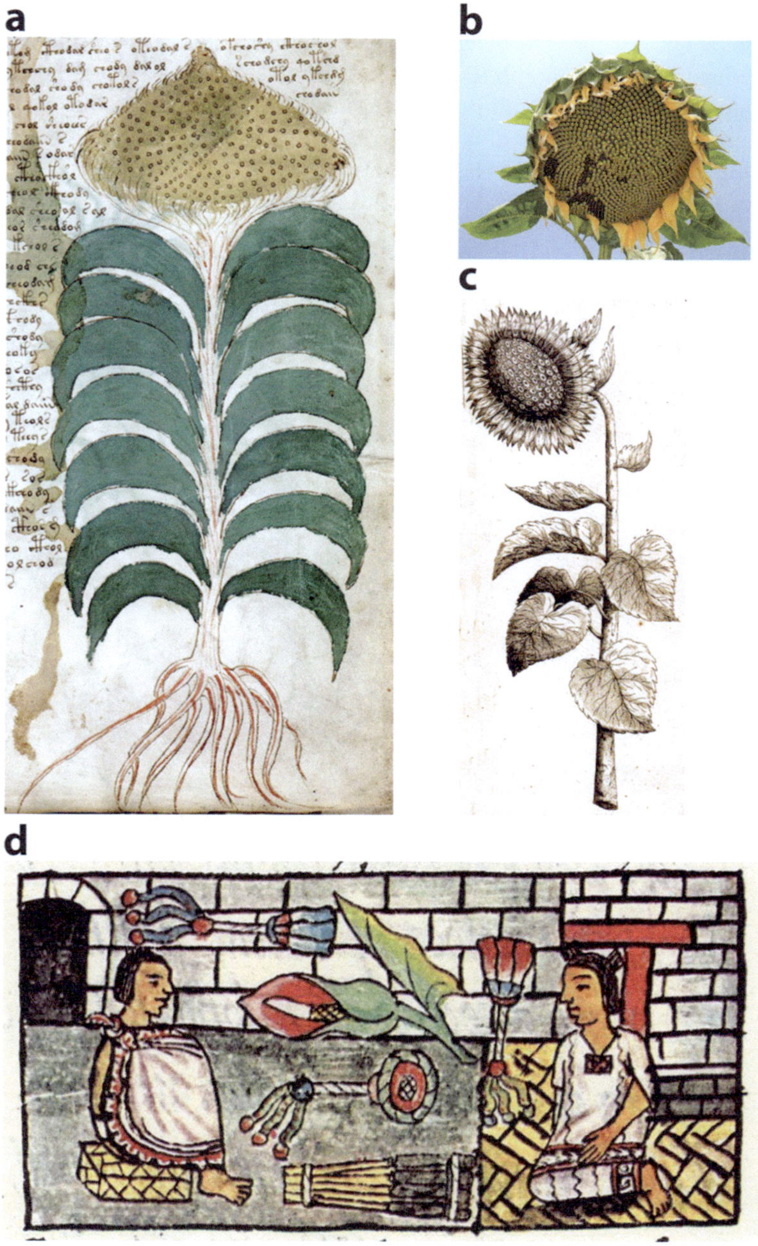

Fig. 4.39 *Helianthus annuus*: (**a**) fol. 93r; (**b**) seedhead of *H. annuus*. (Source: seagul CC0 1.0); (**c**) *Solis flor Peruuinus* of Emanuel Sweerts of 1612 (Bleiler 1976), showing short petioles towards the apex; (**d**) Nahua illustration of sunflower and three other plants; flowers as banquet gifts. (Notice the sunflower in the center of the illustration.). (Source: *Florentine Codex* Book 9; http://www.mexicolore.co.uk/aztecs/flora-and-fauna/sunflower)

Fol. 53v. *Hulsea heterochroma*? (Fig. 4.40)

Description This phytomorph (Fig. 4.40a) has a reddish Asteraceae inflorescence of subfamily Heliantheae, with petiolate hastate green leaves, four per node. The overall growth pattern is that of a shrub or, if created from fragmented dried samples, a vigorous stem. The root is ample but not tuberous. This might fit the variability of *Hulsea heterochroma* A. Gray (Fig. 4.40b).

Distribution Native from Utah to California.

Names Redray alpinegold (English).

Uses Unknown.

Fig. 4.40 *Hulsea heterochroma*: (**a**) fol. 53v; (**b**) flowers. (Source: http://www.calflora.net/recentfieldtrips/lilysprings11parttwoB.html)

Fol. 24v. *Iostephane heterophylla*? (Fig. 4.41)

Description Matching of this phytomorph (Fig. 4.41a) with a terminal actinomorphic floral feature that has seven or eight petals is impossible outside the Asteraceae. If this is considered as an Asteraceae inflorescence of subfamily Asterideae tribe Heliantheae, a match might fall within the variability of *Iostephane heterophylla* (Cav.) Benth., assuming that it was assembled from fragmented specimens in which the leaves were reversed (Fig. 4.41b). All features of this species, from foliage to flowers, are variable (Sandoval et al. 2005). This species might be in Hernández et al. (1651:254) and Hernández (1942:116) as *acuitzehuariracua* (Díaz 1976). In Sahagún's *Florentine Codex*, it is called *tecpatli/tecpalotl* or *tlacopatli/tlacopatle* (Estrada Lugo 1989). Additional Nahuatl names include *chipaoacyztic* and *huichoquachaqua* (Díaz 1976). *Iostephane heterophylla* var. *dicksonii* W. M. Sharp, without the typically serrated petals, is illustrated in Fig. 4.41c.

Distribution Native to Mexico.

Names Aciotze-uariracua, hierba del manzo (Spanish).

Uses Medicinal uses in Mexico are recorded by Martinez (1969).

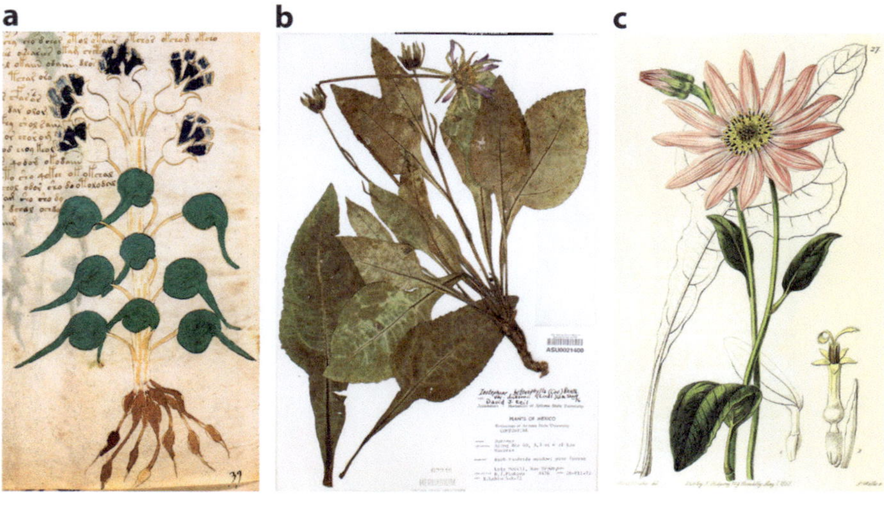

Fig. 4.41 *Iostephane heterophylla*: (**a**) fol 24v; (**b**) *I. heterophylla* herbaria sheet. (Source: Arizona State University Vascular Plant Herbarium); (**c**) *I. heterophylla* var. *dicksonii* as *Echinacea dicksoni*, (Bot. Reg. XXIV, 27 (1838). http://delta-intkey.com/angio/images/brns027.jpg)

Fol. 43v (Left Side). *Isocarpha oppositifolia*? (Fig. 4.42)

Description In this phytomorph (Fig. 4.42a) (as in fol. 30r, shown in Fig. 4.50 below), the trifoliate, entire leaves do not have a counterpart in nature; the artist re-assembled this from dried, fragmented specimens. The roots are ample and have been drawn snake-like, but not really tuberous. The trifoliate leaves aside, the Asteraceae whitish inflorescence is of subfamily Asterideae tribe Eupatorieae and falls within the variability of *Isocarpha oppositifolia* (L.) Cass. Flowers of the var. *acyhranthes* (DC.) D.J. Keil & Stuessy are shown in Fig. 4.42b.

Distribution Native from Texas to Mexico.

Names Rio Grande pearlhead (English).

Uses Ayensu (1981) records for *I. oppositifolia*: "Boil roots with salt for tea for chest colds; use leaves to increase appetite. Tonic known as Manzanilla de la Tierra."

Fig. 4.42 *Isocarpha oppositifolia*?: (**a**) fol. 43v (left side); (**b**) flowers of var. *acyhranthes*. (Source: http://chalk.richmond.edu/flora-kaxil-kiuic/i/isocarpha_oppositifolia_ssp_achyranthes_4914_03s.JPG)

Fol. 34r. *Jungia pringlei*? (Fig. 4.43)

Description This phytomorph (Fig. 4.43a) bears an Asteraceae inflorescence that is whitish to greenish and seems to be subfamily Mutisioideae. The leaves are palmately lobed, probably alternate. Roots are ample but not really tuberous. Species of the Asteraceae with palmate leaves are rare, and no species fits this description exactly, but one species that almost fits is the variability of *Jungia pringlei* Greenm. (Fig. 4.43b). The inflorescences are rather narrow when fresh, but expand considerably upon drying.

Distribution Native to Mexico.

Names Unknown.

Uses Unknown.

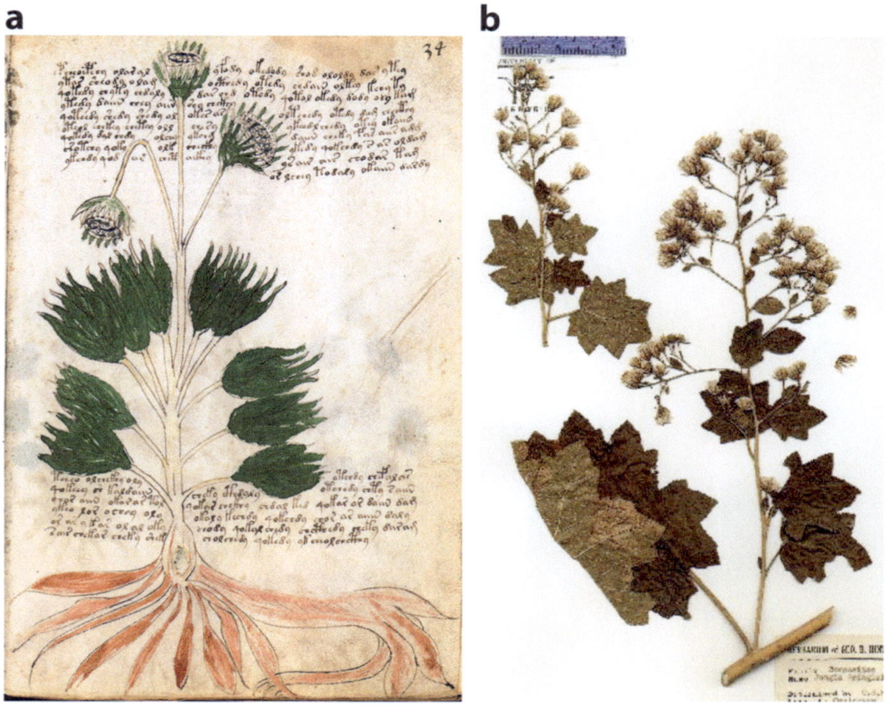

Fig. 4.43 *Jungia pringlei*?: (**a**) fol. 34r; (**b**) *J. pringlei* herbarium sheet. (Source: George B. Hinton Herbarium)

Folio 2r. *Lactuca graminifolia* (Fig. 4.44)

Description The phytomorph (Fig. 4.44a) shows a plant that has linear leaves in groups and an asterid inflorescence. The bracts are prominently tipped in a brownish color, while the petals of the flowers are serrate and off-white. A match might be *Lactuca graminifolia* Michx. (Fig. 4.44b, c), which has linear leaves (Fig. 4.44c) often in groups, and bracts tipped in purple. The petals are serrate and vary from a very light blue-pink to white (Fig. 4.44b).

Distribution Native to Mexico, north to New Jersey, and south to Guatemala.

Names Grassland lettuce (English).

Uses Unknown.

Fig. 4.44 *Lactuca graminifolia*: (**a**) folio 2r; (**b**) *L. graminifolia* flower illustrating serrated petals and purple-tipped bracts. (Source: Mary Keim CC BY-NC-SA 2.0); (**c**) *L. graminifolia* illustrating tufted foliage that sometimes occurs. (Source: Arizona State University Vascular Plant Herbarium CC BY-NC 3.0)

Fol. 5r. *Oclemena acuminata*? (Fig. 4.45)

Description The phytomorph (Fig. 4.45a) has one terminal white Asteraceae inflorescence of subfamily Asterideae tribe Astereae on whorled green leaves with white, somewhat tuberous roots. An absolutely perfect fit is whorled woodland aster, *Oclemena acuminata* (Michx) Greene (Fig. 4.45b), but currently this plant is native only as far south as the mountains of Georgia. If this is correctly identified, it either was raised in a montane woodland setting at a botanic garden or was obtained in trade. An alternative hypothesis is that the distribution of this species has become more restricted within the past five centuries.

Distribution North America, currently as far south as Georgia.

Names Whorled wood aster (English).

Uses Unknown.

Fig. 4.45 *Oclemena acuminata*?: (**a**) fol. 5r; (**b**) *O. acuminata*. (Source: Gerald C. Williamson, USWildflowers.com, all rights reserved)

Fol. 38v, 100v(2)#23. *Onoseris onoseroides* (Fig. 4.46)

Description The phytomorph in Fig. 4.46a) has a single Asteraceae inflorescence, probably subfamily Mutisioideae, with ten bluish ray florets; disk florets of an undetermined number are also bluish. Leaves are hastate, in a bowl-like formation. Roots are ample but not tuberous. This is repeated in the second fold-out (Fig. 4.46b) and may reflect the variability of *Onoseris onoseroides* (Kunth) B.L. Rob., (Fig. 4.46c). To compare with the artistic portrayal of this genus, the similar *O. drakeana* André from Colombia is illustrated in Fig. 4.46d.

Distribution Native from Mexico to Colombia.

Names Unknown.

Uses Breedlove and Laughlin (1993) mention the medicinal value of this species.

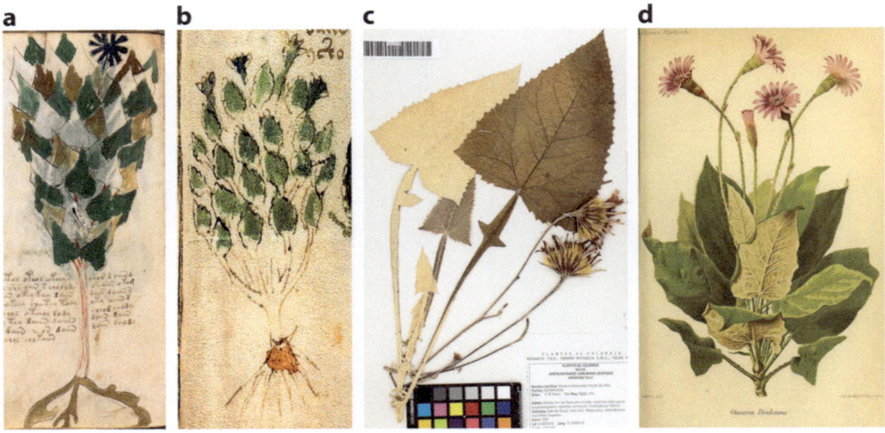

Fig. 4.46 *Onoseris onoseroides*: (**a**) fol. 38v; (**b**) fol. 100v(2)#23; (**c**) *O. onoseroides* herbarium sheet. (Source: Herbario Universidad del Tolima CC BY-NC 2.0); (**d**) illustration of *O. drakeana* André from Colombia (André 1883)

Fol. 31r. *Parthenium confertum* (Fig. 4.47)

Description This phytomorph (Fig. 4.47a) has an Asteraceae inflorescence of subfamily Asterideae tribe Heliantheae with whitish ray and disk florets. The leaves are pinnatifid, but the green wash obscures the lobes. This matches the variability of *Parthenium confertum* A. Gray, native to Mexico (Fig. 4.47b, c).

Distribution Native to Mexico.

Names Gray's feverfew, lyreleaf feverfew (English).

Uses Anti-fungal from the pseudoguaianolides (Mears 1980).

Fig. 4.47 *Parthenium confertum*: (**a**) fol. 31r; (**b**, **c**) *P. confertum*. (Sources: Steven Schwartzman, Lady Bird Johnson Wildflower Center and http://museum2.utep.edu/chih/gardens/plants/NtoQ/partheniumconf.htm)

Fol. 13r. *Petasites* sp., cf. *P. frigidus* var. *palmatus* (Fig. 4.48)

Description Based on the asterid inflorescence, large-cleft orbicular leaves, and relatively large root system, the phytomorph shown in Fig. 4.48a is most probably a *Petasites* sp. of the Asteraceae. The closest match might be *P. frigidus* (L.) Fr. var. *palmatus* (Aiton) Cronquist (Fig. 4.48b).

Distribution Native to North America from Canada to California.

Names Western sweet coltsfoot (English).

Uses *Petasites* spp. are used as antiasthmatics, antispasmodics, and expectorants and in salve or poultice form (Bayer et al. 2006).

Fig. 4.48 *Petasites* sp.: (**a**) fol. 13r; (**b**) *P. frigidus* var. *palmatus*. (Courtesy of Ben Legler)

Fol. 10r. *Plectocephalus americanus* (Fig. 4.49)

Fig. 4.49 *Plectocephalus americanus*: (**a**) fol. 10r; (**b**) *P. americanus* flowers. (Source: James O. Baines, Lady Bird Johnson Wildflower Center); (**c**, **d**) herbarium sheets of *P. americanus*. (Source: Univ of Texas at Austin, Billie L. Turner Plant Resources Center CC0 1.0 and University of Nevada Herbarium CC BY-NC 3.0)

Description This phytomorph (Fig. 4.49a) has a solitary, bluish composite head of subfamily *Carduoideae* borne on a scape (Fig. 4.49b). Leaves are green and slightly serrate. Roots are brown, sometimes displaying tuber-like regions. The flower matches the variability of the Eurasian *Centaurea* cyanus L., cornflower, but the absence of tubers in this species rules it out. This also matches the variability of *Plectocephalus americanus* (Nutt.) D. Don (*Centaurea americana* Nutt.) (Fig. 4.49c, d), which often has thickened roots.

Distribution Native from the central United States to Mexico

Names *Cardo del valle* (Spanish). American basket-flower, American star-thistle, powderpuff thistle, thornless thistle (English).

Uses Ornamental.

Fol. 30r. *Pluchea odorata* (Fig. 4.50)

Description This phytomorph (Fig. 4.50a) has Asteraceae bluish inflorescence of subfamily Asterideae tribe Inuleae. The trifoliate leaves, however, are a conundrum. No species in tribe Inulaeae and only *Koanophyllon tripartitum* B.L. Turner and some specimens of *Eupatorium cannabinum* L. of the tribe Eupatoreae have been reported with trifoliate leaves, and these do not have an entire margin. This leads to the hypothesis that the artist drew these from fragmented, dried specimen and artificially created trifoliate leaves. Ignoring the trifoliate leaves, this falls within the variability of *Pluchea odorata* (L.) Cass. (Fig. 4.50b). A Nahuatl name of this species is *cihuapatli* (Farfán and Elferink 2010).

Distribution Native from Maine to South America.

Names Sweetscent (English).

Uses Ayensu (1981) records for this species: "Leaves in decoctions for flu, fever, pneumonia, hypertension, headache; in eyewash for ophthalmia. Leaf poultice for cold in head…Used for sores and women in labor. Used as an antidote and stomachic. Pains, pulmonary tuberculosis, fever, convulsions, fainting. Hasten afterbirth. Boil for drinking or bathing; they wrap a sheet to cause sweating and eliminate colds." Asprey and Thornton (1953, 1954) record for this species: "Said by Steggerda to be applied to sores and used for women in labour." Alcorn (1984) and Martinez (1969) record similar uses. The medicinal value of this genus is discussed by Moerman (1998). More recent research has centered on the value of this species in treating cancer (e.g., Blaschke et al. 2017; Seelinger et al. 2012). Mayans today uise an infusion of the leaves to treat retention of the after-birth or to hasten menstruation (Roys 1976).

Fig. 4.50 *Pluchea odorata*: (**a**) fol. 30r; (**b**) *P. odorata* inflorescence. (Source: Jim Conrad www.backyardnature.net)

Fol. 27v. *Podachaenium eminens* (Fig. 4.51)

Description The phytomorph (Fig. 4.51a) has an Asteraceae inflorescence of subfamily Asterideae tribe Heliantheae with white ray florets and brownish disk florets. Leaves are elliptic with serrate margins. The stem is stout and woody. Roots are ample. This matches the variability of *Podachaenium eminens* (Lag.) Sch. Bip. ex Sch. Bip. (Fig. 4.51b). The leaves (scented of pineapples) are elliptic towards the apex, but the lower leaves are usually palmately lobed.

Distribution Native from Costa Rica to Mexico.

Names Giant tree daisy (English). This is called *tacote* (Sinaloa) or *tora* (Costa Rica) (Standley 1920–1926).

Uses Standley (1920–1926): states: "In Sinaloa [Mexico] the fragrant leaves are applied as poultices to wounds." This species is anti-inflamatory from the sesquiterpene lactones (Castro et al. 2000).

Fig. 4.51 *Podachaenium eminens*: (**a**) fol. 27v; (**b**) *P. eminens*. (Source: Michael Kartuz)

Fol. 8r. *Prenanthes* sp., cf. *P. trifoliata*? (Fig. 4.52)

Description This phytomorph (Fig. 4.52a) is entirely vegetative. However, the saggitate leaf, subtended on the petiole by a pair of leaflets, is quite distinctive. It is most probably a species of *Prenanthes,* possibly *P. trifoliata* (Cass.) Fernald (Fig. 4.52b), or a related species.

Distribution Native to eastern North America.

Names Gall of the earth (English).

Uses This and related species have been used as medicinal herbs by the Cherokee, Creek, Iroquois, and Seminole Nations (Hutton 2010; Moerman 1998).

Fig. 4.52 *Prenanthes trifoliata*?: (**a**) fol. 8r; (**b**) *P. trifoliata* leaf. (Source: Louis-M. Landry)

Fol. 27r (Left Side). *Pseudelephantopus spicatus* (Fig. 4.53)

Description This phytomorph (Fig. 4.53a) has a bluish Asteraceae inflorescence of subfamily Vernonieae with subsessile, broad, strap-like leaves. This matches the variability of *Pseudelephantopus spicatus* (B. Juss. ex Aubl.) Rohr ex C.F. Baker (Fig. 4.53b). Though normally considered a monospecific genus, the unresolved name *P. spiralis* (Less.) Cronquist is sometimes applied to some variations of the species (Fig. 4.53c).

Distribution Native from Florida to South America.

Names Dog's tongue (English).

Uses Ayensu (1981) records for this species: "Herb teas for heat and fever; poultices on sprains. Leaf juice in eye for ophthalmia. Root teas for diarrhoea and dysentery… Plant tea for colds and heart trouble." This species has been recently documented to have antileishmanial sesquiterpene lactones (Odonne et al. 2010).

Fig. 4.53 *Pseudelephantopus spicatus*: (**a**) fol. 27r; (**b**) *P. spicatus* herbarium sheet. (Source: Wunderlin et al. 2018); (**c**) inflorescense. (Source: Mauricio Mercadante CC BY-NC-SA 2.0)

Fol. 29r. *Pseudognaphalium macounii*? (Fig. 4.54)

Description This phytomorph (Fig. 4.54a) has an Asteraceae inflorescence of subfamily Asterideae tribe Gnaphalieae with whitish phyllaries and brownish disk florets. Leaves are serrate and somewhat elliptic. Roots are ample. This appearance matches the variability of *Pseudognaphalium macounii* (Greene) Kartesz (Fig. 4.54b).

Distribution Native from British Columbia to Mexico.

Names Macoun's cudweed (English).

Uses Species of this genus are sold as *gordolobo* for a variety of ailments.

Fig. 4.54 *Pseudognaphalium macounii*?: (**a**) fol. 29r; (**b**) *P. macounii*. (Source: CC BY-SA 3.0 SEINet Arizona – New Mexico Chapter)

Fol. 14r. *Pseudogynoxys chenopodioides*? (Fig. 4.55)

Description This phytomorph (Fig. 4.55a) has a reddish Asteraceae inflorescence of subfamily Heliantheae, with petiolate, hastate green leaves on a reddish stem. The overall growth pattern is that of a vine or shrub or, if created from fragmented dried samples, a vigorous stem. The root is curiously segmented. This might fit the variability of the vine *Pseudogynoxys chenopodioides* (Kunth) Cabrera (*Senecio confusus* Burtt) (Fig. 4.55b).

Distribution Native of Mexico

Names Mexican flame vine (English).

Uses This species is sometimes reputed to cause contact dermatitis in humans, but under the name *árnica*, it is used to treat skin wounds and sores in animals (Morales and Cruz 2007).

a b

Fig. 4.55 *Pseudogynoxys chenopodioides*: (**a**) fol. 14r; (**b**) *P. chenopodioides* flower. (Source: Dinesh Valke CC BY-SA 2.0)

Fol. 15r. *Rafinesquia neomexicana* (Fig. 4.56)

Description This phytomorph (Fig. 4.56a) is a white Asteraceae of subfamily Cichorioideae. The leaves are deeply serrate, clasping to the stem. A good fit is *Rafinesquia neomexicana* A. Gray (*R. californica* Nutt.) (Fig. 4.56b, c). Older leaves, especially in arid situations, become purple and eventually dried at the tips.

Distribution Native from Oregon to Mexico.

Names Annual California or desert chicory (English).

Uses Unknown.

Fig. 4.56 *Rafinesquia neomexicana*: (**a**) fol. 15r; (**b**) inflorescence. (Source: http://sev.lternet.edu/file/2691); (**c**) shoot. (Source: Barry Breckling CC BY-NC-SA 3.0)

Fol. 40v. *Smallanthus* sp.(Fig. 4.57)

Description This folio contains two phytomorphs of the same plant, vegetative and flowering (Fig. 4.57a). It is quite definitely a member of the Asteraceae, but the genus is less obvious. With bluish petals, reddish involucre, palmately compound leaves, and tuberous roots, this seems to fit a *Smallanthus* sp. It resembles somewhat the leading *Smallanthus* species cultivated today, the edible yacón [*S. sonchifolius* (Poepp.) H. Rob.]. (Fig. 4.57b, c).

Distribution Native to western South America.

Names Peruvian ground apple (English); jicama, yacón (Spanish).

Uses Food and natural sweetener.

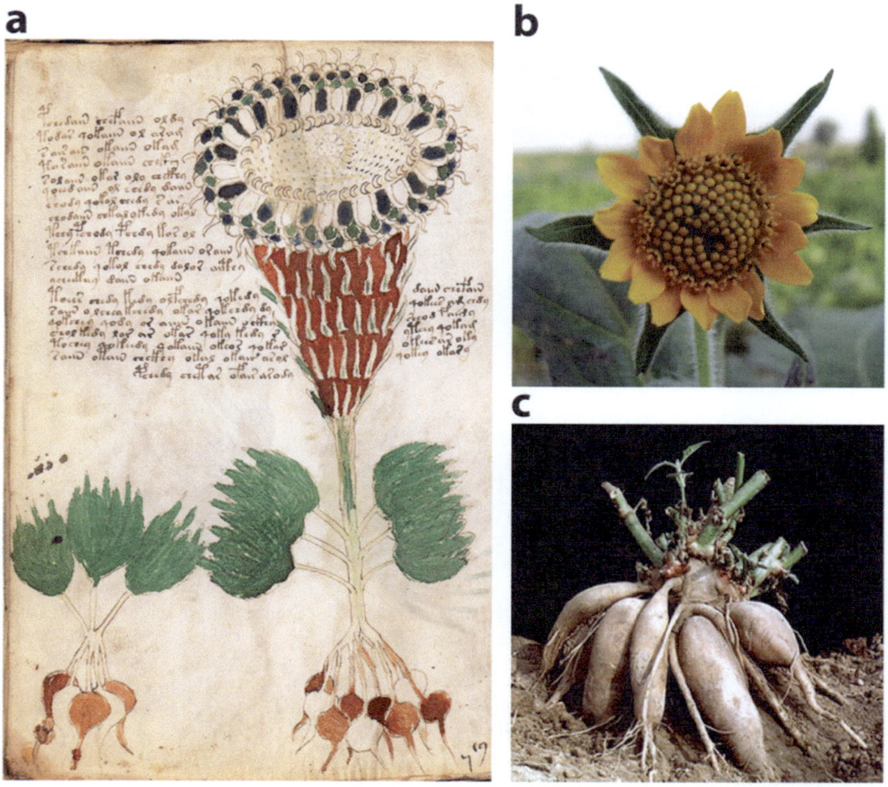

Fig. 4.57 *Smallanthus* sp.: (**a**) fol. 40v; (**b**) inflorescence and (**c**) tuberous roots of *S. sonchifollius* (yacón). (Source: Rob Hille and NusHub, cropsforthefuture.org CC BY-SA 3.0, respectively)

Fol. 39r. *Stenotus acaulis*? (Fig. 4.58)

Description This phytomorph (Fig. 4.58a) is a stoloniferous perennial Asteraceae of subfamily Asterideae tribe Astereae. The leaves are elliptic on an alpine, stoloniferous root. This may match the variability of *Stenotus acaulis* (Nutt.) Nutt. (Fig. 4.58b).

Distribution Native from Saskatchewan to California.

Names Stemless goldenrod (English).

Uses The medicinal value of this genus is discussed by Moerman (1998).

Fig. 4.58 *Stenotus acaulis*: (**a**) fol. 39r; (**b**) *Stenotus acaulis*. (Source: http://science.halleyhosting.com/nature/basin/sun/daisy/stenotus/acaulis/acaulis3.jpg)

Fol. 18r. *Symphyotrichum laeve* var. *geyeri* (Fig. 4.59)

Description This phytomorph (Fig. 4.59a) has 12 separated, ligulate, bluish florets on a branched Asteraceae inflorescence of subfamily Asterideae tribe Astereae. Disk florets are reddish and raised. Involucre is prominent. Green leaves are cuneate, entire, and clasping. Roots are obvious but minimal. This matches the variability of *Symphyotrichum laeve* (L.) Á. Löve & D. Löve var. *geyeri* (A. Gray) G.L. Nesom (Fig. 4.59b, c). Disk florets are initially yellow but change to orange upon drying.

Distribution Native from western Canada to Mexico.

Names Smooth aster (English).

Uses The value of this species is discussed by Moerman (1998) as a drug, food, and dye.

Fig. 4.59 *Symphyotrichum laeve* var. *geyeri*: (**a**) fol 18r; (**b**) flowers; Source: http://botany.cz/foto2/symphyolaeveherb1.jpg); (**c**) leaf. (Source: http://snowbirdpix.com/montana_plant_page_test.php?id=1609)

Fol. 43r. *Townsendia* sp.? (Fig. 4.60)

Description This phytomorph (Fig. 4.60a) bears an Asteraceae inflorescence of subfamily Asterideae tribe Astereae. The disk florets are white; the ray florets are yellow. The habit is that of an alpine perennial plant, such as a *Townsendia* sp. A herbarium sheet of *T. bakeri* Greene is shown for comparison (Fig. 4.60b).

Distribution North America.

Names Genus known as Townsend daisies (English).

Uses The medicinal value of this genus is discussed by Moerman (1998).

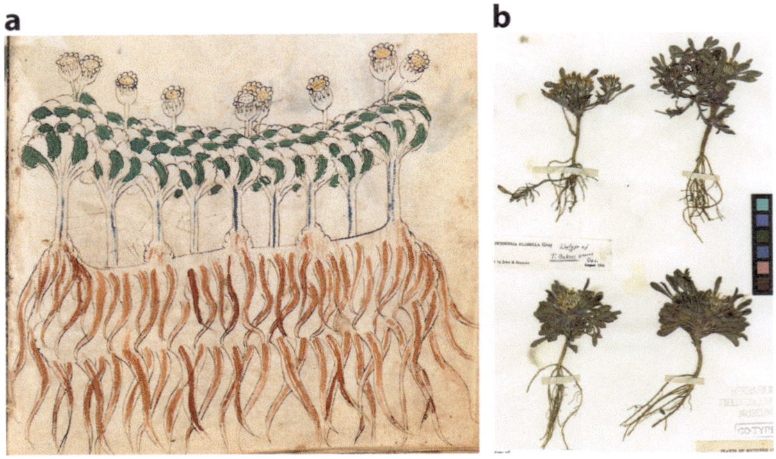

Fig. 4.60 *Townsendia* sp.?: (**a**) fol. 43r; (**b**) *T. bakeri* herbarium sheet. (Source: Field Museum of Natural History CC BY-NC 3.0)

Fol. 20v. *Tragopogon porrifolius* (Fig. 4.61)

Description This phytomorph (Fig. 4.61a) has a whitish involucre with sunken, bluish phyllaries, putting this Asteraceae inflorescence in subfamily Cichorioideae. Leaves are alternate, linear-lanceolate, tapering to a point, with a short petiole on a stout stem. This matches the variability of *Tragopogon porrifolius* L. (Fig. 4.61b). The tap root is banded (Fig. 4.61c). The leaves actually arise from a sheath by a short petiole. The stout stem, milky sap, and leaf attachment of *T. dubius* Scop. is shown in Fig. 4.61d. This was introduced early enough in North America to produce *T.* × *mirus* Ownbey as an amphidiploid hybrid of *T. porrifolius* × *T. dubius*. *Tragopogon* spp., like cornflowers (*Centaurea cyanus* L.), often occur as contaminants in plantings of wheat and rye, both of which were planted in early New Spain (Dunmire 2004).

Distribution Native to Eurasia but introduced from Canada to Mexico. We have no record of its introduction in New Spain.

Names Vipers-grass, goats-beard, salsify, oysterplant (English); *salsifí, barba de cabra, barbón* (Spanish).

Uses McMahon (1806) lists this as a common garden vegetable in Pennsylvania, planted in 1774 by Thomas Jefferson in Virginia (Betts 1985). Under the names "vipers-grass" or "goats-beard," early settlers to New England cultivated *T. porrifolius* and *T. dubius* (Leighton 1970). It has been a medicinal herb since the time of Dioscorides (ca. 40–90 CE) (Janick et al. 2013). Moerman (1998) discusses the medicinal use of this species by the Navajo.

Fig. 4.61 *Tragopogon porrifolius*: (**a**) fol. 20v; (**b**) flowers. (Source: Alice Stöcklin CC BY-SA 4.0); (**c**) taproot. (Source: Women Fitness [www.womenfitness.net]); (**d**) *T. dubius* stem. (Source: http://wimastergardener.org/article/yellow-goatsbeard-tragopogon-spp)

Fol. 17r. *Vernonia marginata* (Fig. 4.62)

Description This phytomorph (Fig. 4.62a) has an Asteraceae inflorescence of subfamily Vernonieae with dark bluish corollas, minimal involucre, and petiolated linear to hastate leaves. Roots are without tubers. A good match is *Vernonia marginata* (Torr.) Raf. (Fig. 4.62b) (Redonda-Martinez 2017).

Distribution Native from Utah to Mexico.

Names Plains ironweed (English).

Uses The medicinal value of this genus is discussed by Moerman (1998).

Fig. 4.62 *Vernonia marginata*: (**a**) fol. 17r; (**b**) *V. marginata*. (Source: Courtesy of Wynn Anderson, University of Texas at El Paso)

Fol. 87r. *Xylorhiza* sp.? cf. *X. wrightii* (Fig. 4.63)

Description This phytomorph (Fig. 4.63a) of Asteraceae subfamily Asterideae tribe Astereae, with an inflorescence painted with blue pigment, long involucre, spatulate leaves, stout stem, and ample roots, resembles an alpine or desert aster that is a subshrub. This resembles a *Xylorhiza* sp., possibly *X. wrightii* (A. Gray) Greene (Watson 1977) (Fig. 4.63b)

Distribution Native from Texas to Mexico.

Names Big bend woody aster (English).

Uses Unknown.

Fig. 4.63 *Xylorhiza wrightii*?: (**a**) fol. 87r; (**b**) *X. wrightii*. (Source: W.D. and Dolphia Bransford, Lady Bird Johnson Wildflower Center)

Boraginaceae/Hydrophyllaceae + Namaceae

Fol. 47v. *Cynoglossum grande* (Fig. 4.64)

Description This phytomorph (Fig. 4.64a) has terminal blue flowers of six to seven petals, with a raised white center, prominent cauline leaves, broadly elliptic basal leaves, and broad, branched brown roots. This matches the variability of *Cynoglossum grande* Douglas ex Lehm, except that this species has only five petals and the cauline leaves are smaller and closer to the base (Fig. 4.64b).

Distribution Native from British Columbia to California. Found in shady areas in woodland and chaparral.

Names Pacific hound's tongue (English)

Uses Grated root used as a dressing on burns and scalds; root used for stomach aches and venereal disease among Native Americans (Moerman 2009).

Fig. 4.64 *Cynoglossum grande*: (**a**) fol. 47v; (**b**) *C. grande*. (Source: Eugene Zelenko, CC BY-SA 2.0)

Fol. 56r. *Phacelia campanularia* (Fig. 4.65)

Description With blue flowers in a scorpioid cyme, dentate leaves, and overlapping leaf-like basal scales (Fig. 4.65a), the phytomorph is a good match for *Phacelia campanularia* A. Gray, California bluebell (Fig. 4.65b, c), recently placed in the Hydrophyllaceae.

Distribution A California native.

Names Desertbells, desert bluebells, desert scorpionweed, desert Canterbury bells (English).

Uses Ornamental, poisonous to mammals.

Fig. 4.65 *Phacelia campanularia*: (**a**) fol. 56r; (**b**, **c**) *P. campanularia* inflorescence. (Source: courtesy of Chez Brungraber and GeorgeWilliams, respectively)

Fol. 39v. *Phacelia crenulata* (Fig. 4.66)

Description This phytomorph (Fig. 4.66a) has bluish flowers in a cyme with deeply pinnately lobed green leaves on broad, brown, branched roots. This closely matches *Phacelia crenulata* Torr. (Fig. 4.66b), recently placed in the Hydrophyllaceae.

Distribution Native from Colorado to Mexico.

Names Notch-leaf scorpionweed, notch-leaved phacelia, cleft-leaf wild heliotrope (English).

Uses Moerman (2009) records that it is used by the Hopi for injury in animals, especially horses.

Fig. 4.66 *Phacelia crenulata*: (**a**) fol. 39v; (**b**) *P. crenulata*. (Courtesy of Tom Chester)

Fol. 51v. *Phacelia integrifolia* (Fig. 4.67)

Description The phytomorph (Fig. 4.67a) has blue flowers on a scorpioid cyme and narrowly elliptic, crenate leaves that curl. The roots are brown and branched. This matches fairly well the variability of *Phacelia integrifolia* Torr. (Fig. 4.67b, c), recently placed in the Hydrophyllaceae.

Distribution Native from Utah and Kansas south to Mexico.

Names Gypsum phacelia (English).

Uses Unknown.

Fig. 4.67 *Phacelia integrifolia*: (**a**) fol. 51v; (**b**) inflorescence and (**c**) shoot of *P. integrifolia*. (Source: Nieland and Finley 2009 and Western New Mexico University Department of Natural Sciences and the Dale A. Zimmerman Herbarium, respectively)

Fol. 26r. *Wigandia urens* (Fig. 4.68)

Description This phytomorph (Fig. 4.68a) bears what can be interpreted as bluish flowers on a scorpiod cyme with leaves that are green, crenate, and obtuse. This matches *Wigandia urens* (Ruiz & Pav.) Kunth very well (Fig. 4.68b), recently placed in the Namaceae.

Distribution Shrub found from Mexico and south to Peru.

Names Wide/broad nettle (English); in Nahuatl, called *chichicaztle* (Díaz 1976); also might match *patlāhua-ctzītzicāztli* on fol. 47r in the *Codex Cruz-Badianus* (Emmart 1940; Díaz 1976; Cruz and Badiano 1991; Gates 2000; Clayton et al. 2009; de Ávila Blomberg 2012).

Uses Unknown.

Fig. 4.68 *Wigandia urens*: (**a**) fol. 26r; (**b**) *W. urens.* (Source: Dick Culbert, CC-BY 2.0)

Brassicaceae

Fol. 90v. *Caulanthus heterophyllus* (Fig. 4.69)

Description This phytomorph (Fig. 4.69a) might be *Caulanthus heterophyllus* (Nutt.) Payson, an annual species. The flowers of *C. heterophyllus* are four-petaled, white with a purple streak down the center, with four protruding dark purple anthers (Fig. 4.69b). Leaves vary from dentate to lobed but are typically clasping, not petiolate (Fig. 4.69c).

Distribution Native to California and Baja California (Al-Shehbaz 2012).

Names San Diego wild cabbage, San Diego jewelflower (English).

Uses Unknown.

Fig. 4.69 *Caulanthus heterophyllus*: (**a**) fol. 90v; (**b**) inflorescence and (**c**) shoots of *C. heterophyllus*. (Source: National Park Service and Anthony J. Valois, respectively)

Bromeliaceae

Fol. 7r. *Guzmania lingulata*? (Fig. 4.70)

Description This phytomorph (Fig. 4.70a) seems to be entirely vegetative, with white to green lanceolate leaves in a rosette, acute at the tips and very slightly and remotely serrate, on a white stem. Roots are bifurcated. Actually, this bears resemblance to the colored, cupped bracts of the very variable *Guzmania lingulata* (L.) Mez. (Fig. 4.70b). The flowering bracts vary from red to orange and yellow to white.

Distribution Native to rain forests from southern Mexico to South America and the West Indies.

Names Scarlet star (English).

Uses The medicinal value of this species is limited but noted (Belotto et al. 2014).

Fig. 4.70 *Guzmania lingulata*?: (**a**) fol. 7r; (**b**) *G. lingulata*. (Source: http://sinaiplants.com/index.php/2013-12-14-05-10-26/bromeliad)

Cabombaceae

Fol. 49v. *Brasenia schreberi* (Fig. 4.71)

Description This phytomorph (Fig. 4.71a) bears one six-petalled flower, painted bluish, which sprigs from a ring of orbicular peltate leaves with entire margins. Remotely, they arise from a rhizome or corm. This matches very well the water shield, *Brasenia schreberi* J.F. Gmel. (Fig. 4.71b).

Distribution Native to not only North and South America but also Eurasia and Africa.

Names Water shield (English).

Uses The polysaccharide gel covering of the leaves has been documented to lower cholesterol (Kim et al. 2015).

Fig. 4.71 *Brasenia schreberi*: (**a**) fol.49v; (**b**) *B. schreberi*. (Source: https://alchetron.com/Brasenia-1963638-W)

Cactaceae

Fol. 99v#10. *Cylindropuntia* sp.? (Fig. 4.72)

Description This herbal root is labelled 8aɯ̀c, *chollă*, Spanish for skull or cholla cactus, which may refer to a *Cylindropuntia* sp. (Fig. 4.72a). Some cacti roots are tuberous when fresh, such as the roots of *Opuntia chaffeyi* Britton & Rose (Fig. 4.72b), a related species.

Distribution Cactaceae are native to the New World, except for a few outliers in Madagascar.

Names Cholla (Spanish and English).

Uses *Cylindropuntia imbricata* (Haw.) F. M. Knuth [*Opuntia imbricata* (Haw.) DC.], a cholla cactus, is used in Mexico as a "Tea taken to treat asthma. Tea used against diarrhea (plant contains tannins)." (González Stuart 2004).

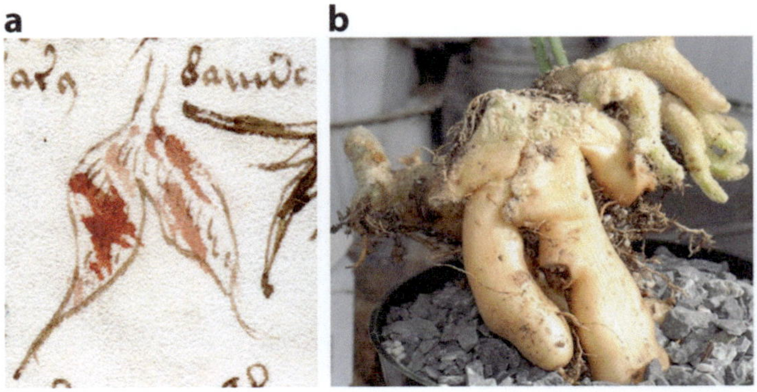

Fig. 4.72 *Cylindropuntia* sp.: (**a**) Fol. 99v#10; (**b**) *Opuntia chaffey* roots. (Source: Roberts and Hill 2009)

Fol. 100v(3)#7, 101v(1)#9. *Hylocereus undatus* (Fig. 4.73)

Description These phytomorphs (Figs. 4.73a, b), which resemble an epiphytic cactus, are most probably *Hylocereus undatus* (Haw.) Britton and Rose, the night-blooming cereus dragon fruit, or pitaya. Though today the segments are long and bear pink fruit that is commercially marketed (Fig. 4.73c), in areas of deprived moisture and nutrition (*i.e.,* when grown as an epiphyte) the segments are much shorter, with prominent glochids (Fig. 4.73d).

Distribution Native to the New World tropics.

Names Dragonfruit (English). Estrada Lugo (1989) identifies the *teonochtli/teō-nōchtli* of fol. 17v in the *Codex Cruz-Badianus* as this species. The name accompanying the second phytomorph, **o8o?**, transliterates to *āchāe*, of unknown translation.

Uses Useful in treating Type II diabetes (Song et al. 2016).

Fig. 4.73 *Hylocereus undatus*: (**a**) fol. 100v(3)#7; (**b**) fol. 101v(1)#9; (**c**) *H. undatus*. (Source: Bùi Thụy Đào Nguyên CC BY-SA 3.0); (**d**) *H. undatus* grown as an epiphyte. (Source: Rhett Butler, mongabay.com)

Fol. 100r#8. *Opuntia* sp., cf. *O. ficus-indica* (Fig. 4.74)

Description This phytomorph (Fig. 4.74a) has the shape of a prickly pear cactus pad or fruit, with areoles bearing leaf primordia and tiny fruits on the top edge, *i.e.*, *Opuntia* sp., quite possibly *Opuntia ficus-indica* (L.) Mill., *O. megacantha* Salm-Dyck, or *O. streptacantha* Lem. (Dressler 1953). Today, *Opuntia ficus-indica* is widely cultivated, but it is apparently native to central Mexico.

Distribution Native to Central Mexico and the Caribbean region, but presently widely cultivated.

Names Prickly pear, cactus pear (English); called by the Nahuatl name *anochtli* and *tlatoc nochtli/tla-tōc-nōchtli* in the *Codex Cruz-Badianus* (Emmart 1940; Dressler 1953; Cruz and Badiano 1991; Gates 2000; Zepeda and White 2008; Clayton et al. 2009; de Ávila Blomberg 2012). The name accompanying this phytomorph, , *nāshtli*, probably refers to the period pronunciation of the primary name *nochtli*.

Uses Edible fruit and leaves (cladodes). *Nopalea cochenillifera* (L.) Salm-Dyck [*O. cochenillifera* (L.) Mill.] is also widely cultivated for the insect that is the source for cochineal (Standley 1920–1926:863).

Fig. 4.74 *Opuntia* sp.: (**a**) fol. 100r#8; (**b**) cladode (pad) of *O. ficus-indica* with fruits. (Source: Public domain)

Fol. 100v(3)#6. *Pseudorhipsalis ramulosa* (Fig. 4.75)

Description This phytomorph (Fig. 4.75a), which resembles an epiphytic cactus, is awash with blue pigment. A good fit is *Pseudorhipsalis ramulosa* (Salm-Dyck) Bertholott), which is prominently pink in sunlight (Fig. 4.75b).

Distribution Native from Mexico to Peru.

Names Red mistletoe cactus, red rhipsalis (English).

Uses Unknown.

Fig. 4.75 *Pseudorhipsalis ramulosa*: (**a**) fol. 100v(3)#6; (**b**) *P. ramulosa*. (Source: TopTropicals.com)

Fol. 28v. *Selenicereus grandiflorus*? (Fig. 4.76)

Description Figure 4.76a is perhaps the most alien-looking phytomorph in the *Voynich Codex*. This seems to show a large white flower, perhaps unfinished, attached to a tree or shrub stem. A comparable plant from reality would be a large white blossom of an epiphytic cactus, such as *Selenicereus grandlflorus* (L.) Britton & Rose (Fig. 4.76b).

Distribution Native to the Caribbean, Mexico, and Nicaragua.

Names Night-blooming cereus (English), *organillo* or *reina de la noche* (Spanish).

Uses Martinez (1969) discusses this species as valuable in rheumatism and diseases of the heart.

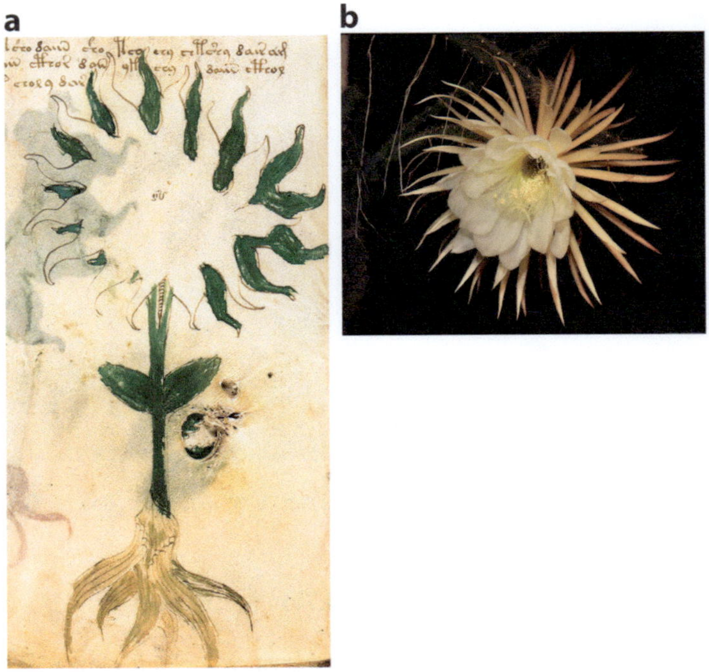

Fig. 4.76 *Selenicereus grandiflorus*?: (**a**) fol. 28v; (**b**) *S. grandiflorus*. (Source: dé.wé. CC BY-SA 2.0)

Campanulaceae

Fol. 56v. *Centropogon* sp.? (Fig. 4.77)

Description The shape of the terminal flower in this phytomorph (Fig. 4.77a) indicates pollination by birds (ornithophily), most likely the Campanulaceae. Again, it has probably been re-assembled from dried, fragmented specimens, so the sessile, deeply serrate leaves and tuberous roots may not be accurate. Otherwise, this bears resemblance to a *Centropogon* sp. (Fig. 4.77b).

Distribution Genus is native to Central and South America.

Names Unknown.

Uses Unknown.

Fig. 4.77 *Centropogon* sp.?: (**a**) fol. 56v; (**b**) *Centropogon* sp. (Source: Herbaria of Northwestern Mexico CC BY-SA)

Caryophyllaceae

Fol. 24r. *Silene* sp. cf. *S. menziesii* Infected with *Microbotryum violaceum* (Fig. 4.78)

Description This phytomorph (Fig. 4.78a) is probably a *Silene* sp., but the crudeness of the image prevents accurate designation of a species. This phytomorph might possibly be based, in part, on the variable species *Silene menziesii* Hook. (Fig. 4.78b). The white flowers are a good match, even showing the typical infection with the fungus *Microbotryum violaceum* (Pers.) G. Deml & Oberw., anther smut fungus, which turns the anthers purple. However, the leaves are shown as hastate, and *C. menziesii* has attenuate leaf bases. Is this another case of disparity of the leaves between reality and portrayal, or is there another species of *Silene* that is a better match to the illustration?

Distribution Native from Alaska to California and New Mexico (Morton 2005).

Names Menzies' campion or Menzies' catchfly (English).

Uses Unknown.

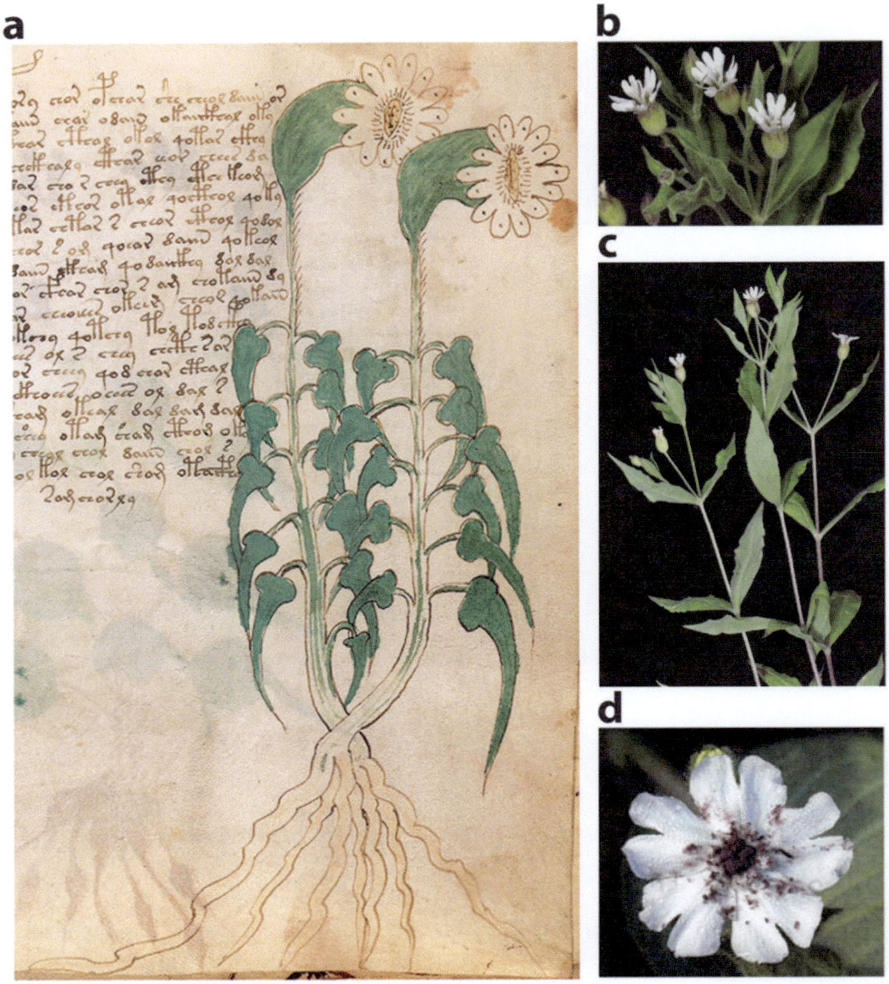

Fig. 4.78 *Silene* sp.: (**a**) fol. 24r; (**b**, **c**) stems and inflorescence of *S. menziesii*. (Source: Robert L. Carr); (**d**) flower infected with *Microbotryum violaceum*. (Source: Michael Hood)

Convolvulaceae

Fol. 1v + Fol. 101v(2)#4. *Ipomoea arborescens* (Fig. 4.79)

Description The phytomorph (Fig. 4.79a) shows a single leafy shoot with a single terminal flower bud arising from a thick caudex. The flower bud has whitish sepals and brownish petals; leaves are alternate, cordate, petiolate, green on the adaxial surface, and tan on the abaxial surface; coarse roots emanate from the basal caudex. This phytomorph is repeated on fol. 101v(2)#4 (Fig. 4.79b). This illustration is overwhelmingly similar in style and substance to *xiuhamolli*/*xiuhhamolli* (soap plant) found in fol. 9r (Plate 11) in the 1552 Mexican *Codex Cruz-Badianus*, which has been identified as *Ipomoea murucoides* Roem. & Schult. (Emmart 1940; Cruz and Badiano 1991; Gates 2000; Clayton et al. 2009). Reko (1947) and Bye and Linares (2013) also identify this as an *Ipomoea* sp. The flower bud of *I. murucoides* (Fig. 4.79c) is similar. Phytomorphs in the *Codex Cruz-Badianus* and in the *Voynich Codex* have a large, broad, gray to whitish basal caudex with ridged bark. However, the leaves in the *Voynich Codex* are cordate rather than attenuate as observed in the *Cruz-Badianus Codex*. The phytomorph in the *Voynich Codex* must then be *Ipomoea arborescens* (Humb. & Bonpl. ex Willd.) G. Don (Fig. 4.79d), found from northern to southern Mexico (Standley 1920–1926:1205) and commonly known in Nahuatl as *quauhtzahuat* (Ocaranza 2011). Additional botanical characters of both species are discussed by Austin et al. (2005), McPherson (1981) and Standley (1920–1926). Curiously, McPherson described the bases of the leaves of *I. murucoides* as truncate, whereas Standley described the bases as rounded or obtuse, and they would be better described as cuneate on all herbarium sheets of this species that we have seen. Leaves of *I. murucoides* are described by McPherson as variously pubescent, while the leaves of *I. arborescens* are usually tomentose, especially on the lower surface, rendering the abaxial surface grey-green and the adaxial surface green. The tomentose abaxial surface often turns brownish green on drying, which is similar to that of the phytomorph in the *Voynich Codex*.

Distribution Native to Mexico, originating in in the Sonoran Desert.

Names Morning glory tree (English); additional Spanish names for *I. arborescens* include "'*Palo blanco*' (Sonora, Sinaloa); '*palo del muerto*,' '*casahuate*,' 'quauhzahuatl,' [Nahuatl] 'casahuate blanco' (Morelos); '*palo santo*' (Sonora); '*palo bobo*' (Morelos, El Salvador); '*tutumushte*,' '*siete pellejos*,' '*siete camisas*' (El Salvador)." (Standley 1920–1926:1205).

Uses The seeds of *Ipomoea tricolor* Cav. (*I. violacea* auct.) (Witters 1975) and the related species *Argyreia nervosa* (Burm.f.) Bojer, a tropical wood rose, contain d-lysergic acid amide (LSA), similar to LSD (Borsutsky et al. 2002; Miller 1970). Species of *Ipomoea* are known for their resin glycosides and are used to counter several diseases (Pereda-Miranda et al. 2010; Batres et al. 2012; Meira et al. 2012). The ashes of the arborescent *Ipomoea* species, *I. murucoides* and *I. arborescens*, are used to prepare soap and are also used in hair and skin care (Batres et al. 2012; Standley 1920–1926:1205). Medicinal uses in Mexico are recorded by Martinez (1969).

Fig. 4.79 *Ipomoea arborescens*: (**a**) fol. 1v; (**b**) *I. murucoides* from *Codex Cruz-Badianus* plate 11; (**c**) bud of *I. murucoides*. (Source: Kevin C. Nixon); (**d**); flower, fruits, and leaves of *I. arborescens*. (Source: Tony Rodd, CC BY-NC-SA 2.0); (**e**) fol. 101v(2)#4

Fol. 57r. *Ipomoea nil* (Fig. 4.80)

Description The phytomorph (Fig. 4.80a) has a terminal dark blue flower with a white edge, acute petals, and elongated calyx lobes. Leaves are lobed peltate on an herbaceous bush. Roots are brown and branched. This might match the variability of *Ipomoea nil* (L.) Roth (Fig. 4.80b, c). The species is extremely variable, from vine to herbaceous bush, with floral colors from blue to pink to white, but often with a distinctive white edge. Leaves are often hastate but vary to palmately lobed.

Distribution Native from Mexico to Argentina.

Names Picotee morning glory, ivy morning glory, Japanese morning glory (English). A Nahuatl name is *acaquitl mecatl* (Díaz 1976).

Uses The seeds of *Ipomoea tricolor* Cav. (*I. violacea* auct.) (Witters 1975) and the related species *Argyreia nervosa* (Burm.f.) Bojer, a tropical wood rose, contain d-lysergic acid amide (LSA), similar to LSD (Borsutsky et al. 2002; Miller 1970). Species of *Ipomoea* are known for their resin glycosides and are used to counter several diseases (Pereda-Miranda et al. 2010; Batres et al. 2012; Meira et al. 2012).

Fig. 4.80 *Ipomoea nil*: (**a**) fol. 57r; (**b**, **c**) flowers and leaves of *I. nil*. (Source: Rare and Exotic Seeds and Image by Ghost32 at ghost32writer.com, respectively)

Fol. 89v#12. *Ipomoea plummerae* var. *cuneifolia* (Fig. 4.81)

Description This phytomorph (Fig. 4.81a) appears to be a desert Convolvulaceae, and *Ipomoea plummerae* A. Gray var. *cuneifolia* (A. Gray) J. F. Mabr. matches rather well (Fig. 4.81b). A photo of the bulb of this taxon could not be located, but Fig. 4.81c shows *I. plummerae* var. *plummerae*.

Distribution Native to the Gila Wilderness in Arizona and New Mexico.

Names Huachaua morning glory (English). The word accompanying this phytomorph, 8a22co89, is transliterated as *chonts/tzāchi*. This may be derived from the Nahuatl *chonetl*, demon, + *zachi*, to give a little bit more (Herrera 2004; Karttunen 1983; Siméon 2010).

Uses The seeds of *Ipomoea tricolor* Cav. (*I. violacea* auct.) (Witters 1975) and the related species *Argyreia nervosa* (Burm.f.) Bojer, a tropical wood rose, contain d-lysergic acid amide (LSA), similar to LSD (Borsutsky et al. 2002; Miller 1970). Species of *Ipomoea* are known for their resin glycosides and are used to counter several diseases (Pereda-Miranda et al. 2010; Batres et al. 2012; Meira et al. 2012).

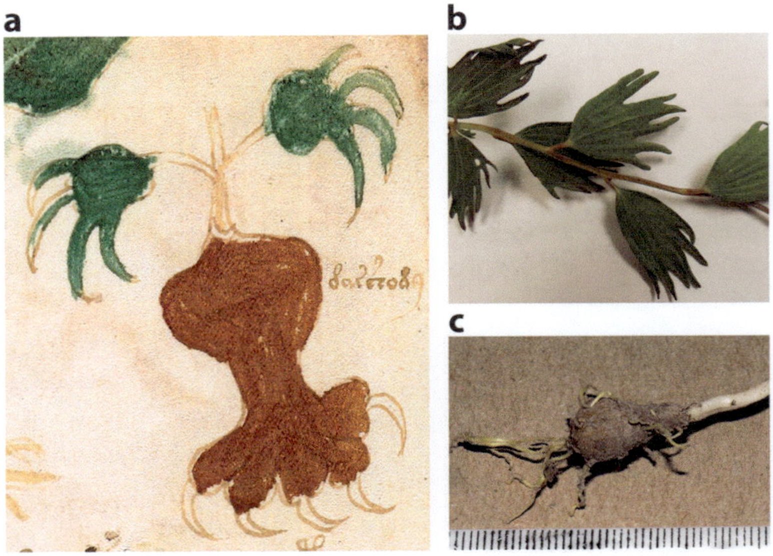

Fig. 4.81 *Ipomoea plummerae* var. *cuneifolia*: (**a**) fol. 89v#12; (**b**) *I. plummerae* A. Gray var. *cuneifolia*. (Source: Western New Mexico University Department of Natural Sciences and the Dale A. Zimmerman Herbarium); (**c**) *I. plummerae* var. *plummerae* bulb. (Source: Russ Kleinman)

Fol. 32v, fol. 101v(3)#2. *Ipomoea pubescens* (Fig. 4.82)

Description Figure 4.82a, b show two phytomorphs. The blue flowers, deeply lobed leaves, and tuberous roots are all good fits for (most probably) *Ipomoea pubescens* Lam. (Fig. 4.82c).

Distribution Native from Arizona and New Mexico to Mexico, and also in Bolivia, Peru, and Argentina.

Names Silky morning glory (English).

Uses The seeds of *Ipomoea tricolor* Cav. (*I. violacea* auct.) (Witters 1975) and the related species *Argyreia nervosa* (Burm.f.) Bojer, a tropical wood rose, contain d-lysergic acid amide (LSA), similar to LSD (Borsutsky et al. 2002; Miller 1970). Species of *Ipomoea* are known for their resin glycosides and are used to counter several diseases (Pereda-Miranda et al. 2010; Batres et al. 2012; Meira et al. 2012).

Fig. 4.82 *Ipomoea pubescens*: (**a**) fol 32v; (**b**) fol. 101v(3)#2; (**c**) flower, leaf, and rhizome of *I. pubescens*. (Source: Apostolou Starvos)

Crassulaceae

Fol. 25v. *Dudleya ingens* (Fig. 4.83)

Description This phytomorph (Fig. 4.83a) is a rosette of lanceolate green leaves. Roots are ample but not tuberous. This is most probably a *Dudleya* sp., probably *D. ingens* Johans. (Fig. 4.83b). Although not recognized as a myrmecophyte (a plant associated with ants) (Díaz-Castelazo et al. 2004; Rico-Gray 1993), under cultivation, some *Dudleya* species are subject to severe ant infestations, which carry mealy bugs. This would explain the carnivorous Mexican horned lizard, *Phrynosoma taurus* Dugès, 1873, nibbling at the lower corner of the phytomorph.

Distribution Endemic to Baja California.

Names Baja liveforever (English).

Uses Unknown.

Fig. 4.83 *Dudleya ingens*: (**a**) fol. 25v; (**b**) *D. ingens*. (Source: https://davesgarden.com/guides/pf/go/94671)

Fol. 47r. *Echeveria* sp. (Fig. 4.84)

Description This phytomorph (Fig. 4.84a) is a rosette of elliptic, finely crenate leaves, along with an identical offset. Roots are ample but not tuberous. This bears an uncanny resemblance to an *Echeveria* sp. No species of this genus of succulents are truly crenate, but some are ruffled on the margin, such as the very variable *E. gibbiflora* DC. (Fig. 4.84b).

Distribution *E. gibbiflora* is native from Mexico to Guatemala.

Names *E. gibbiflora* is callecd *oreja de burro* (Spanish).

Uses Unknown.

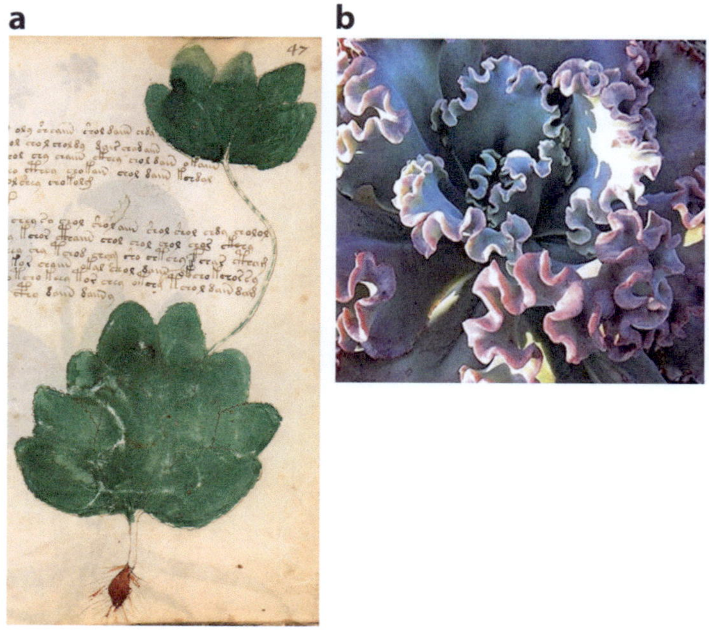

Fig. 4.84 *Echeveria* sp.: (**a**) fol. 47r; (**b**) *E. gibbiflora*. (Source: http://www.myseeds.co/echeveria-gibbiflora-ruffled-leaves-succulent-seeds)

Dioscoreaceae

Fol. 17v. *Dioscorea composita* (Fig. 4.85)

Description Most probably this phytomorph (Fig. 4.85a) is *Dioscorea composita* Hemsl. (Fig. 4.85b). The roots (Fig. 4.85c) are quite often segmented, as shown in the *Voynich Codex*. The flowers are yellow when fresh but rust-colored upon drying, and they are borne on a vine, fitting rather well, but the phytomorph is shown with leaves more hastate than *D. composita* normally exhibits.

Distribution Native from southern Mexico to Costa Rica.

Names *Barbasco* (Spanish).

Uses Medicinal uses in Mexico are recorded by Martinez (1969). This is a source of cortisone precursors (Correll et al. 1955).

Fig. 4.85 *Dioscorea composita*: (**a**) fol. 17v; (**b**) leaves and inflorescences of *D. composita*. (Source: Abisaí García Mendoza, Institute of Biology, National Autonomous University of Mexico); (**c**) segmented roots of *D. composita*. (Source: Ryan Somma, CC BY-SA 2.0)

Fol. 96v. *Dioscorea mexicana* (Fig. 4.86)

Description The rust-colored flowers and sagittate leaves of the phytomorph (Fig. 4.86a) fit rather well for *Dioscorea mexicana* Schweidw. (Fig. 4.86b, c).

Distribution This vine is native from northern to southern Mexico to Panama.

Names *Cabeza de negro* (Spanish), Mexican yam (English).

Uses Medicinal uses in Mexico are recorded by Martinez (1969). This is a source of cortisone precursors (Correll et al. 1955).

Fig. 4.86 *Dioscorea mexicana*: (**a**) fol. 96v; (**b**) inflorescence and (**c**) rhizome of *D. mexicana*. (Source: Michael Charters and The Smithsonian, respectively)

Fol. 99r#28. *Dioscorea* sp., cf. *D. remotiflora* (Fig. 4.87)

Description This phytomorph (Fig. 4.87a) is most probably *Dioscorea remotiflora* Kunth (Fig. 4.87b). The large, dark root resembles a paddle or bat (Fig. 4.87c, d). The rust-colored flowers and cordate leaves on a vine also match.

Distribution Native from northern to southern Mexico.

Names Mountain yam (English). The name accompanying this phytomorph, *tlācânoni*, is Nahuatl for bat or paddle, which matches the long, tuberous root.

Uses Unknown.

Fig. 4.87 *Dioscorea remotiflora*: (**a**) fol. 99r#28; (**b**) herbarium sheet of *D. remotiflora* (F1405679). (Source: The Field Museum); (**c**, **d**) bat- or paddle-like roots of *D. remotiflora*. (Source: Ignacio Garcia Ruiz, PhytoImages.siu.edu)

Ericaceae

Fol. 94v(2). *Macleania insignis*? (Fig. 4.88)

Description This phytomorph (Fig. 4.88a) bears an overwhelming similarity to the Ericaceae, with swollen corolla and opposite leaves on a shrub with ample roots. The corolla is painted with blue and red paints. This might be *Macleania insignis* M. Martens & Galeotti (Fig. 4.88b), which can form caudiciform lignotubers with edible fruit on arching stems.

Distribution Native to cloud forests of Mexico, Costa Rica, and Colombia.

Names Unknown.

Uses Unknown.

Fig. 4.88 *Macleania insignis*: (**a**) fol. 94v(2); (**b**) *M. insignis*. (Source: https://farm1.staticflickr.com/390/19231071609_30b076d1d1_b.jpg)

Fol. 94v(3). *Vaccinium caespitosum*? (Fig. 4.89)

Description This phytomorph (Fig. 4.89a) is most probably an alpine, stoloniferous shrub with elliptic, acute leaves. The campanulate flowers, painted with blue pigment, resemble the Ericaceae, and a guess might be *Vaccinium caespitosum* Michx. (Fig. 4.89b).

Distribution Native from California to Mexico.

Names Sierra bilberry (English).

Uses Moerman (1998) discusses the culinary value of this species for the Alaska Native, Gosiute, and Hanaksiaia.

Fig. 4.89 *Vaccinium caespitosum*?: (**a**) fol. 94v(3); (**b**) *V. caespitosum*. (Source: Jane S. Richardson CC BY 3.0)

Euphorbiaceae

Fol. 6v. *Cnidoscolus texanus* (Fig. 4.90)

Description This phytomorph (Fig. 4.90a) has palmately compound leaves and trichomes on the fruit and matches a *Cnidoscolus* sp. Both *C. chayamansa* McVaugh and *C. aconitifolius* (Mill.) I.M. Johnst. are called *chaya* and widely cultivated from Mexico to Nicaragua; the leaves are matches (Ross-Ibarra and Molina-Cruz 2002). However, these cultivated species have relatively smooth fruits, and a closer correspondence would be *C. texanus* (Müll. Arg.) Small (Fig. 4.90b, c), which has fruits that are coated with trichomes.

Distribution Native to the United States (Texas, Arkansas, Kansas, Lousiana, Oklahoma) and northern Mexico.

Names Texas bullnettle, treadsoftly (English), *mala mulher* (Spanish).

Uses Causes intense pain upon contact (Pollard 1986).

Fig. 4.90 *Cnidoscolus* sp.: (**a**) fol. 6v; (**b**) fruits of *C. texanus* with trichomes. (Source: Carl Fabre, Lady Bird Johnson Wildflower Center); (**c**) leaves and flower of *C. texanus*. (Source: Tiana Rehman, Botanical Research Institute of Texas)

Fol. 20r. *Euphorbia chamaesula*? (Fig. 4.91)

Description This phytomorph (Fig. 4.91a) is similar to both fols. 4r and 20r; while it is overwhelmingly similar to a *Euphorbia* sp. (*Chamaesyce* sp.), the inflorescence is a simple, non-foliar raceme, unlike the other species in this genus. The simplest explanation, agreeing with other phytomorphs, is that it was incorrectly re-assembled from fragmented dried specimens. If the arrangement of the inflorescence is ignored, this might be *E. chamaesula* Boiss. (Fig. 4.91b).

Distribution Native from Arizona and New Mexico to Mexico.

Names Mountain spurge (English).

Uses The medicinal value of this genus is discussed by Moerman (1998).

Fig. 4.91 *Euphorbia chamaesula*?: (**a**) fol. 20r; (**b**) *E. chamaesula*. (Source: SEINet Arizona-New Mexico Chapter CC BY-SA)

Fol. 4r. *Euphorbia glyptosperma*? (Fig. 4.92)

Description This phytomorph (Fig. 4.92a) has small, green-to-red entire leaves. The terminal capsules are quite characteristic of a *Euphorbia* sp. (*Chamaesyce* sp.). This might be *E. glyptosperma* Engelm. [*Chamaesyce glyptosperma* (Engelm.) Small] (Fig. 4.92b).

Distribution Native to Quebec, west to British Columbia, south to New Mexico.

Names Ridge-seed spurge (English).

Uses The medicinal value of this genus is discussed by Moerman (1998).

a b

Fig. 4.92 *Euphorbia glyptosperma*?: (**a**) fol. 4r; (**b**) *E. glyptosperma*. (Source: Photo ©Al Schneider, http://www.swcoloradowildflowers.com)

Fol. 7v. *Euphorbia nutans*? (Fig. 4.93)

Description This phytomorph (Fig. 4.93a) is almost entirely vegetative, with serrate, elliptic, primarily opposite leaves with an acute tip, with color light green to green to orange. The limited inflorescence of three flowers has a pale calyx and darker corolla. The roots are small. The distinguishing character, however, is the brown-to-black spot in the center of the leaves. This might match a member of the Euphorbiaceae, such as *Euphorbia nutans* Lag. [*Chamaesyce nutans* (Lag.) Small] (Fig. 4.93b). The wide geographical range of this species accounts for its high variability, with leaf tips ranging from obtuse to acute and variation in spots.

Distribution Native from the eastern United States to Venezuela.

Names Nodding spurge (English).

Uses The medicinal value of this genus is discussed by Moerman (1998).

Fig. 4.93 *Euphorbia nutans*?: (**a**) fol. 7v; (**b**) *E. nutans*. (Source: Dr. James Van Kley, Stephen F. Austin State University, Nacogdoches, Texas)

Fol. 21r. *Euphorbia thymifolia* (Fig. 4.94)

Description The spreading growth pattern, tiny green to reddish leaves, and reddish axillary flowers of this phytomorph (Fig. 4.94a) are a good fit for *Euphorbia thymifolia* L. [*Chamaesyce thymifolia* (L.) Millsp.] (Fig. 4.94b).

Distribution Native to the tropics in Africa, Asia, and the Americas (Florida to Argentina).

Names *Golondrina* (Spanish); chickenweed (English).

Uses Popular medicinal herb; leaves, seeds and fresh juice of the whole plant are used in worm infections, in bowel complaints, and in many more diseases therapeutically (Mali and Panchal 2013).

Fig. 4.94 *Euphorbia thymifolia*: (**a**) fol. 21r; (**b**) *E. thymifolia*. (Source: Forest and Kim Starr, CC BY-NC-SA 3.0)

Fol. 5v. *Jatropha cathartica* (Fig. 4.95)

Description The appropriate identification for this phytomorph (Fig. 4.95a) is most probably *Jatropha cathartica* Terán & Berland., *jicamilla* (Fig. 4.95b). The palmately compound dentate leaves, red flowers, and tuberous roots are similar. Another similar species is *J. podagrica* Hook., native from southern Mexico to Nicaragua, but the leaflets are typically broader and not as deeply cut as *J. cathartica.*

Distribution Native from Texas to northern Mexico.

Names *Jicamilla* (Spanish), Berlandier's nettlespurge (English).

Uses Cathartic and poisonous, as the specific name implies.

Fig. 4.95 *Jatropha cathartica*: (**a**) fol. 5v; (**b**) *J. cathartica*. (Source: Frank Vincentz, CC BY-NC-SA 3.0)

Fol. 36r. *Jatropha gossypifolia* (Fig. 4.96)

Description This phytomorph (Fig. 4.96a) has small, inferior eight-petaled flowers that are painted bluish. Leaves are palmately lobed, on a small shrub. The roots are substantial. This matches the variability of *Jatropha gossypifolia* L. (Fig. 4.96b).

Distribution Native to Mexico, Caribbean to South America, and (curiously) India.

Names Names include *Frailecillo* (Costa Rica, Cuba, Venezuela, Colombia); *frailejón, purge de frail* (Colombia); *túatúa* (Venezuela, Santo Domingo, Puerto Rico, Cuba); *San Juan del Cobre* (Cuba); and *higuereta cimarrona* (Puerto Rico) (Standley 1920–1926); bellyache bush, black physicnut, and cotton-leaf physicnut (English).

Uses Standley (1920–1926) records for Mexico: "The seeds are eaten by doves and domestic fowl. They contain much oil and have drastic purgative and emetic properties. A decoction of the leaves is employed as a blood purifier and for venereal diseases, and is administered as an emetic for pains in the stomach. The root has some repute as an antidote for snake bites and as a remedy for the poison of manchineel (*Hippomane mancinella*)." Ayensu (1981) records for this species: "Leaf teas for diarrhoea, heat, venereal diseases, indigestion. Leaf baths for rash, bewitchment; poultices for sores, bich, piles…Decoction of boiled leaves as purgative. Leaves steeped in cold water for cooling off. Diarrhoea: leaves boiled and tea drunk; also for fever and for prickly heat in babies. Infant diarrhoea: leaf heated in water. Leaf decoction: stomach ache, laxative, throat inflammation, venereal sores in men, diabetes, gall bladder, bathing wounds. Roots with leaves of *Crescentia cujete* and *Croton flavens* for colds. Root decoction: sore throat, diabetes. Root boiled, salad oil added for stomach ache and bloody stool...Sores: mashed fresh leaves or powdered baked leaves applied. Cutting teeth: fruits crushed, boiled, and given to infant...Sap from stem for cuts and wounds. Clean scraped stem as rectal suppository for baby when bowels have not moved. Asthma: flowering tips boiled with 7 raw coffee beans. Tea for laxative; purge, constipation. Seeds purgative. Leaf tea to wash sores and ulcers." Asprey and Thornton (1953, 1954) record for this species: "Most of the old writers speak of the use of the leaves of this plant, either in decoction or boiled like spinach, as a purgative remedy for 'dry belly-ache.' Beckwith says it is used to prepare tea for constipation, the part used not being specified but it is probably the leaves, as given in the old remedies. The seeds are said to be purgative, the oil being similar to castor oil." Martinez (1969) also records medicinal uses for this species, including "*para males venéreos*." The medicinal value of this species is also discussed by Félix-Silva et al. (2014).

Fig. 4.96 *Jatropha gossypifolia*: (**a**) fol. 36r; (**b**) *J. gossypifolia*. (Source: Sue Carnahan, CC BY-SA)

Fol. 93v. *Manihot rubricaulis* (Fig. 4.97)

Description The stout, thickened roots, palmately compound leaves, and reddish fruits of this phytomorph (Fig. 4.97a) all fit the genus *Manihot*. This phytomorph is most probably *Manihot rubricaulis* I. M. Johnst. (Fig. 4.97b). This close relative to the cassava, *M. esculenta* Crantz, has thinner, more deeply lobed leaves; it also bears tuberous roots (Hancock 2012).

Distribution Native to northern Mexico.

Names *Manihot rubricaulis* is perhaps discussed in fol. 43v of the *Codex Cruz-Badianus* as *yamanquipatlis* (gentle or weak medicine) (Emmart 1940; Cruz and Badian 1991; Gates 2000; Clayton et al. 2009). It is also perhaps illustrated as *ytzticpatli* in Hernández et al. (1651:116).

Uses Potential source of cold tolerance for cassava.

Fig. 4.97 *Manihot rubricaulis*: (**a**) fol. 93v; (**b**) leaf of *M. rubricaulis*. (Source: Sky Jacobs, wildsonora.com); (**c**) tubers of *M. esculenta*. (Source: David Monniaux, CC BY-SA 3.0)

Fol. 16r. *Manihot pringlei* (Fig. 4.98)

Description This phytomorph (Fig. 4.98a) has deeply lobed leaves, probably compound. The inflorescence is leafy and reddish. The roots are swollen. This is similar to fol. 93v, *Manihot rubricaulis,* but probably *M. pringlei* S. Watson (Fig. 4.98b)

Distribution Native to Mexico.

Names *Yuca*, *matorral* (Tamaulipas, Mexico) (Standley 1920–1926).

Uses Unknown

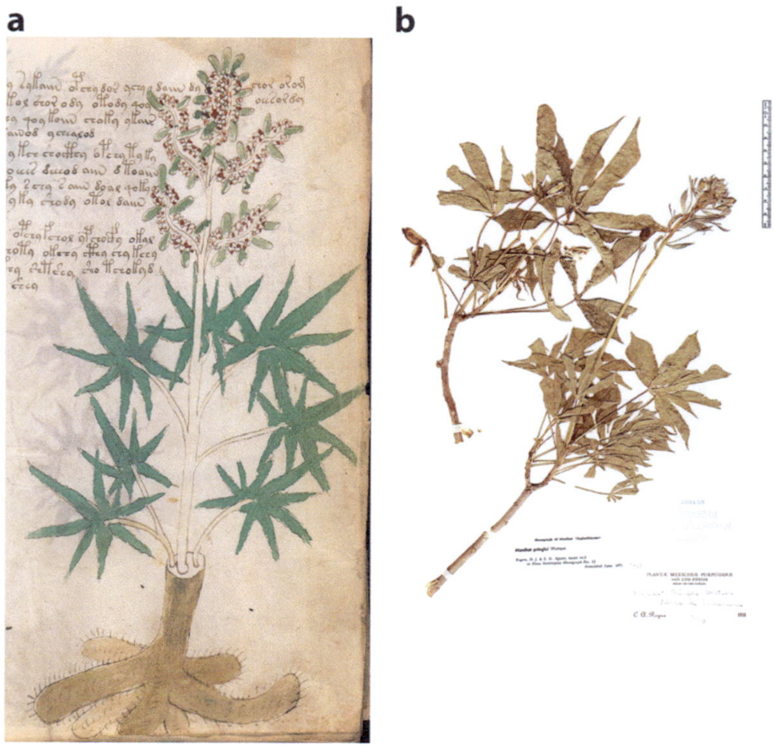

Fig. 4.98 *Manihot pringlei*: (***a***) fol. 16r; (**b**) *M. pringlei*. (Source: The Field Museum, CC-BY-NC)

Fabaceae

Fol. 88r#11. *Lupinus* sp., cf. *L. montanus* (Fig. 4.99)

Description This phytomorph (Fig. 4.99a) displays stylized compound peltate leaves and callus-like, nitrogen-fixing root nodules on one side of the roots. Many members of the Fabaceae have nitrogen-fixing nodules, but none have leaves exactly as depicted. Although the image is crudely rendered, it may possibly be *Lupinus montanus* Humb., Bonpl. & Kunth (Fig. 4.99b). The compound peltate leaves and soft, callus-like, nitrogen-fixing root nodules on one side of the roots (Fig. 4.99c) are all typical of this species.

Distribution Mexico and Central America.

Names *Garbancillo* (Spanish). The name accompanying this phytomorph, *āhuocâchā*, might be derived from *agua* (water) + *cacha* (Nahuatl), callus; the nitrogen-fixing root nodules do resemble watery calluses.

Uses This lupine is noted to contain alkaloids (Dunn and Harmon 1977; Ruiz-López et al. 2010).

Fig. 4.99 *Lupinus* sp.: (**a**) fol. 88r#7; (**b**) *L. montanus*. (Source: Harry Douwes); (**c**) root nodules on *L. montanus*. (Source: Therrell et al. 2006)

Geraniaceae

Fol. 101r#1. *Geranium tenue*? (Fig. 4.100)

Description This phytomorph (Fig. 4.100a) bears alternate, deeply lobed, palmate leaves. The roots indicate a rhizome. This may be *Geranium tenue* Hanks (Fig. 4.100b).

Distribution Native to Mexico.

Names The name accompanying this phytomorph, , might be transliterated as *noshe ocai* or *noche icai,* Nahuatl for *induces night* (Herrera 2004; Karttunen 1983; Siméon 2010).

Uses Unknown.

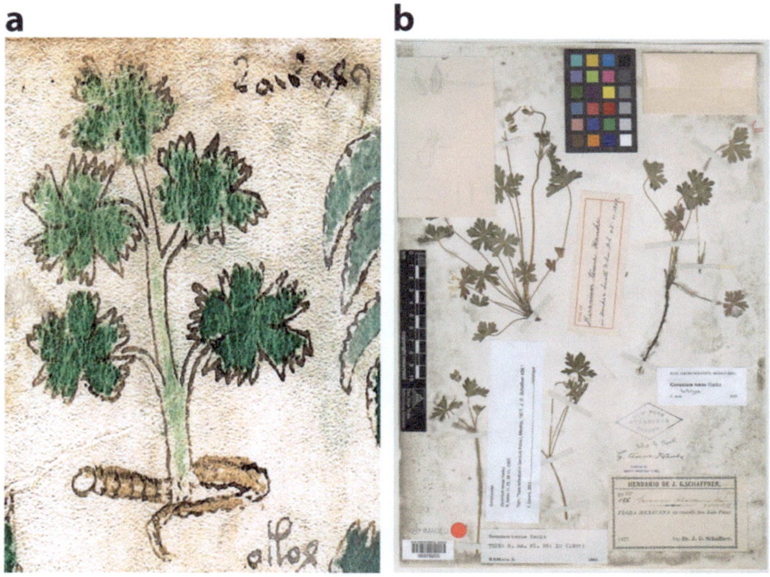

Fig. 4.100 *Gernanium tenue*?: (**a**) fol. 101r#1; (**b**) *G. tenue* herbarium sheet. (Source: SEINet Arizona-New Mexico Chapter, CC BY-SA)

Gesneriaceae

Fol. 55r. *Diastema hispidum* (Fig. 4.101)

Description This phytomorph (Fig. 4.101a) has six petals that are white and bluish, and a long corolla. Leaves are green and deeply lobed. Multiple stems arise from a rhizomatous base, with many brown roots. This might possibly be a species of *Geranium*, but the swollen fruits seem incongruous with this genus; a better match might be *Diastema hispidum* (DC.) Fritsch. (Fig. 4.101b, c).

Distribution Native from Nicaragua to Peru.

Names Unknown.

Uses Unknown.

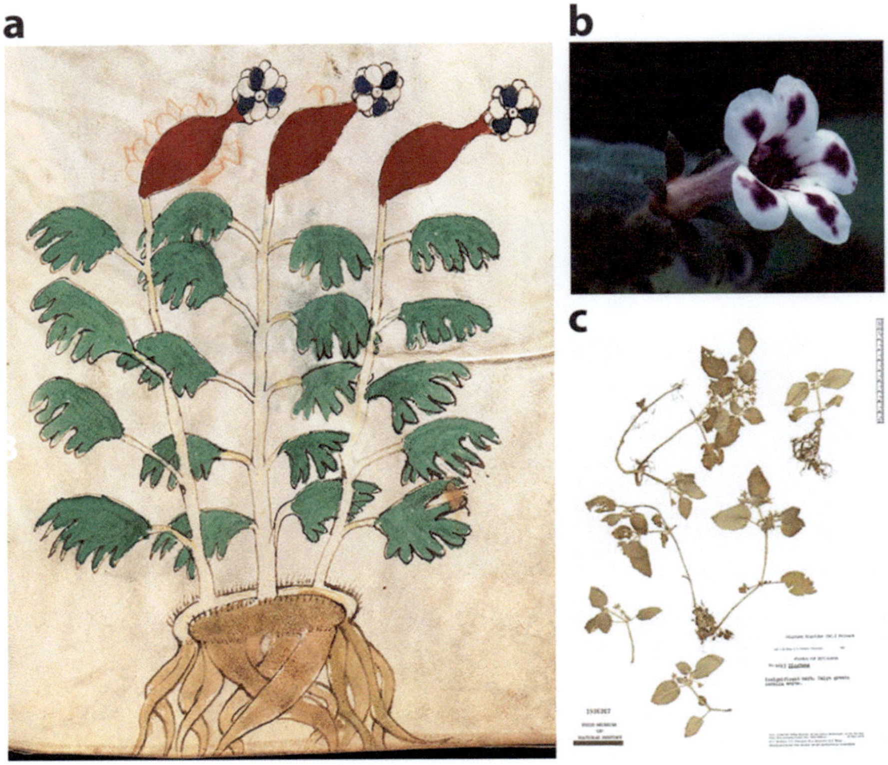

Fig. 4.101 *Diastema hispidum*: (**a**) fol. 55r; (**b**) flower of *D. hispidum*. (Source: Leslie Brothers); (**c**) herbarium sheet of *D. hispidum* (F1836367), showing thin rhizomatous roots. (Source: The Field Museum, CC-BY-NC)

Grossulariaceae

Fol. 37v. *Ribes cereum* (Fig. 4.102)

Description This phytomorph (Fig. 4.102a) has opposite orbicular, peltate leaves with a crenate margin. Worldwide, except for *Hydrocotyle* and *Tropaeolum* spp., there are no species that match the foliage of this phytomorph, and when the inferior ovary and eight petals are added to the description, nothing matches. Obviously, artistic license was taken, and it might have been created from dried specimens. The only species that comes close to this description is *Ribes cereum* Douglas (Fig. 4.102b), the flowers of which range from white to pink. Fruit is typically red but blackens upon drying. Leaves are crenate and lightly lobed, with the cordate bases often overlapping on drying. This is also a shrub with substantial roots.

Distribution Ranges from British Columbia to California, east to Oklahoma.

Names Wax currant (English).

Uses Moerman (1998) discusses the culinary and medicinal value of this species by the Navajo and Hopi.

Fig. 4.102 *Ribes cereum*: (**a**) fol. 37v; (**b**) *R. cereum*. (Source: http://www.wildflowerchild.info/large/wax_currant2.jpg)

Fol. 23r. *Ribes malvaceum* (Fig. 4.103)

Description This phytomorph (Fig. 4.103a) is most probably *Ribes malvaceum* Sm. (Fig. 4.103b, c). This woody, stoloniferous shrub has purple-magenta flowers and palmately lobed leaves.

Distribution Native from California to Baja Norte, Mexico (Standley 1920–1926:316).

Names Chaparral currant (English).

Uses Unknown.

Fig. 4.103 *Ribes malvaceum*: (**a**) fol. 23r; (**b**) flower and (**c**) shoots of *R. malvaceum*. (Source: Stan Shebs, CC BY-SA 3.0 and Toedrifter, CC BY-SA 3.0)

Fol. 95v. *Ribes montigenum*? (Fig. 4.104)

Distribution This phytomorph (Fig. 4.104a) is most probably a stoloniferous, alpine shrub. The flowers, painted with a blue pigment, superficially could be Asteraceae, but the general gestalt of the whole plant is Grossulariaceae. A guess might be *Ribes montigenum* McClatchie (Fig. 4.104b).

Distribution Native from Canada to California, Arizona, and New Mexico.

Names Mountain gooseberry (English).

Uses The medicinal value of this genus is discussed by Moerman (1998).

Fig. 4.104 *Ribes montigenum*?: (**a**) fol. 95v; (**b**) *R. montigenum*. (Source: PhytoImages.siu.edu)

Lamiaceae

Fol. 45v. *Hyptis albida* (Fig. 4.105)

Description The gray leaves, blue flowers, and stout root in this phytomorph (Fig. 4.105a) are all good fits for most probably *Hyptis albida* Kunth (Fig. 4.105b, c), now sometimes classified as *Condea albida* (Kunth) Harley and J.F.B. Pastore.

Distribution This shrub is native from Sonora and Chihuahua to San Luis Potosí, Guanajuato, and Guerrero.

Names *Orégano* (Spanish); desert lavender (English).

Uses Standley (1920–1926:1275) relates that "The leaves are sometimes used for flavoring food. In Sinaloa they are employed as a remedy for ear-ache, and in Guerrero a decoction of the plant is used in fomentations to relieve rheumatic pains."

Fig. 4.105 *Hyptis albida*: (**a**) fol. 45v; (**b, c**) inflorescence of *H. albida*. (Source: Benjamin Wilder, Desert Laboratory on Tumamoc Hill, University of Arizona and Jim Conrad, respectively)

Fol. 32r. *Ocimum campechianum* (Fig. 4.106)

Description This phytomorph (Fig. 4.103a) is most probably *Ocimum campechianum* Mill. (*O. micranthum* Willd.) (Fig. 4.106b). The terminal inflorescence, bluish flowers, and ovate leaves are all good fits (Standley 1920–1926:1272).

Distribution This suffrutescent plant is native from Florida to Argentina; in Mexico it is found from Sinaloa to Tamaulipas, Yucatán, and Colima (Standley and Williams 1973:269).

Names Wild sweet basil, least basil (English).

Uses Standley (1920–1926:1272) relates "In El Salvador bunches of the leaves of this plant are put in the ears as a remedy for earache."

Fig. 4.106 *Ocimum campechianum*: (**a**) fol. 32r; (**b**) inflorescence and leaves of *O. campechianum*. (Source: Roger L. Hammer)

Fol. 45r. *Salvia cacaliifolia* (Fig. 4.107)

Description In this phytomorph (Fig. 4.107a), blue flowers are in a tripartite inflorescence, and distantly dentate deltoid-hastate leaves are quite characteristic of *Salvia cacaliifolia* Benth. (Fig. 4.107b).

Distribution Native from Mexico (Chiapas) to Guatemala and Honduras (Standley and Williams 1973:278).

Names Guatamalan blue sage (English).

Uses Ornamental.

Fig. 4.107 *Salvia cacaliifolia*: (**a**) fol. 45r; (**b**) inflorescence and leaves of *S. cacaliifolia*. (Source: Ashwood Nurseries Ltd)

Fol. 100r#5. *Scutellaria mexicana* (Fig. 4.108)

Description This phytomorph (Fig. 4.108a) shows three flowers that match to *Scutellaria mexicana* (Torr.) A.J. Paton (*Salazaria mexicana* Torr.) (Fig. 4.108b, c). Inflated bladder-like calyces vary in color, depending upon maturity, from green to white to magenta, with a dark blue and white corolla emerging from it (Standley 1920–1926:1271).

Distribution This shrub is native from Utah to Mexico (Baja California, Chihuahua, and Coahuila).

Names The name accompanying this phytomorph is , which transliterates to *noe, moe-choll-chi*. This includes the Spanish *cholla*, skeleton or skull, plus the Nahuatl root word for owl, *chi* (*chi:chtli* or *chi:cuahtli* is owl, *chicuatli* is the barn or screech owl). The calyces do resemble the skull of the great horned owl (Fig. 4.108d). This species also seems to match the description of *tenamaznanapoloa* (carrying triplets?) of Hernández et al. (1651:129) (alias *tenamazton* or *tlalamatl*).

Uses Unknown.

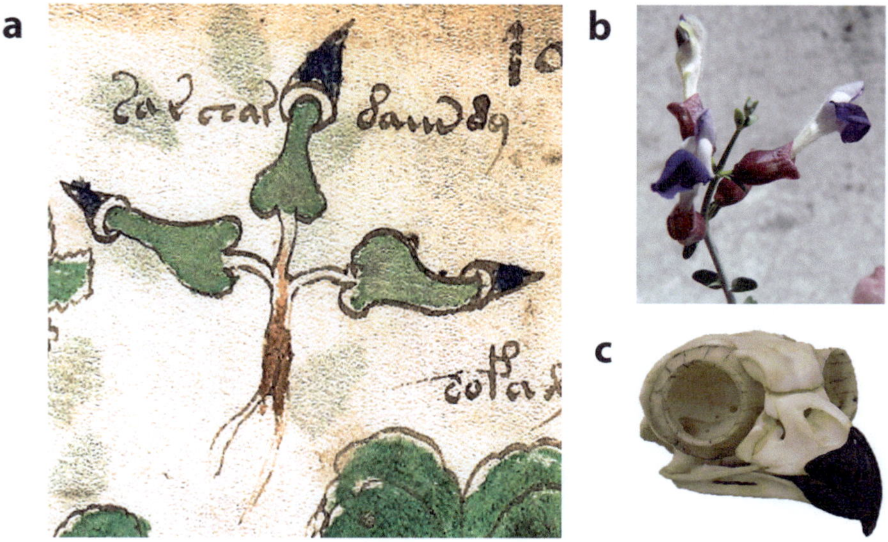

Fig. 4.108 *Scutellaria mexicana*: (**a**) fol. 100r#5; (**b**) inflorescence of *S. mexicana.* (Courtesy of Wynn Anderson); (**c**) skull of the great horned owl. (Source: http://www.connecticutvalleybiological.com/great-horned-owl-skull-bubo-virginianus-p-13913.html)

Fol. 52r. *Teucrium laciniatum* (Fig. 4.109)

Description This phytomorph (Fig. 4.109a), apparently in bud, whitish with bluish lines, bears resemblance to the Lamiaceae, with a terminal inflorescence on a woody subshrub. Leaves are four per node and palmately lobed at the tips. Roots are rhizomatous and ample, not tuberous. Assuming that this was assembled from dried specimens, it matches the variability of *Teucrium laciniatum* Torr. (Fig. 4.109b). Flowers are white with purple lines in the throat.

Distribution Native from Colorado to Mexico.

Names Lacy germander (English).

Uses Unknown.

Fig. 4.109 *Teucrium laciniatum*: (**a**) fol. 52r; (**b**) *T. laciniatum*. (Source: Richard Spellenberg, CC BY-NC-SA 3.0)

Liliaceae

Fol. 55v. *Clintonia andrewsiana* (Fig. 4.110)

Description The flowers of this phytomorph (Fig. 4.110a), apparently of the Liliaceae, are painted with blue pigment. Leaves are broadly elliptic and basal, margin entire. Roots are ample but not tuberous or bulbous. This matches the genus *Clintonia* very well, probably *C. andrewsiana* Torr. (Fig. 4.110b).

Distribution Native from Oregon to California.

Names Red bead lily or western bluebead-lily (English).

Uses Moerman (1998) says: "Pomo *Poison* Plant considered poisonous."

Fig. 4.110 *Clintonia andrewsiana*: (**a**) fol. 55v; (**b**) *C. andrewsiana*. (Source: David A. Hofmann)

Malvaceae

Fol. 102r#11. *Chiranthodendron pentadactylon* (Fig. 4.111)

Description This phytomorph (Fig. 4.111a) is very curious, looking more like a very dark, blue-black flag than a possible plant part. However, this is often the appearance of the five-parted stamens of *Chiranthodendron pentadactylon* Larreat. (*C. platanoides* Bonpl.) when they are pressed and dried. When fresh, the stamens are a brilliant vermillion (Fig. 4.111b), but they turn blue-black when improperly dried and/or aged (Fig. 4.111c), and the five-parted, hand-like stamens can assume a flag-like figure when pressed.

Distribution This species typically grows in wet areas in the mountains of Oaxaca and Guatemala but is widely planted in the Valley of Mexico (Standley 1920–1926).

Names This is called *macpalxochi quahuitl* in Hernández et al. (1651:383, 459) (Reko 1947; Safford 1912). It may be the same as *teonacaztli/ueinacaztli* (sacred ear) in pl. 420 and 686 and *ueinacaztli/ueynacaztli* in Sahagún's *Florentine Codex* (Estrada Lugo 1989). Additional Nahuatl names are *mapasúchil*, *mapilxochitl*, and *teyacua* (Díaz 1976); *mapasúchil* is derived from the Nahuatl *macpal-xochitl*, "hand flower" (Standley 1920–1926). Monkey's hand tree (English). The name accompanying this phytomorph, ollo?occ9, transliterates to *ātlāeāăăi*, with an unknown meaning.

Uses Medicinal uses in Mexico are recorded by Martinez (1969). It has been documented to have antiprotozoal, antibacterial, and antidiarrheal properties and to be antisecretory (Calzada et al. 2017; Velázquez et al. 2009).

Fig. 4.111 *Chiranthodendron pentadactylon*: (**a**) fol. 102r#11; (**b**) five-parted stamens of *C. pentadactylon*. (Source: Jan Conayne); (**c**) dried shoot of *C. pentadactylon* showing leaves and flower with protruding stamens. (Source: George B. Hinton Herbarium, CC BY 4.0)

Fol. 19v. *Melochia tomentosa* (Fig. 4.112)

Description This phytomorph (Fig. 4.112a) has abundant green, serrate, hastate leaves on a shrubby plant with obvious but minimal roots. Flowers are towards the summit and painted bluish, with two lobes of the actinomorphic corolla visible. The best match is *Melochia tomentosa* L. (Fig. 4.112b).

Distribution Ranges from Texas to Brazil and the West Indies and is very variable (Goldberg 1967).

Names Teabush (English); Spanish names include *malvisco* (Puerto Rico); *bretónica* (Puerto Rico, Venezuela); and *malva* and *varita de San José* (Nicaragua) (Standley 1920–1926).

Uses Ayensu (1981) records for this species: "Tea for colds; boil roots for decoction for back pain." Asprey and Thornton (1953, 1954) record for this species: "Used to make tea for colds." Encarcion Dimayuga et al. (1987) also record its use for colds in Baja California Sur. Mayans today use an infusion of the leaves to hasten delivery (Roys 1976).

Fig. 4.112 *Melochia tomentosa*: (**a**) fol. 19v; (**b**) inflorescence. (Source: Joseph A. Marcus, Lady Bird Johnson Wildflower Center)

Marantaceae

Fol. 42v. *Calathea* sp., cf. *C. loeseneri* (Fig. 4.113)

Description The phytomorph inflorescence (Fig. 4.113a) is a crude representation of a *Calathea* sp., probably allied to *C. loeseneri* J.F. Macbr. (Fig. 4.113b). Many species of *Calathea* were recently transferred to the genus *Goeppertia*, and a synonym of this species is now *G. loeseneri* (J.F. Macbr.) Borschs. & Suárez (Borschenius et al. 2012). The crudeness of the illustration, coupled with inadequate surveys of the genus *Calathea/Goeppertia* in Mexico, prevent precise identification.

Distribution Native to tropical America, including Peru, Brazil, Colombia, Ecuador, and Bolivia.

Names Genus called calathea or prayer plant (English).

Uses *Calathea loesneri* yields a blue dye; in a related species in Peru (*C. standleyi* J. F. Macbr.), this is used to ward off dental caries (Graham et al. 2004).

Fig. 4.113 *Calathea* sp.: (**a**) fol. 42v; (**b**) inflorescence and leaves of *C. loeseneri*. (Source: Milan Kořínek)

Menyanthaceae

Fol. 2v. *Nymphoides aquatic*a. (Fig. 4.114)

Description This phytomorph (Fig. 4.114a) has a rounded, notched leaf, four white petals with a crenate margin, and a thick rhizome. It closely resembles *Nymphoides aquatica* (J.F. Gmel.) (Fig. 4.114b), with five wide, crenulated petals. The horizontal rhizome bears thick, unbranched, adventitious roots that in young plants resemble a bunch of bananas (Richards et al. 2010), earning it the popular name, *banana plant*. Another possibility is *N. indica* (L.) Kuntze, but the petals are narrower and fimbriated in this species (Fig. 4.114c).

Distribution *Nymphoides aquatica* is native to the southern United States, from New Jersey to Texas; *N. indica* is native to southern Mexico (Campeche, Chiapas, Oaxaca, Quintana Roo, Tabasco, Veracruz, Yucatan) but also Asia, Africa, and Australia.

Names *Nymphoides aquatica*: banana plant, banana lily, big floating heart; *N. indica*: banana plant, robust marshwort, and water snowflake (English).

Uses Aquarium plant.

Fig. 4.114 *Nymphoides aquatica*: (a) fol. 2v; (b) *N. aquatica*. (Source: Center for Aquatic and Invasive plants, University of Florida, Gainesville); (c) *N. indica*. (Source: Edward Jaser, CC BY-SA 3.0)

Moraceae

Fol. 36v. *Dorstenia contrajerva* (Fig. 4.115)

Description The inflorescence (Fig. 4.115a), appearing like a split-open fig, is quite distinct and matches most probably a *Dorstenia* sp., likely the very variable *D. contrajerva* L. (Fig. 4.115b). Leaves for this species vary "in spirals, rosulate or spaced; lamina broadly ovate to cordiform to subhastate, pinnately to subpalmately or subpedately, variously lobed to parted with 3–8 lobes at each side or subentire" (Berg 2001).

Distribution Native from Mexico to Peru.

Names The Nahuatl name is *tozpatli* or *tuzpatli* (Díaz 1976); snakewort (English).

Uses The specific epithet *contrajerva*, the Latinized form of Spanish *contrahierba*, suggests its use for treating poisoning and venomous bites and stings. Medicinal uses in Mexico are recorded by Martinez (1969).

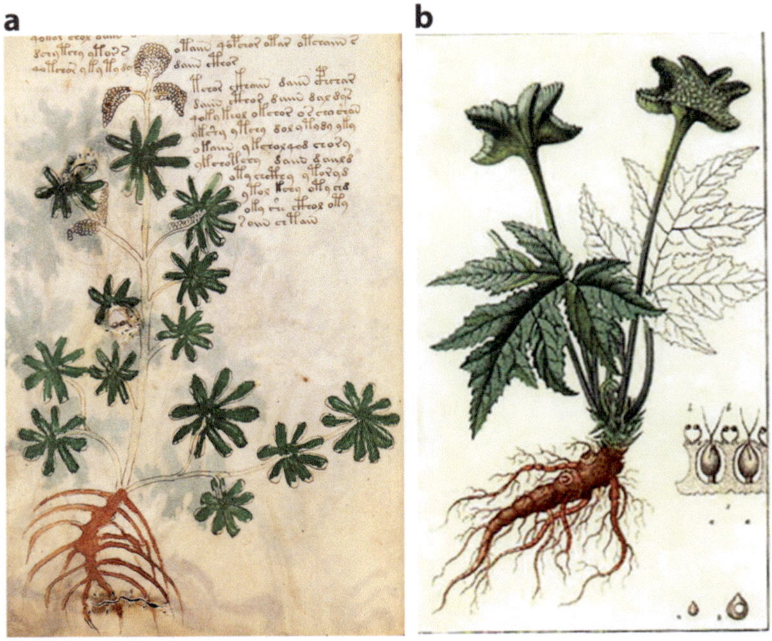

Fig. 4.115 *Dorstenia contrajerva*: (**a**) fol. 36v; (**b**) botanical image of *D. contrajerva* by Pierre Turpin from Chaumeton (1830:131)

Fol. 14v. *Dorstenia lindeniana* (Fig. 4.116)

Description This phytomorph (Fig. 4.116a) is overwhelmingly similar to fol. 36v, which has been identified as a *Dorstenia* sp. However, this phytomorph has serrated simple leaves, not deeply lobed compound leaves. This might fall within the variability of *D. lindeniana* Bureau in DC (Fig. 4.116b).

Distribution Native to Mexico. Guatemala, and Belize (Berg 2001).

Names Unknown.

Uses Used medicinally by the Maya for skin problems, being anti-inflammatory (Hitziger 2016; Walshe-Roussel 2014.). Further chemistry and pharmacology are explored by Ngadjui and Abegaz (2003).

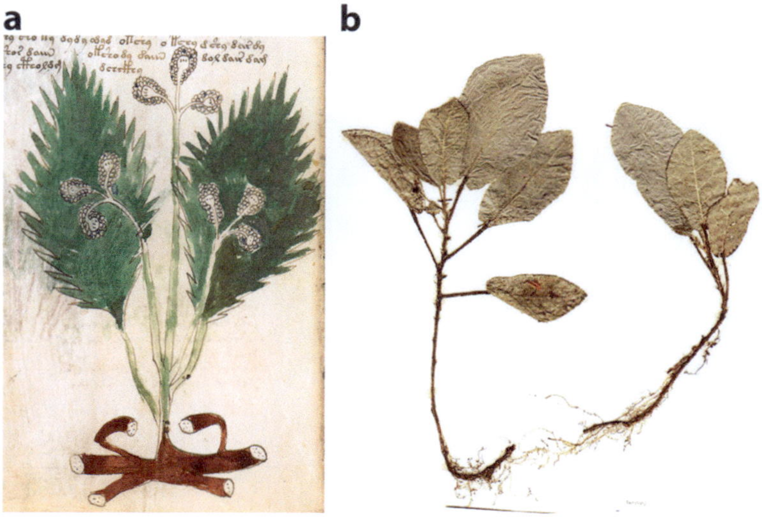

Fig. 4.116 *Dorstenia lindeniana*: (**a**) fol. 14v; (**b**) *D. lindeniana* shoots. (Source: The Field Museum)

Nelumbonaceae

Fol. 101(2)#3. *Nelumbo lutea* (Fig. 4.117)

Description This phytomorph (Fig. 4.117a) has a very slightly crenate margin on the leaf, peltate attachment of the petiole, and a substantial root with aerenchyma. With the peltate attachment and aerenchyma, this can match only a *Nelumbo* sp., such as *N. lutea* (Willd.) Pers. (Fig. 4.117b).

Distribution Native from Canada to Honduras and the Caribbean.

Names American lotus (English).

Uses Often a food plant (Swan 2010).

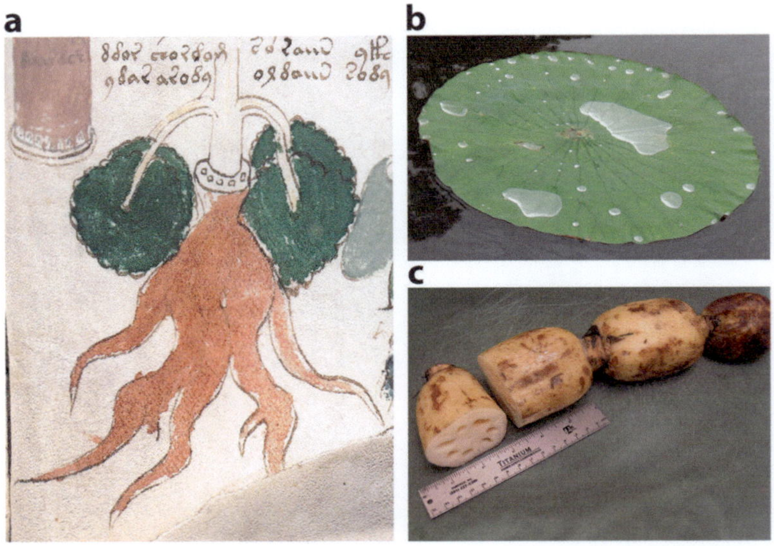

Fig. 4.117 *Nelumbo lutea*: (**a**) fol. 101(2)#3; (**b**) *N. lutea* leaf. (Source: john clintoneasywildflowers.com); (**c**) *N. lutea* root. (Source: Mark "Merriwether" Vorderbruggen, Ph.D., www.foragingtexas.com)

Nymphaeaceae

Fol. 89v(1)#6. *Nymphaea ampla* (Fig. 4.118)

Description This phytomorph shows a leaf that is orbicular and deeply notched, with attachment of the petiole at the center. The margin is hairy or deeply serrated. Roots are ample. This might match *Nymphaea ampla* (Salisb.) DC. (Fig. 4.118a) (Conard 1905; Wiersema et al. 2008).

Distribution Mayan lowlands in southern Mexico.

Names The name accompanying this phytomorph, [Voynich script], transliterates to *āhu/guācacholl,* which may be derived from the Spanish *agua* + Nahuatl *cacçolli*, or watery hardness (Herrera 2004; Karttunen 1983; Siméon 2010), perhaps alluding to its use in treating kidney stones. White water lily (English).

Uses This is rich in psychoactive apomorphine alkaloids and currently is being investigated to treat erectile dysfunction (Bertol et al. 2004; Dobkin de Rios 1974; Díaz 1977; Emboden 1979, 1981a, b, 1982a, b, 1983; McDonald and Stross 2012; Rands 1953).

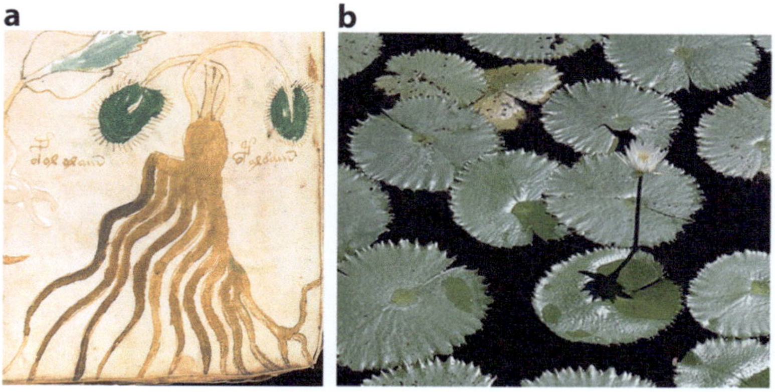

Fig. 4.118 *Nymphaea ampla*: (**a**) fol. 89v(1)#6; (**b**) *N. ampla*. (Source: Reinhard Jahn CC BY-SA 2.0 DE)

Nyctaginaceae

Fol. 33r. *Allionia incarnata* (Fig. 4.119)

Description This has many petals united into a greenish corolla, backed by a swollen, whitish calyx, with sagittate, green leaves, and swollen, branched brown roots (Fig. 4.119a). This matches the wide variability of *Allionia incarnata* L. (Fig. 4.119b, c). Curiously, the ends of the two main roots in the phytomorph in the *Voynich Codex* have a face, and the roots of *Allionia incarnata* do bear bumps with indentations that could be interpreted as tiny faces.

Distribution Native from Utah to Mexico.

Names *Hierba de la hormiga* (ant herb), *hierba del golpe* (knock herb) (Spanish); four o'clock, trailing windmills (English).

Uses Moerman (2009) reports that *A. incarnata* is used by the Ramah Navajo as a dermatological aid, a cold infusion used as a lotion for swellings.

Fig. 4.119 *Allionia incarnata*: (**a**) fol. 33r; (**b**) inflorescence and leaves of *A. incarnata*. (Source: Campbell and Lynn Loughmiller, Lady Bird Johnson Wildflower Center); (**c**) herbarium specimen of *A. incarnata* (DES00067664) showing the swollen, knobby roots. (Source: Desert Botanical Garden Herbarium Collection)

Onagraceae

Fol. 51r. *Fuchsia thymifolia* (Fig. 4.120)

Description The phytomorph (Fig. 4.120a) has four red petals and four pale sepals backed by a corolla tube and a swollen ovary; the leaves are green and deeply serrate; the roots are brown and tuberous. This may be *Fuchsia thymifolia* Kunth (Fig. 4.120b).

Distribution Native from Mexico to Guatemala.

Names Thyme-leaved fuchsia (English). This species may also be in Sahagún's *Florentine Codex* as *tlacozazalic* (Estrada Lugo 1989).

Uses Unknown.

Fig. 4.120 *Fuchsia thymifolia*: (**a**) fol. 51r; (**b**) leaves and flowers of *F. thymifolia*. (Source: Todd Boland)

Passifloraceae

Fol. 101v(1)#15. *Passiflora coriacea* (Fig. 4.121)

Description This phytomorph (Fig. 4.121a) has unique, obdeltoid leaves with three dark blue berries. The most obvious choice is the bat-leaved passion flower, *Passiflora coriacea* Juss. (Fig. 4.121b), probably assembled from dried fragments.

Distribution Native from Mexico to Bolivia.

Names Wild sweet calabash, bat-leaved passion flower (English). This is the same species discussed by Francisco Hernández under the Nahuatl name *tzinacanatlapatli* (Hernández et al. 1651:435).

Uses Roys (1976) records from an unpublished manuscript on Mayan herbal medicine (*Yerbas y Hechicherias del Yucatan*), "The root of the *escorzonera*, which in this language they call *xik-zoɔ* or in ours, sorrel, is crushed, and its juice or milk is applied with some cotton to the eye. It is a tried remedy." Further names and medicinal uses by the Huastecs are recorded by Alcorn (1984).

Fig. 4.121 *Passiflora coriacea*: (**a**) fol. 101v(1)#15; (**b**) *P. coriacea*. (Source: https://www.ebay.com/itm/passiflora-coriacea-collectors-bat-wing-rare-passion-flower-5-seeds-/321520705556)

Fol. 23v. *Passiflora* Subgenus *Decaloba,* cf. *P. morifolia* (Fig. 4.122)

Description From the flower alone, this is definitely a *Passiflora* sp. of subgenus *Decaloba* (Fig. 4.122a). *Passiflora* is primarily a New World genus, with a few species also occurring in Australia and Southeast Asia, but none in Europe. The prominent corona with filaments of the genus *Passiflora* is very distinctive and cannot be confused with any other genus. The paired petiolar glands in the upper third of the leaf, blue tints in the flower, and dentate leaves that are deeply cordate only seem to match the variability of *P. morifolia* Mast. (Killip 1938) (Fig. 4.122b, c), although the artist has made the leaves slightly more orbicular than is normal in mature foliage. Young plants (*i.e.,* root suckers), however, often exhibit juvenile leaves that are orbicular, entire leaves.

Distribution *Passiflora* is primarily a New World genus with a few species in Australasia.

Names Blue sweet calabash, woodland passion flower (English).

Uses Unknown.

Fig. 4.122 *Passiflora* sp.: (**a**) fol 23v; (**b**) flower of *P. morifolia*. (Source: Anke and Ralf Schlosser); (**c**) leaf showing petiolar gland of *P. morifolia*. (Source: Hans B., CC BY-SA 3.0)

Penthoraceae

Fol. 30v. *Penthorum sedoides* (Fig. 4.123)

Description The cymose inflorescence, dentate leaves, and stolons of this phytomorph (Fig. 4.123a) match *Penthorum sedoides* L. (Fig. 4.123b, c) quite well. The artist has apparently illustrated this in very early bud (or glossed over the details of the flowers), however, because prominent pistils emerge later and are very obvious in fruit, often turning rosy.

Distribution Native from Canada to Texas.

Names Ditch stonecrop (English).

Uses Unknown.

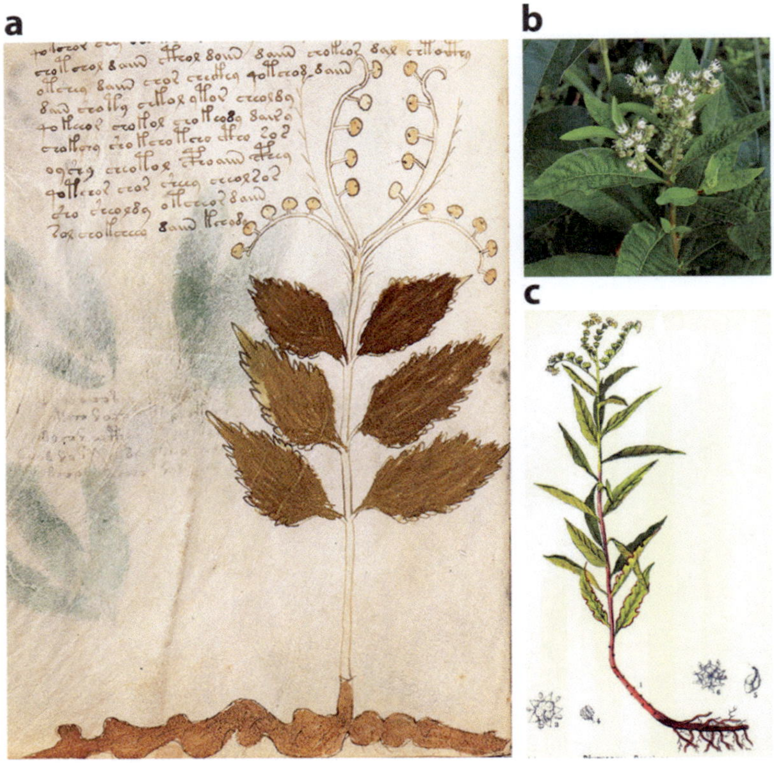

Fig. 4.123 *Penthorum sedoides*: (**a**) fol. 30v; (**b**) inflorescence and leaves of *P. sedoides*. (Source: Fritz Flohr Reynolds, CC BY-SA 3.0); (**c**) botanical painting of *P. sedoides* from Millspaugh (1892 1:t. 57)

Poaceae

Fol. 101v(1)#4. *Zea* sp.? (Fig. 4.124)

Description This phytomorph (Fig. 4.124a) has a broad green blade and a terminal brownish inflorescence. This matches the tassels of a *Zea* sp., and the absence of ears makes it likely to be Zea *mays* L. subsp. *mexicana* (Schrad.) H.H. Iltis, teosinte, a progenitor of maize (Fig. 4.124b).

Distribution Native to Mexico.

Names *Maiz* (Spanish), maize (English), *cintli* (Nahuatl); Mexican teosinte (English), *teocintli* (Nahuatl). The name accompanying this phytomorph, 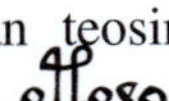, is transliterated as *ātlācai* and may be derived from *ātlacui,* Nahuatl, to draw or carry water (Herrera 2004; Karttunen 1983; Siméon 2010).

Uses Maize is widely cultivated as a grain and is a medicinal herb among the Huastecs (Alcorn 1984).

Fig. 4.124 *Zea* sp.?: (**a**) fol. 101v(1)#4; (**b**) teosinte tassel. (Source: Jan Thomas Johansson)

Polemoniaceae

Fol. 4v. *Cobaea* sp., cf. *C. biaurita* (Fig. 4.125)

Description With a basally woody stem, pinnately compound elliptic leaves, campanulate corolla, segmented calyx, and exserted style (Fig. 4.125a), the best match would be a *Cobaea* sp., most probably *C. biaurita* Standl. (Fig. 4.125b, c), which is closely related to the cultivated *C. scandens* Cav. and has elliptic leaflets with acute to acuminate apices and flowers that emerge cream-colored but later mature to purple (Standley 1914; Prather 1999).

Distribution Native to Chiapas and Oaxaca, Mexico.

Names Unknown.

Uses Unknown.

Fig. 4.125 *Cobaea* sp.: (**a**) fol 4v; (**b**) flower of *C. scandens*. (Source: Michael Wolf, CC BY-SA 3.0); (**c**) herbarium sheet of *C. biaurita* (F1662840). (Source: The Field Museum)

Polygonaceae

Fol. 43v (Right Side). *Neomillspaughia emarginata*? (Fig. 4.126)

Description This phytomorph (Fig. 4.126a) has a tripartite inflorescence of the family Polygonaceae, mostly white. The leaves curl around the stems. This was probably assembled from fragmented dried specimens in which the leaves had curled, but it might be *Neomillspaughia emarginata* (H. Gross) S.F. Blake (Fig. 4.126b).

Distribution Native to Mexico.

Names Unknown.

Uses Unknown.

Fig. 4.126 *Neomillspaughia emarginata*?: (**a**) fol. 43v, right side; (**b**) *N. emarginata*. (Source: http://www.cicy.mx/sitios/flora%20digital/ficha_virtual.php?especie=1987)

Ranunculaceae

Fol. 95r. *Actaea rubra* f. *neglecta* (Fig. 4.127)

Description This phytomorph (Fig. 4.127a) has crenate, pinnately compound leaves and noticeably white, globose fruits in a raceme, fitting quite definitely an *Actaea* sp., probably the white-fruited *Actaea rubra* (Aiton) Willd. f. *neglecta* (Gillman) B.L. Rob (Fig. 4.127b).

Distribution *Actaea rubra* is native to Eurasia and to North America from Canada to New Mexico, but f. *neglecta* is more common in North America (Compton et al. 1998).

Names Baneberry (English).

Uses The common name *baneberry* suggests that it is poisonous.

Fig. 4.127 *Actaea rubra* f. *neglecta*: (**a**) fol. 95r; (**b**) fruits and leaves of *Actaea rubra* f. *neglecta*. (Courtesy of Donald Cameron)

Fol. 44r. *Anemone narcissiflora* (Fig. 4.128)

Description This phytomorph (Fig. 4.128a) bears six white petals subtended by green bracts and a ring of linear leaves. The roots are ample but not tuberous. It bears an uncanny resemblance to *Anemone narcissiflora* L. (Fig. 4.128b).

Distribution Ranges from Eurasia and northwest North America to Colorado, with at least 12 varieties.

Names Narcissus anemone (English).

Uses Moerman (1998) records culinary and medicinal value by the Alaska natives and Aleut, especially for hemorrhage.

Fig. 4.128 *Anemone narcissiflora*: (**a**) fol. 44r; (**b**) *A. narcissiflora*. (Source: Σ64, CC BY-SA 3.0)

Fol. 52r. *Anemone patens* (Fig. 4.129)

Description This phytomorph (Fig. 4.129a) has a terminal, pubescent blue flower with many linear bracts; leaves are deeply laciniate; roots are brown, long, and tuberous. This matches very nicely the variability of *Anemone patens* L., (Fig. 4.129b, c).

Distribution Circumboreal, south to Texas and New Mexico.

Names Cutleaf anemone, pasqueflower, prairie crocus (English).

Uses Unknown.

Fig. 4.129 *Anemone patens*: (**a**) fol. 52r; (**b**, **c**) flowers and leaves of *Anemone patens*. (Source: Tom Koerner, U.S. Fish and Wildlife Service and John Richards, respectively)

Fol. 29v. *Anemone tuberosa* (Fig. 4.130)

Description This phytomorph (Fig. 4.130a) has blue-green, hairy flower buds with a multi-petaled, bluish corolla; basal leaves are deeply divided; roots are tuberous. This matches *Anemone tuberosa* Rydb. (Fig. 4.130b).

Distribution Native from Utah to Mexico.

Names Desert anemone (English)

Uses Unknown.

Fig. 4.130 *Anemone tuberosa*: (**a**) fol. 29v; (**b**) *A. tuberosa*. (Source: 7Song)

Rosaceae

Fol. 42r (Right Side). *Rubus trivialis*? (Fig. 4.131)

Description This phytomorph (Fig. 4.131a) is entirely vegetative, but the trifoliate leaves with crenate margin in orange with spatulate leaflets matches a senescent leaf of a *Rubus* sp., possibly one of dewberries, such as *R. trivialis* Michx. (Fig. 4.131b).

Distribution Native from Virginia to Texas.

Names Dewberry (English).

Uses This was a food plant of the Indians in the Florida Panhandle (Stockdale and Bryenton 1978).

Fig. 4.131 *Rubus trivialis*?: (**a**) fol. 42r (right side); (**b**) *R. trivialis*. (Source: Melody Lytle, Lady Bird Johnson Wildflower Center)

Rubiaceae

Fol. 13v. *Randia aculeata* (Fig. 4.132)

Description This phytomorph (Fig. 4.132a) has an unusual corolla with only three white lobes (one seems to be hidden to the rear), with a slightly exserted stigma on an inferior bluish ovary. Leaves are entire, dense, and green to coppery. Except for the corolla lobes, this fits rather well the Rubiaceae, most probably a *Randia* sp., *R. aculeata* L. (Fig. 4.132b, c). This species usually has five corolla lobes but can have fewer.

Distribution Native from the Caribbean to Mexico and Colombia.

Names White indigoberry (English); Spanish names include *escontria*, *maíz tostado* (Colombia), *agalla de costa*, *espinoso*, *pitajoni bravo*, *pitajoni espinoso*, *yamaguey*, *yamaguey de costa* (Cuba); *crucilla*, *crucilla de la costa*, *papachilla* (Sinaloa, Mexico); *cruceto*, *crucilla* (Tamaulipas, Mexico); *crucete*, *espino cruz* (Veracruz, Mexico); *cambrón*, *escambrón*, *palo de cotorra*, *tintillo* (Puerto Rico) (Standley 1920–1926). A Nahuatl name is *tempixquiztli* (Díaz 1976).

Uses Ayensu (1981) records for this species: "Boil leaves for tea for fever in bones; bathe with cooled decoction for sores and infections. With *Cassia bahamensis* use roots of leaves for tea to help expel childbirth. Boil with *Tamarindus indica* as tea for fever and chills."

Fig. 4.132 *Randia aculeata*: (**a**) fol. 13v; (**b**) *R. aculeata* flowers with four corolla lobes. (Source: Sam Fraser-Smith, CC BY 2.0); (**c**) flowers with five corolla lobes. (Source: Scott Zona, CC BY 2.0)

Salicaceae

Fol. 100r#11. *Salix* sp.? (Fig. 4.133)

Description The phytomorph (Fig. 4.133a) has the appearance of a *Salix* sp.

Distribution *Salix humboldtiana* Willd. is a species widespread in Mexico (Fig. 4.133b).

Names The word accompanying this, , transliterates as *nāhuoha*, which may be a phonetic version of the Nahuatl *nahua,* to dance (Herrera 2004; Karttunen 1983; Siméon 2010), perhaps referring to the dance of the leaves in wind. *Salix lasiolepis* Benth. or *S. prardoxa* Kunth is known as *quetzala[h]uexotl, quetzalhuexotl, quetzalauexotl,* or *quetzalāhuexōtl* after fol. 57v of the

Codex Cruz-Badianus (de Ávila Blomberg 2012; Díaz 1976; Linares and Bye 2013; Ortiz de Montellano 1990; Reko 1947).

Uses Willowleaf tea, prepared with *S. taxifolia* Kunth, is still medicinal in Mexico today (Martinez 1969).

Fig. 4.133 *Salix* sp.?: (**a**) fol. 100r#11; (**b**) *S. humboldtiana*. (Source: http://www.conservamosica.org/conica/wp-content/uploads/2016/05/Salix_humboldtiana-Diapo2.jpg)

Saxifragaceae

Fol. 49r. *Lithophragma affine* (Fig. 4.134)

Description This phytomorph (Fig. 4.134a) has blue flowers with fringed petals and red calyces; leaves are numerous and borne tightly together on long petioles; roots are tuberous. This matches the variability of *Lithophragma affine* A. Gray (Fig. 4.134b, c), which intergrades with *L. parviflorum* (Hook.) Torr. & A. Gray (Park and Elvander 2012).

Distribution Native from Oregon to Baja California.

Names San Francisco woodland star (English).

Uses Unknown.

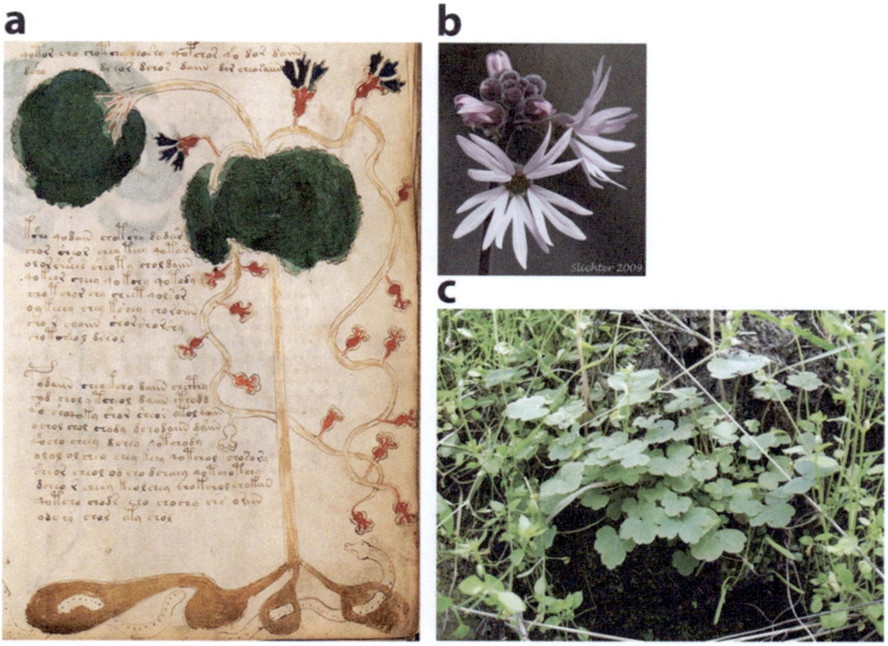

Fig. 4.134 *Lithophragma affine*: (**a**) fol. 49r; (**b**, **c**) inflorescence and leaves of *L. affine*, respectively. (Source: Paul Slichter and ©Rebecca Snyder www.fallbrooksource.com, respectively)

Solanaceae

Fol. 101r#3 and Fol. 101v(1)#2. *Capsicum annuum* (Fig. 4.135)

Description O'Neill (1944) identified a *Capsicum* sp. in the *Voynich Codex*. However, in the text he said fol. 101v, but the illustration provided was from fol. 101r. We agree that the shape and color of the fruits of both phytomorphs (Fig. 4.135a, b) agree with the genus *Capsicum*. Phytomorph #3 on fol. 101r has erect green fruits with depressed stem attachments and a forked primary root. Phytomorph #2 on fol. 101v has pendant red fruits and a forked primary root. Both fall within the wide variation of *C. annuum* L. (Fig. 4.135b, c).

Distribution The common red capsicum pepper originated in Mesoamerica and was introduced to Eurasia and Africa by the early sixteenth century.

Names Capsicum pepper, chili pepper, hot and sweet pepper (English); *chili* (Nahuatl). The name accompanying the phytomorph in fol. 101r#3, oHax, *atloca,* might be derived from the Nahuatl, *atococ*, drowned, perhaps alluding to its use to induce thirst. The name accompanying fol. 101v(1)#2, oHax, *atloca*, has not been translated. Nahuatl names, including those in Sahagún's *Florentine Codex*, are *ahui-*

yac tlatancuaye, *aji*, *chictecitl,* and *chile* (Díaz 1976; Estrada Lugo 1989). In Hernández et al. (1651:135–137) and Hernández (1942:430,431,433,434), it is *chilli* (Farfán and Elferink 2010; Ocaranza 2011).

Uses Widely used as food, and the hot (piquant) types, as a spice.

Fig. 4.135 *Capsicum annuum*: (**a**) fol. 101#3; (**b**) fol. 101v(1)#2; (**c**, **d**) green and red fruit of *C. annuum*. (Source: Consell Comarca Baix Empordà, CC BY 2.0 and H. Zell, CC BY-SA 3.0, respectively)

Fol. 21v. *Lycianthes moziniana*? (Fig. 4.136)

Description This phytomorph (Fig. 4.136a) displays a foliar organ unknown in nature. The logical solution is that the artist created artificial foliage to accommodate fragmented dried leaves, alternating adaxial and abaxial surfaces into a telescoping foliar organ; this idea indicates that we also cannot trust the opposite leaf arrangement. The flower, actinomorphic with five petals, darker bluish on the ribs, with an inferior ovary and reddish stamens, fits a *Lycianthes* sp. well, *e.g.*, *L. moziniana* (Dun.) Bitter. (Fig. 4.136b).

Distribution Native to Mexico, endemic to the highland plateau and Oaxana.

Names *Tlanochtle* (Nahuatl).

Uses Edible fruit used for ceremonial purposes (Lindsay 1994; Williams 1993).

Fig. 4.136 Fol. 21v. *Lycianthes moziniana*?: (**a**) fol. 21v; (**b**) *L. moziniana* flower. (Source: Courtesy of plantae mexicanae tenorianae, Pedro Tenorio-Lezama)

Urticaceae

Fol. 25r. *Urtica* sp., cf. *U. chamaedryoides* (Fig. 4.137)

Description This phytomorph (Fig. 4.137a) was first postulated by the Rev. Hugh O'Neill (1944) to be a member of the Urticaceae, or nettle family. The best match, because of the dentate, lanceolate leaves and reddish inflorescences, seems to be *Urtica chamaedryoides* Pursh, heart-leaf nettle (Fig. 4.137b).

Distribution Native from Canada to northern Mexico.

Names *Urtica* and the closely related genus *Urera* also occur in the *Codex Cruz-Badianus* and Hernández. *Urtica chamaedryoides* may be *atzitzicaztli/atzizicaztli/ā-tzītzicāztli* (water nettle) of fol. 16 of the Cruz-Badianus, also discussed in Hernández et al. (1651:356) and Hernández (1942:384) (Bye and Linares 2013; de Ávila Blomberg 2012; Díaz 1976; Farfán and Elferink 2010; Guerra 1952; Miranda and Valdés 1964; Reko 1947; Valdés Gutiérrez et al. 1992). Additional Nahuatl names are *chichicastli/chichicastle/tsitsikaxtli* (Díaz 1976).

Uses Unknown.

Fig. 4.137 *Urtica* sp.: (**a**) fol. 25v; (**b**) *U. chamaedryoides*. (Source: Steve Baskauf, CC BY-NC-SA 3.0)

Valerianaceae/Caprifoliaceae

Fol. 65r. *Valeriana albonervata* (Fig. 4.138)

Description This phytomorph (Fig. 4.138a) bears palmate or cleft-lobed leaves; inflorescence; and taproots that are napiform to fusiform, often forked. It is a good match for *Valeriana albonervata* B.L. Rob. (Fig. 4.138b), recently placed in the Caprifoliaceae.

Distribution Native to the Sierra Madre of Mexico, abundant in the Sierras of Tamaulipas and Nuevo Leon (Meyer 1951).

Names Unknown.

Uses Medicinal uses of the genus in Mexico are recorded by Martinez (1969).

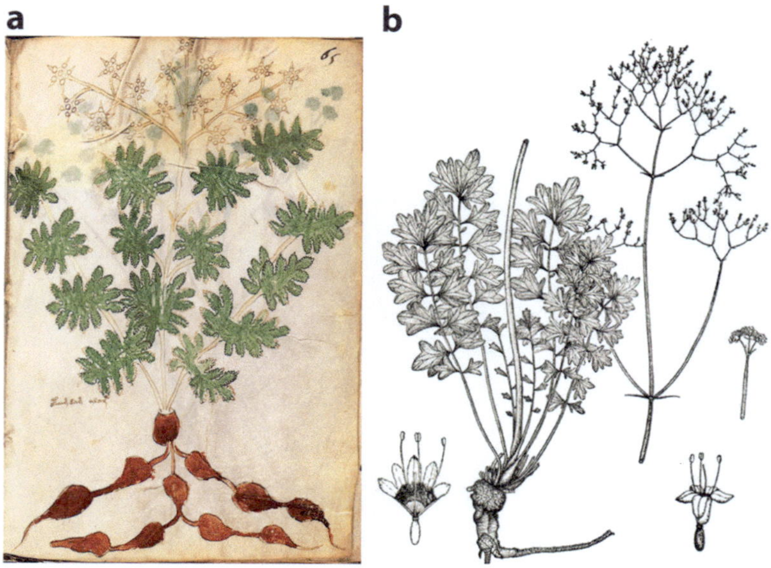

Fig. 4.138 *Valeriana albonervata*: (**a**) fol. 65r; (**b**) botanical drawing of *V. albonervata* from Meyer (1951)

Verbenaceae

Fol. 11r. *Citharexylum flabellifolium* (Fig. 4.139)

Description This phytomorph (Fig. 4.139a) has rotund to orbicular, crenate, gray to green leaves on a dense shrub and lobed, actinomorphic, bluish corollas. This matches *Citharexylum flabellifolium* S. Watson (Fig. 4.139b).

Distribution Native to Mexico, from Baja California Sur to Sonora.

Names *Citharexylum spinosum* L. (*C. fruticosum* L.) is called *penda* in Taino (García Bidó 2014). Florida fiddlewood (English). Spanish names in Mexico for this genus include *comida de cuervo* (Durango); *chachalaca* (Michoacán); *naranjilo, tepesi, roble, roble amarillo* (Veracruz) (Standley 1920–1926).

Uses Landscape plants.

Fig. 4.139 *Citharexylum flabellifolium*: (**a**) fol. 11r; (**b**) flower. (Source: Sue Carnahan, SEINet Arizona-New Mexico Chapter, CC BY-SA); (**c**) shoot. (Source: Sue Carnahan, SEINet Arizona-New Mexico Chapter, CC BY-SA)

Fol. 94r. *Duranta erecta* (*D. repens*) (Fig. 4.140)

Description This phytomorph (Fig. 4.140a) has terminal black fruits with a white terminal knob, arranged in a corymb or umbel. Leaves are obcordate and crenate. The roots are brown, thick and branched. Except for the shape of the inflorescence (Fig. 4.140b), this matches perfectly a dried specimen of the repent form of *Duranta erecta* L. (*D. repens* L.) (Fig. 4.140c). The fruits are golden yellow when fresh and are borne on a panicle, but when dried, they often quickly fall apart and turn black. Because the color of the fruits matches the color of a dried specimen, the sample presumably would have been fragmented when the artist prepared this illustration, and the artist was likely unaware of the color of the mature fruit on a living branch.

Distribution The shrub is native from Florida and Texas south to Argentina, but it is widely cultivated around the world in semitropical and tropical areas.

Names Golden dewdrops (English).

Uses Landscape plant, popular in the subtropics around the world.

Fig. 4.140 *Duranta erecta*: (*D. repens*) (**a**) fol. 94r; (**b**) fruit cluster of *D. erecta*. (Source: Hari Krishnan, CC BY-SA 3.0); (**c**) herbarium sheet of *D. erecta* (F1721033). (Source: The Field Museum, CC-BY-NC)

Fol. 10v. *Petrea volubilis* (Fig. 4.141)

Description This phytomorph (Fig. 4.141a) has both bluish calyces and corollas subtended by entire, petiolated leaves, the inflorescence weeping or vining. The inferior ovary is prominent below. This matches the variability of *Petrea volubilis* L. (Fig. 4.141b).

Distribution Vine is native to Mexico and South America.

Names Queen's wreath (English). Spanish and Mayan names include *jazmín azul*, *chaparrito* (Colombia); *choreque* (Costa Rica); *adelfa*, *flor de Jesús*, *lengua de vaca*, *adolfina* (El Salvador); *opp tzimin* (Yucatán, Mexico); *flor de Santa Mariá* (Tabasco, Mexico); *totopostillo*, *soltero* (Oaxaca, Mexico); *bejuco de caballo* (Yucatán, Veracruz, Mexico); *raspa-sombrero*, *jazmín*, *coamecate azul* (Veracruz, Mexico); *hoja chigüe* (Nicaragua); *buirá* (Panama); (Standley 1920–1926).

Uses Widely planted as a horticultural subject around the tropics, and a renowned medicinal herb. The Huastecs use this as a medicinal herb (Alcorn 1984); at Huasteca Potosina, Mexico, infusions of the leaves are used for kidney stones, rheumatism, diarrhea, and urinary infections under the name *raspasombrero* (Josabad Alonso-Castro et al. 2012). It has also been shown to be antidiabetic, anticancer, and antioxidant (Abdel-Hady et al. 2011; Sharma and Vijayvergia 2013).

Fig. 4.141 *Petrea volubilis*: (**a**) fol. 10v (**b**) inflorescence. (Source: J.M. Garg, CC-BY-SA-3.0,2.5, 2.0,1.0 GFDL); (**c**) dried shoot. (Source: Forest & Kim Starr, CC-BY-3.0)

Violaceae

Fol. 9v. *Viola bicolor* (*V. rafinesquei*) (Fig. 4.142)

Description This phytomorph (Fig. 4.142a) clearly shows linear, terminal stipular lobes, as in the North American native *V. bicolor* Pursh (*V. rafinesquei* Greene) (Fig. 4.142b)—not spatulate, as in the Eurasian *V. tricolor* L. Also, this phytomorph matches the blue flowers of *V. bicolor,* not the tricolored ones of *V. tricolor*; *V. bicolor* flowers are uniformly cream to blue, whereas those of *V. tricolor* usually have two purple upper petals and three cream to yellow lower petals. *Viola bicolor* is native from New Jersey to Texas and west to Arizona, with a center of diversity in eastern Texas (Shinners 1961; Russell 1965).

Distribution North America. The delineation of *V. bicolor* as native to North America and not introduced from elsewhere was only first fully elucidated by Shinners in 1961. Prior to 1961, *V. bicolor* was considered doubtfully native and was classified as a variety of the Eurasian *V. kitaibeliana* Schult. [var. *rafinesquii* (Greene) Fernald] and was often confused with *V. tricolor* and *V. arvensis* Murray (Fernald 1938). The North American *V. bicolor,* besides being seasonally dimorphic, with fertile and cleistogamous flowers, differs from the three other related species of Eurasia by "roundish, almost entire basal leaves, and by pectinate, palmately divided stipules. The petals of its open flowers are twice as long as the sepals, slightly shorter than the petals of *V. tricolor* but longer than those of *V. arvensis* and *V. kitaibeliana*." (Clausen et al. 1964). Thus, because *V. bicolor* was only known as a North American endemic since 1961, any attempt to propose a forgery prior to 1912 (Barlow 1986) will have to explain this anomaly.

Names American field pansy, wild pansy (English).

Uses Moerman (2009) records the uses of *Viola bicolor* among Native Americans as an analgesic, antidiarrheal, blood medicine, cold medicine, dermatological aid, respiratory aid, and tonic.

Fig. 4.142 *Viola bicolor*: (**a**) fol. 9v; (**b**) flowers and leaves of *V. bicolor*. (Source: Daniel Reed)

Zingiberaceae

Fol. 101v(1)#13. *Renealmia alpinia* (Fig. 4.143)

Description This phytomorph (Fig. 4.143a) is apparently a member of the Zingiberaceae, and *Renealmia alpinia* (Rottb.) Maas (Fig. 4.143b) fits rather well.

Distribution This species is native from Mexico to South America and the Caribbean.

Names In Oaxaca, the ripe fruits of different species of *Renealmia* are used in cooking as *cherimole* and *huilimole* (Kennedy 2010).

Uses Culinary herb. Therapeutic against snakebite. It has also been used to treat wounds, malignant tumors, epilepsy, and fungal infections (Gómez-Betancur and Benjumea 2014).

Fig. 4.143 *Renealmia alpinia*: (**a**) fol. 101v(1)#13; (**b**) R. alpine. (Source: https://i.ebayimg.com/images/g/XR4AAMXQWzNSm3UD/s-l500.jpg)

Hybrid Phytomorphs

Anemiaceae + Convolvulaceae

Fol. 19r. Anemiaceae: *Anemia mexicana*? + Convolvulaceae: *Ipomoea pubescens* (Fig. 4.144)

Description This phytomorph (Fig. 4.144a) combines fern fronds with an angiosperm flower. The morning glory blossom bears similarities to fol. 32v, *Ipomoea pubescens* Lam., silky morning-glory (Fig. 4.144b, see also Fig. 4.82). The foliage is that of a pinnate-pinnatifid, dimorphic fern frond with a terminal four-lobed morning-glory blossom. The roots are obvious but minimal. The fern frond with a branched fertile frond arising from the sterile frond simply does not match any fern on earth, but a guess might be an *Anemia* sp., probably *A. mexicana* Klotzsch (Fig. 4.144c).

Distribution *I. pubescens:* native from Texas to Mexico, also in Bolivia, Peru, and Argentina. *A. mexicana*: native from Texas to Mexico (Mickel and Smith 2004).

Names *I. pubescens*: silky morning glory (English). *A. mexicana*: Mexican flowering fern (English).

Uses Unknown.

Fig. 4.144 Anemiaceae: *Anemia mexicana*? + Convolvulaceae: *Ipomoea pubescens*: (**a**) fol. 19r; (**b**) *Ipomoea pubescens* flower. (Source: Apostolou Starvos); (**c**) *Anemia mexicana* foliage. (Source: Casa Flora, Inc.)

Araceae + Caryophyllaceae

Fol. 44v. Araceae: *Monstera deliciosa?* + Caryophyllaceae: *Silene* sp.? (Fig. 4.145)

Description The phytomorph (Fig. 4.145a), unlike any plant found in nature, bears a consistently oval pinnatifid leaf that is also crenate, on a woody shrub or vine. The closest match is *Monstera deliciosa* Liebm., Mexican breadfruit (Fig. 4.145b). This phytomorph is also combined with a denticidal capsule of the Caryophyllaceae, probably *Silene antirrhina* L. (Fig. 4.145c).

Distribution *M. deliciosa*: Native to tropical forest of southern Mexico, south to Panama.
Silene antirrhina: Widespread throughout North America and parts of South America.

Names *M. deliciosa*: American breadfruit (English). *S. antirrhina:* sleepy silene, sleepy catchfly (English).

Uses *M. deliciosa*: Ayensu (1981) records: "Root as internal alexiteric. Used against snake bite." Breedlove and Laughlin (1993) also discuss the medicinal value of *M. deliciosa.* All parts of the plant are poisonous except ripe fruit, which can be consumed. *S. antirrhina*: unknown.

Fig. 4.145 Araceae: *Monstera deliciosa*? + Caryophyllaceae: *Silene* sp.?: (**a**) Fol. 44v; (**b**) *M. deliciosa* fronds. (Source: http://wildlifeofhawaii.com/flowers/1438/monstera-deliciosa-monstera/); (**c**) *Silene antirrhina* flowers. (Source: Stan Shebs, CC BY-SA 3.0)

Arecaceae +?

Fol. 38r. + fol. 101v(1)#1 + fol 100r#6. Arecaceae: *Chamaedorea tuerckheimii* + ? (Fig. 4.146)

Description The phytomorph shown in Fig. 4.146a is a single palm frond (Johnson 1998). The five "apostrophes" are probably mottling due to mites. The five inflorescences close to the roots may actually be pedicels, re-assembled from dried, fragmented specimens, but the family is unknown. The roots are ample but not enlarged. This might match the variability of leaf of *Chamaedorea tuerckheimii* (Dammer) Burret, (Fig. 4.146b, c). Though not hybrid phytomorphs, the putative *C. tuerckheimii* from fol. 101v(1)#1 and fol. 100r#6 are also included here for comparison.

Distribution Native from Mexico (states of Chiapas, Tabasco, and Veracruz) to Guatemala in montane rainforests and cloud forests; very susceptible to mites (Hodel 1992).

Names Potato-chip palm. The name accompanying fol. 101v(1)#1, which is likely to be a *Chamaedorea* sp. (Fig. 4.146d), 8oℓℓo?, *chātlāe,* might be an orthographic variation of the Nahuatl *chāntli,* vagina (Herrera 2004; Karttunen 1983; Siméon 2010. The phytomorph on fol. 100r#6 (Fig. 4.146e) is not cleft but is probably a species of *Chamaedora;* the name, o2a?o, (*ānoeā,* Spanish), is a medical term meaning without intelligence, from the French *anoie* (de Echegaray 1889).

Uses In Oaxaca and El Salvador, the buds of *tepejilote* (*C. tepejilote* Liebm.) are used in cooking (Chizmar Fernández 2009; Kennedy 2010).

Fig. 4.146 Arecaceae +? *Chamaedorea tuerckheimii*: (**a**) fol. 38r; (**b**) *C. tuerckheimii* leaves. (Source; http://www.htbg.com/Arecaceae/PAL-010-11-4-134/); (**c**) *C. tuerckheimii* cleft leaf. (Source: https://www.rarepalmseeds.com/pix/ChaVer.shtml); (**d**) fol. 101v(1)#1; (**e**) fol. 100r#6

Asteraceae + Apiaceae

Fol. 31v. Asteraceae: *Elephantopus mollis* + Apiaceae: *Lomatium dissectum*? (Fig. 4.147)

Description This phytomorph (Fig. 4.147a) was obviously agglomerated from dried, fragmented plant parts. The bluish Asteraceae of subfamily Vernonieae has two green bracts and associated pinnately lobed foliage (Fig. 4.147b, c), a dead giveaway of the genus *Elephantopus*. This falls within the variability of *E. mollis* Kunth. The inflorescence from the Apiaceae is unknown buts bears resemblance to Fol. 41v, (*see* Fig. 4.15), *Lomatium dissectum* (Fig. 4.147d).

Distribution *E. mollis*: native from Mexico to South America but now naturalized worldwide. *L. dissectium*: native British Columbia to Mexico.

Names *E. mollis*: elephantsfoot (English). *L. dissectum*: biscuit root (English).

Uses *E. mollis:* Ayensu (1981) records: "Tea for colds and back pains with rum or *Desmodium* sp. and *Commelina* sp., vulnerary; tonic, diaphoretic." Kabiru and Por (2013) discuss the traditional uses of *E. mollis* in the tropics. Recently, *E. mollis* has been shown to be useful in treating cancer (Kuete et al. 2017; Lee et al. 1980; Ooi et al. 2014) and as an antimicrobial (Ragasa et al. 2009). *L. dissectum*: Moerman (1998) discusses the medicinal value among American Indians for a multitude of symptoms.

Fig. 4.147 Asteraceae: *Elephantopus mollis* + Apiaceae: *Lomatium dissectum*?: (**a**) fol. 31v; (**b**) *E. mollis* inflorescence. (Source: Ong Kwan Han @ natureloveyou.sg); (**c**) *E. mollis* leaves. (Source: Ong Kwan Han @ natureloveyou.sg); (**d**) *L. dissectum* inflorescence. (Source: iDigBio, CC BY)

Asteraceae + *Asteraceae*

Fol. 94v(1). Asteraceae: *Acourtia* sp.? + *Dahlia* sp.? (Fig. 4.148)

Description This phytomorph (Fig. 4.148a) combines an Asteraceae inflorescence, possibly subfamily *Mutisioideae,* possibly an *Acourtia* sp. for inflorescence and leaves (Fig. 4.148b, see Fig. 4.25) with the tuberous roots of a *Dahlia* sp. (Fig. 4.148c, see Fig. 4.36). Because the elliptic leaves are arranged unnaturally, this was probably re-created from dried, fragmented specimens.

Distribution Native to Mexico.

Names *Acourtia* sp.: desert peonies (English). *Dahlia* sp.: dahlia (English), *dalia* (Spanish), *acocotli* (Nahuatl).

Uses *Acourtia* sp.: Decoction of *A. thurberi* is antidiabetic and antihyperalgesic (Martínez et al. 2017). *Dahlia* sp.: Still used in gastrointestinal diseases by the Highland Maya of Chiapas, Mexico (Berlin and Berlin 1996) and was also used for many medicinal and nutritive properties by the Aztecs (Whitley 1985).

Fig. 4.148 Asteraceae: *Acourtia* sp.? + *Dahlia* sp.?: (**a**) fol. 94v(1); (**b**) leaves of *Acourtia* sp. (Source: Nickrent, D.L., Costea, M., Barcelona, J.F., Pelser, P.B. & Nixon, K. (2006 onwards) PhytoImages. Available from: http://www.phytoimages.siu.edu); (**c**) roots of *Dahlia* sp. (Source: F.D. Richards, CC BY-SA 2.0)

Fol. 54r. Asteraceae: *Calea zacatechichi*? + *Dahlia* sp.? (Fig. 4.149)

Description This phytomorph (Fig. 4.149a) is a conundrum. No member of the Asteraceae except *Eupatorium* and *Silphium* has perfoliate leaves, but the inflorescence is obviously neither of these genera. This was most likely assembled from dried, fragmented specimens, possibly of *Calea zacatechichi* Schltdl. (*C. terniflora* Kunth *pro parte*) (Fig. 4.149b, c), combined with what looks like the tuberous roots of a *Dahlia* sp. (Fig. 4.144d; *see* Fig. 4.36).

Fig. 4.149 Asteraceae: *Calea zacatechichi*? + *Dahlia* sp.: (**a**) fol. 54r; (**b**) *Calea* sp. dried specimen. (Source: http://dreamstudies.org/2011/09/27/getting-to-know-calea-zacatechichi-aka-the-dream-herb/); (**c**) *C. zacatechichi* inflorescence. (Source: Pineali, GNU Free Documentation License 2.1); (**d**) *Dahlia* sp. roots. (Source: F.D. Richards, CC BY-SA 2.0)

Distribution *C. ternifolia:* native to Mexico and Central America. *Dahlia* sp.: native to Mexico.

Names *C. zacatechichi*: Mexican dream herb, bittergrass, Mexican calea (English); *Dahlia* sp.: dahlia (English), *dalia* (Spanish), *acocotli* (Nahuatl). *Calea zacatechichi* may be the same as *ahuapatli/ahuapahtli/ahoapatli* of Hernández et al. (1651:133) and Hernández (1942:49) (Díaz 1976) or *chichicxihuitl* of fol. 32r of the *Codex Cruz-Badianus* (Díaz 1976; Miranda and Valdés 1964). Additional Nahuatl names might be *xaxachichitl, xikin, zacachichi/zacachichic, and zacatechichi* (Díaz 1976).

Uses *Calea* sp.: Martinez (1969) records extensive medicinal information for *C. ternifolia*. The medicinal value of *C. ternifolia* is also discussed by Aguilar (2010) and Breedlove and Laughlin (1993). Ayensu (1981) records for *C. jamaicensis* (L.) L.: "Cold remedy; fever and colds: head in steam of boiling plant." Asprey and Thornton (1953, 1954) record for this species:" cold remedy." González Stuart (2004) records: "Tea from the leaves promotes appetite and is also used to treat fever and stomach ailments, such as colic, and diarrhea."

Dahlia sp.: Still used in gastrointestinal diseases by the Highland Maya of Chiapas, Mexico (Berlin and Berlin 1996) and was also used for many medicinal and nutritive properties by the Aztecs (Whitley 1985).

Fol. 33v. Asteraceae: *Psacalium* sp. cf. *P. peltigerum*? + cf. *Pippenalia* sp.? cf. *P. delphinifolia* (Fig. 4.150)

Description This phytomorph (Fig. 4.150a) has lobed, peltate leaves and fleshy, round subterranean tubers. The inflorescence is characteristic of the tribe Heliantheae, and the "achenes" or cypselae are round and naked, a rare feature in the Asteraceae family. This illustration is a conundrum. The leaves and tubers suggest *Psacalium* sp., possibly *P. peltigerum* (B.L. Rob. & Seaton) Rydb. (Fig. 4.150b), but the large flower suggests a *Pippenalia* sp., possibly *P. delphinifolia* (Rydb.) McVaugh (Fig. 4.150c).

Distribution Both *Psacalium* and *Pippenalia a*re Mexican genera.

Names Indianbush (English) for *Psacallium* sp.

Uses Unknown.

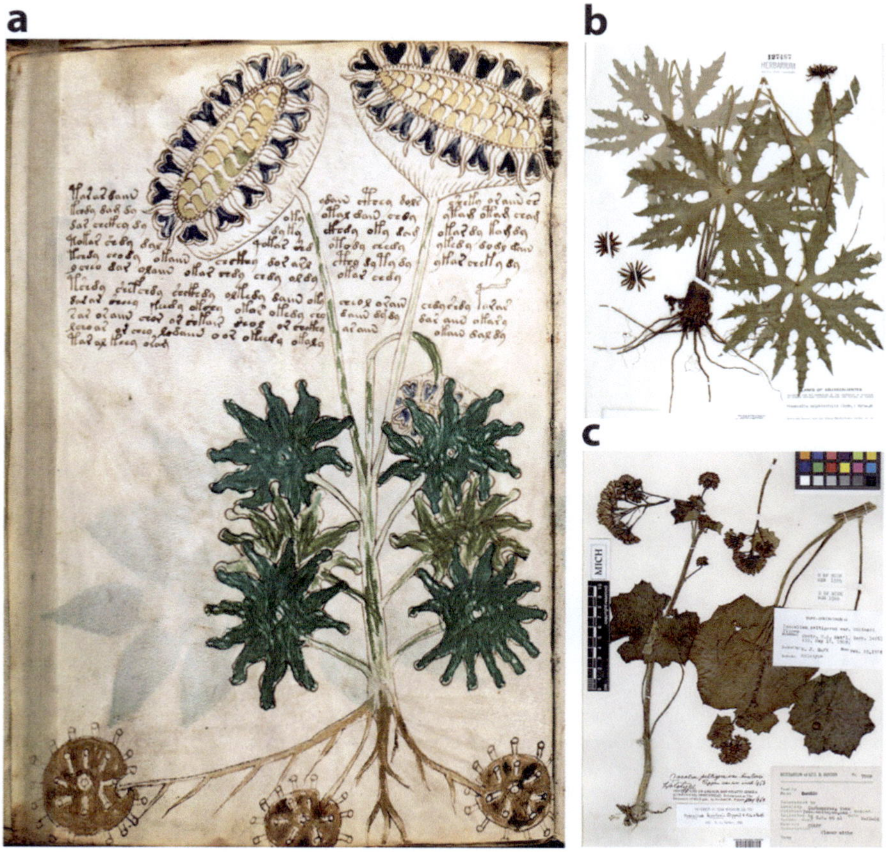

Fig. 4.150 Asteraceae: *Psacalium* sp.? + *Pippenalia* sp.?: (**a**) fol. 33v; (**b**) herbarium sheet of *Pippenalia delphinifolia* (ASU0029020). (Source: Arizona State University Herbarium); (**c**) herbarium sheet of *Psacalium peltigerum* var. *hintonii* R. W. Pippen (MICH1107637). (Source: University of Michigan Library Digital Collections. University of Michigan Herbarium Vascular Plant Type Collection with Specimen Images)

Fol. 65v. Asteraceae: *Chloracantha spinosa*? + *Stebbinsoseris heterocarpa*? (Fig. 4.151)

Description This phytomorph (Fig. 4.151a) consists of two Asteraceae. The primary one seems to be *Chloracantha spinosa* (Benth.) G. L. Nesom, subfamily Asterideae, tribe Astereae, in which the disk florets are rendered green instead of yellow (Fig. 4.151b). Grafted upon this is one bluish inflorescence of what seems to be *Stebbinsoseris heterocarpa* (Nutt.) K.L. Chambers, subfamily Cichorioideae (Fig. 4.151c).

Distribution Both native from California to Mexico.

Names *C. spinosa*: spiny chroracanta (English). *S. heterocarpa*: grassland silver-puffs (English).

Uses Unknown

Fig. 4.151 Asteraceae: *Chloracantha spinosa*? + *Stebbinsoseris heterocarpa*?: (**a**) fol. 65v; (**b**) *C. spinosa* inflorescence. (Source: http://www.smmflowers.org/bloom/species/Stebbinsoseris_heterocarpa.htm); (**c**) *S. heterocarpa* inflorescence. (Source: Peggy Romfh, Lady Bird Johnson Wildflower Center)

Asteraceae + Convolvulaceae

Fol. 40r. Asteraceae: *Lygodesmia* sp.? + Convolvulaceae: *Ipomoea plummerae* var. *cuneifolia*? (Fig. 4.152)

Description This phytomorph (Fig. 4.152a) was probably re-assembled from dried, fragmented specimens, probably more than one species. The twelve florets of this Asteraceae subfamily Cichorioideae, painted bluish, notably have crenate tips. The bifurcated stem, which has no natural counterpoint, bears palmate, deeply lobed leaves. Roots are ample but not tuberous. The closest match for the flowers is a *Lygodesmia* sp., *L. texana* (Torr. & A. Gray) Greene ex Small (Fig. 4.152b). The foliage seems to be a desert Convolvulaceae, *Ipomoea plummerae* A. Gray var. *cuneifolia* (A. Gray) J. F. Macbr. (Fig. 4.152c).

Distribution *Lygodesmia* sp.: native from Oklahoma, Texas, and New Mexico to Mexico. *I. plummeria*: Native to Nevada, New Mexico, Texas, and Mexico.

Names *Lycodesmia* sp.: Texas skeletonweed (English). *I. plummerae:* Huachuca Mountain morning glory (English).

Uses *Lycodesmia* sp.: The medicinal value is discussed by Moerman (1998). *I. plummerae*: Unknown.

Fig. 4.152 Asteraceae: *Lygodesmia* sp.? cf. *L. texana*+ Convolvulaceae: *Ipomoea plummerae* var. *cuneifolia*?: (**a**) fol. 40r; (**b**, **c**) *L. texana*. (Source: https://www.flickr.com/photos/20238838@N07/7185662826 & Robert L. Stone, Lady Bird Johnson Wildflower Center); (**d**) *I. plummerae* foliage. (Source: Western New Mexico University Department of Natural Sciences and the Dale A. Zimmerman Herbarium)

Asteraceae + Dioscoreaceae

Fol. 54v. Asteraceae: *Cyrtocymura scorpioides*? + Dioscoreaceae: *Dioscorea composita*? (Fig. 4.153)

Description This phytomorph (Fig. 4.153a) unites what may be a *Cyrtocymura* sp., possibly *C. scorpioides* (Lam.) H. Rob., subfamily Asteriodeae, tribe Vernonieae [*Vernonia scorpioides* (Lam.) Pers.] (Fig. 4.153b) with a *Dioscorea* sp., possibly *D. composita* Hemsl., seen in fol. 17v and Fig. 4.85 (Fig. 4.153c).

Distribution *C. scorpioides*: South America, Central America, Mexico. *D. composita*: native from southern Mexico to Costa Rica.

Names *C. scorpioides*: *erva-preá*, *assa peixe* (Spanish). *D. composita*: *barbasco* (Spanish).

Uses *C. scorpioides*: cytotoxic to tumor cells (Buskuhl et al. 2009; Pagno et al. 2006) as well as a treatment for skin inflammation (Pollo et al. 2013; Rauha et al. 2011). *D. composita*: used for both rheumatism and as a contraceptive in Mexico (Argueta Villamar et al. 2009).

Fig. 4.153 Asteraceae: Cyrtocymura scorpioides? + Dioscoreaceae: Dioscorea composita?: (**a**) fol. 54v; (**b**) C. scorpiodes flower. (Source: Michael Nee, C. V. Starr Virtual Herbarium, https://www.nybg.org/botany/nee/ambo/Checklist/images-ast/Cyrtocymura_scorp.jpg); (**c**) D. composita root. (Source: Syntex Pharmaceuticals Ltd., CC BY 4.0)

Asteraceae? + Euphorbiaceae?

Fol. 96r. Asteraceae: *Emilia fosbergii*? + Euphorbiaceae: *Manihot rubricaulis*? (Fig. 4.154)

Description This phytomorph (Fig. 4.154a) resembles the bird-pollinated flower (ornithophily) of an Asteraceae species grafted onto a member of the Euphorbiaceae. Unfortunately, the flower does not match any Asteraceae in Mesoamerica (Vogel 2015). Perhaps this is *Emilia fosbergii* Nicolson (Fig. 4.154b), similar to fol. 3v (*see* Fig. 4.37), grafted onto *Manihot rubricaulis* I. M. Johnst. of fol. 93v (*see* Fig. 4.97) (Fig. 4.154b). The tubers appear to be those of *Manihot esculenta* Crantz (Fig. 4.154d).

Distribution *E. fosbergii*: Semitropics of the world, including South America. *M. rubricaulis*: Mexico; native to South America, perhaps originating in Brazil.

Names *E. fosbergii*: Florida tasselflower, Cupid's shaving brush (English). *Manihot esculenta*: *mandioca* (Spanish), cassava (English).

Uses *E. fosbergii*: ornamental, treatment for high blood pressure. *M. rubricaulis*: tuberous roots are edible after treatment.

Fig. 4.154 Asteraceae: *Emilia fosbergii*? + Euphorbiaceae: *Manihot rubricaulis*?: (**a**) fol. 96r; (**b**) *E. fosbergii* flowers. (Source: Richard A. Howard, USDA Natural Resources Conservation Service); (**c**) *M. rubricaulis* foliage. (Source: Sue Carnahan, SEINet Arizona-New Mexico Chapter, CC BY-SA); (**d**) *M. esculenta* roots. (Source: David Monniaux, CC BY-SA)

Asteraceae + Ranunculaceae

Fol. 26v. Asteraceae: *Cyrtocymura scorpioides*? + Ranunculaceae: *Thalictrum* sp.? (Fig. 4.155)

Description This phytomorph (Fig. 4.155a) unites what may be a *Cyrtocymura* sp., possibly *C. scorpioides* (Lam.) H. Rob., subfamily Asteriodeae, tribe Vernonieae, with foliage that is overwhelmingly similar to fol. 89v(3) (*see* Fig. 4.156b). However, there is no extensive tuberous root, as in fol. 89v(3), so this might be another species of *Thalictrum*, such as *T. hernandezii* Tausch (Fig. 4.155c).

Distribution *C. scorpioides*: South America, Central America, Mexico. *T. hernandezii*: Mexico.

Names *C. scorpioides*: *erva-preá*, *assa peixe* (Spanish); *T. hernandezii*: Hernández's meadow rue, (English), cocoztic, cozticpatli, coztic patli acatlanensi/cozticpatli acatlanense (Nahuatl).

Uses *C. scorpioides*: toxic to tumor cells (Buskuhl et al. 2009; Pagno et al. 2006) as well as a treatment for skin inflammation (Pollo et al. 2013; Rauha et al. 2011). *T. hernandezii*: Antimalarial (Martinez 1969).

Fig. 4.155 Asteraceae: *Cyrtocymura scorpioides*? + Ranunculaceae: *Thalictrum hernandezii*?: (**a**) fol. 26v; (**b**) *C. scorpioides* flower. (Source: https://www.nybg.org/botany/nee/ambo/Checklist/images-ast/Cyrtocymura_scorp.jpg); (**c**) *T. hernandezii*. (Source: SEINet, Arizona-New Mexico Chapter)

Fol. 89v(3). Asteraceae: *Cyrtocymura scorpioides*? + Ranunculaceae: *Thalictrum fendleri*? (Fig. 4.156)

Description This phytomorph (Fig. 4.156a) unites what may be a *Cyrtocymura* sp., possibly *C. scorpioides* (Lam.) H. Rob., subfamily Asteriodeae, tribe Vernonieae (Fig. 4.156b), with a *Thalictrum* sp., possibly *T. fendleri* Engelm. ex A. Gray (Fig. 4.156c), which can have a tuberous roots, as in the phytomorph.

Distribution *C. scorpioides*: South America, Central America, Mexico. *T. fendleri*: native to western United States and northern Mexico.

Names *C. scorpioides*: *erva-preá, assa peixe* (Spanish). *T. fendleri*: Fendler's meadowrue (English).

Uses *C. scorpioides*: toxic to tumor cells (Buskuhl et al. 2009; Pagno et al. 2006) as well as a treatment for skin inflammation (Pollo et al. 2013; Rauha et al. 2011). Moerman (1998) discusses the value of *T. fendleri* among Indian tribes as a drug, dye, and other.

Fig. 4.156 Asteraceae: *Cyrtocymura scorpioides*? + Ranunculaceae: *Thalictrum fendleri*?: (**a**) fol. 89v(3); (**b**) *C. scorpioides* flower. (Source: https://www.nybg.org/botany/nee/ambo/Checklist/images-ast/Cyrtocymura_scorp.jpg); (**c**) *T. fendleri* foliage. (Source: Photo ©Al Schneider, www.swcoloradowildflowers.com); (**d**) *T. fendleri*. (Source: SEINet, Arizona-New Mexico Chapter)

Cactaceae + Grossulariaceae

Fol. 18v. Cactaceae: *Carnegiea gigantea* + Grossulariaceae: *Ribes malvaceum*? (Fig. 4.157)

Description The terminal flower of this phytomorph (Fig. 4.157a) has a scaly tube with many petals, along with many hairs that may be stamens. The green leaves are palmately lobed and simple, on a stoloniferous shrub. The flower is overwhelmingly that of *Carnegiea gigantea* (Engelm.) Britton & Rose (Fig. 4.157b), saguaro, native to the Sonoran desert; the foliage matches that of fol. 23r, *Ribes malvaceum* Sm. (Fig. 4.157c).

Distribution *C. gigantea:* Native to the Sonoran desert in the southwestern United States and Mexico. *R. malvaceum*: Native from California to Baja Norte, Mexico.

Names *C. gigantea*: *saguaro* (Spanish, English). *Ribes malvaceum*: chaparral currant (English).

Uses Moerman (1998) discusses the culinary and medicinal value of *C. gigantea* among the Pima, Apache, Chiricahua, and Mescalero, and the medicinal value of *R. malvaceum* among the Luiseño and Cahuilla. Standley (1920–1926) lists many edible uses of the fruit and seeds of saguaro but is silent on the medicinal value of the flowers.

Fig. 4.157 Cactaceae: *Carnegiea gigantea* + Grossulariaceae: *Ribes malvaceum?*: (**a**) fol. 18v. (**b**) *C. gigantea* flowers. (Source: http://www.fireflyforest.com/flowers/339/carnegiea-gigantea-saguaro/); (**c**) R. malvaceum foliage. (Source: Toedrifter, CC BY-SA 3.0).

Convolvulaceae + ?

Fol. 66v. Convolvulaceae: *Ipomoea nil* + ? (Fig. 4.158)

Description The two flowers of this phytomorph (Fig. 4.158a) resemble that of fol. 57r, i.e., *Ipomoea nil* (L.) Roth. The foliage might be a representation of abaxial and adaxial surfaces of rotund leaves, but it is difficult to identify. An alternative interpretation is that this may be *I. pes-caprae* (L.) R.Br., but the flowers in *I. pes-caprae* are larger and not picotee.

Distribution *I. nil*: Native to Mexico to Argentina; *I. pes-caprae*: pantropical.

Names *I. nil*: Picotee morning glory, ivy morning glory, Japanese morning glory (English); *I. pes-caprae*: bay hops (English), *bejuco de playa* (Spanish).

Uses Ornamental.

Fig. 4.158 Convolvulaceae: *Ipomoea nil*? +?: (**a**) fol. 66v; (**b**) *Ipomoea nil* flowers and leaves. (Source: Courtesy of Rare and Exotic Seeds); (**c**) *I. pes-caprae*. (Source: Primejyothi, CC BY-SA 3.0)

Cucurbitaceae + Dioscoreaceae?

Fol. 9r. Cucurbitaceae: *Ibervillea sonorae* + Dioscoreaceae: *Dioscorea* sp.? (Fig. 4.159)

Description This phytomorph (Fig. 4.159a) bears the resemblance of a "hybrid" phytomorph from dried specimens, i.e., *Ibervillea sonorae* (S. Watson) Greene (Fig. 4.157b) "grafted" on a *Dioscorea* sp. such as *D. mexicana* Scheidw. (Fig. 4.159c).

Distribution *I. sonorae*: Sonoran zone into Baja California, Mexico. *Dioscorea* spp.: native and cultivated throughout Mexico.

Names *I. sonorae*: *guarequi* (Mexico), *guareke, wereke, limoncillo, melon coyote* (Spanish); big root, coyote melon, cowpie plant (English). *Dioscorea mexicana*: *cabeza de negro* (Spanish), Mexican yam (English).

Uses *I. sonorae*: used in Mexico for various ailments. "The dried root is used to make a decoction to treat diabetes. Pulverized root is applied to skin as an antiseptic. Avoid in pregnancy and lactation." (González Stuart 2004). *Dioscorea spp.*: discussed by Moerman (1998).

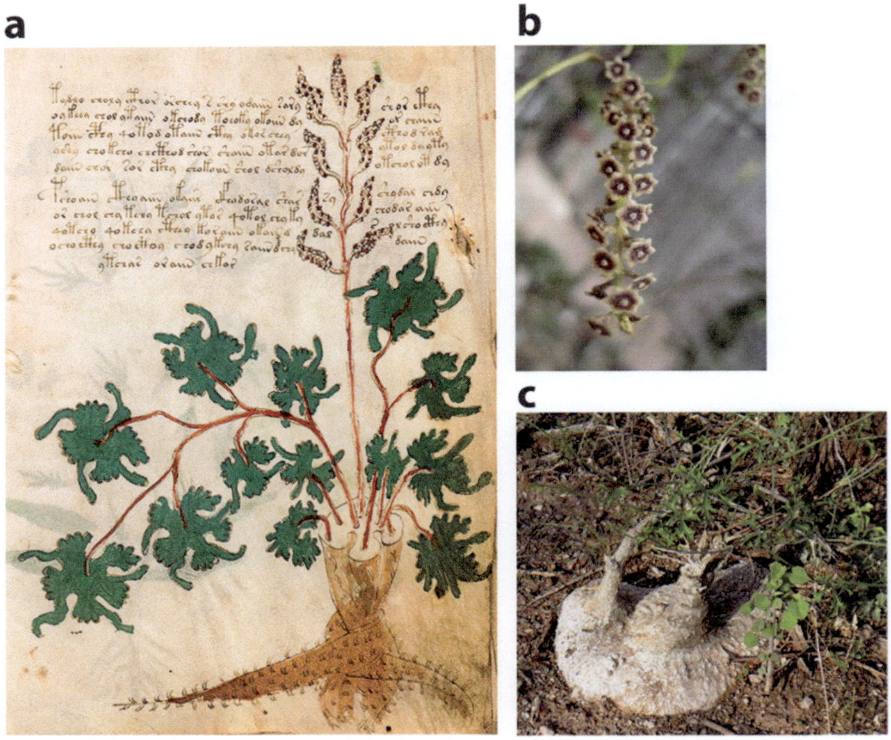

Fig. 4.159 Cucurbitaceae: *Ibervillea sonorae* + Dioscoreaceae: *Dioscorea* sp.?: (**a**) fol. 9r; (**b**) *I. sonorae* inflorescence. (Source: http://www.bihrmann.com/caudiciforms/subs/dio-mex-sub.asp); (**c**) *D. mexicana* root. (Source: T.R. Van Devender)

Fabaceae + ?

Fol. 28r. Fabaceae: *Styphnolobium burseroides*? + ? (Fig. 4.160)

Description This phytomorph (Fig. 4.160a) seemingly was created from fragmented, dried specimens. It represents a legume in the middle of leaves. The three loments, ridged from drying, seem to be the fruit of a *Parkinsonia, Sophora,* or

Styphnolobium sp. It might fall within the variability of the shrub *Styphnolobium burseroides* M. Sousa, Rudd & González Medrano, Francisco (Fig. 4.160b, c). The foliage remains unknown.

Distribution Native to Mexico.

Names Necklace pod (English).

Uses Unknown.

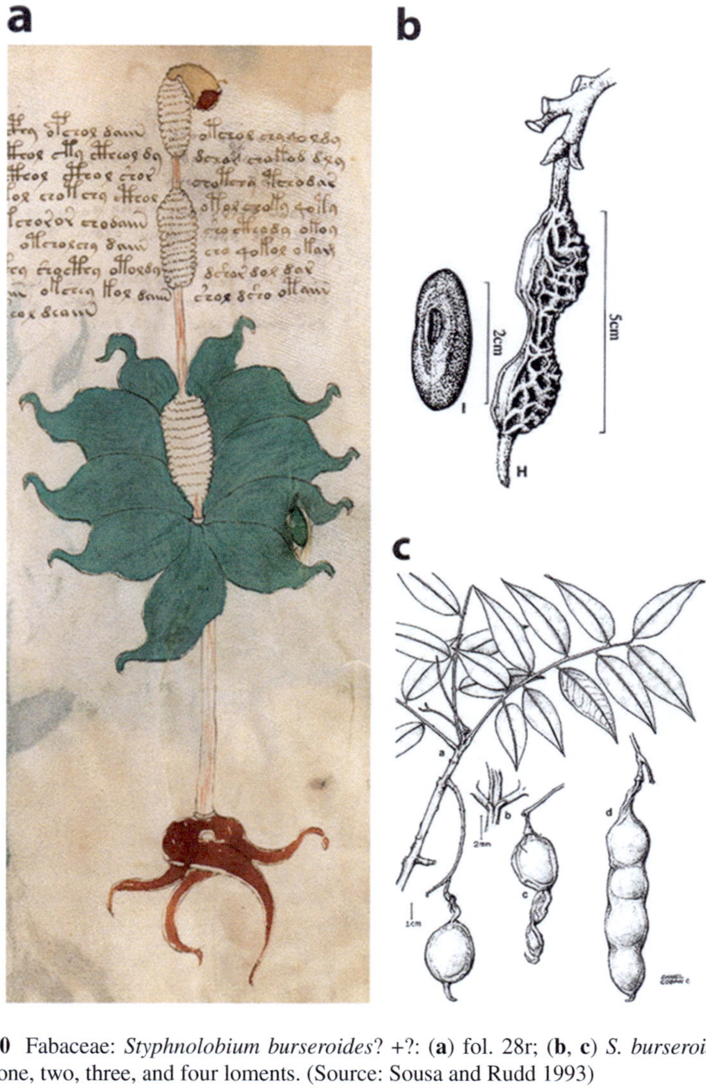

Fig. 4.160 Fabaceae: *Styphnolobium burseroides*? +?: (**a**) fol. 28r; (**b**, **c**) *S. burseroides* fruits, showing one, two, three, and four loments. (Source: Sousa and Rudd 1993)

Osmundaceae + Asteraceae

Fol. 90r. Osmundaceae: *Osmunda regalis* var. *spectabilis* + Asteraceae: *Vernonia* sp. (Fig. 4.161)

Description This phytomorph (Fig. 4.161a) has an Asteraceae subfamily Asterideae, tribe Vernonieae, inflorescence, but no species in the Asteraceae has twice pinnate foliage. This probably represents an inflorescence of a *Vernonia* sp. such as *Vernonia joyaliae* B. Turner grafted onto the frond of *Osmunda regalis* L. var. *spectabilis* (Willd.) A. Gray (Fig. 4.161c; *see* Fig. 4.6).

Distribution *O. regalis var. spectabilis*: native North and South America; *V. joyaliae:* native to Mexico.

Names *O. regalis*: royal fern (English). *V. joyaliae*: unknown.

Uses The medicinal value of *O. regalis* by the Iroquois is discussed by Moerman (1998); he also discusses the culinary and medicinal value of *Vernonia* spp.

Fig. 4.161 Osmundaceae: *Osmunda regalis* var. *spectabilis* + Asteraceae: *Vernonia* sp.: (**a**) fol. 90r; (**b**) *V. joyaliae*. (Source: SEINet, Arizona-New Mexico Chapter); (**c**) *O. regalis* var. *spectabilis* fronds. (Source: Online Virtual Flora of Wisconsin. 2018. http://wisflora.herbarium.wisc.edu. Accessed on June 06)

Papaveraceae + Moraceae

Fol. 89v(2). Papaveraceae: *Bocconia frutescens* + Moraceae: *Ficus pertusa*? (Fig. 4.162)

Description The foliage of this phytomorph (Fig. 4.162a) resembles *Bocconia frutescens* (Fig. 4.162b, *see* fol. 6r) except that it is paired with two *Ficus* fruits, probably *F. pertusa* L.f. (Fig. 4.162c).

Distribution *B. frutescens*: Native from Mexico to South America. *F. pertusa*: Native from Mexico to Brazil and Jamaica.

Names *B. frutescens*: *gordolobo, llorasangre, tabaquillo palo amarillo, palo de toro, pan cimarrón* (Spanish); plume poppy, tree poppy, tree celadine (English). *F. pertusa*: *bibosi palomo, mata-pau* (Spanish). *Bocconia frutescens* also has Nahuatl names of *cococ patli/cococpatli, iztecauhticmixitl* and *uitzocuitlapilxiuitl* (Díaz 1976; Estrada Lugo 1989); it may be the same as *cococ xihuitl* (stinging herb) in fol. 30v in the *Codex Cruz-Badianus* (Farfán & Elferink 2010; Miranda and Valdés 1964; Ortiz de Montellano 1975, 1990; Reko 1947). *Ficus pertusa* also has Nahuatl names of *amate de madrono, amatillo, amazquitl/amezquite, coamichin,* and *nacapuli* (Díaz 1976).

Fig. 4.162 Papaveraceae: *Bocconia frutescens* + Moraceae: *Ficus pertusa*?: (**a**) fol. 89v(2); (**b**) *B. frutescens* foliage. (Source: JMK, CC BY-SA 3.0); (**c**) *F. pertusa* foliage and fruits (Ibarra-Manríquez and Cornejo-Tenorio 2013)

Uses *B. frutescens:* Ornamental; in Mexico used to treat skin conditions and respiratory tract infections; extracts have antimicrobial action against thuman pathogens, *Escherichia coli*, *Staphyloccus aureu*s, and *Pseudomonas aeruginosa*; dye plant. *F. perusa*: In Mexico, used for toothache and inflammation of the breasts of women who breastfeed (lactifuge) (Argueta Villamar et al. 2009).

Papaveraceae + Rubiaceae?

Fol. 6r. Papaveraceae: *Bocconia frutescens* + Rubiaceae: *Ferdinandusa* sp.? (Fig. 4.163)

Description This phytomorph (Fig. 4.163a) bears the appearance of a "hybrid" phytomorph from dried specimens, *i.e.,* pinnate leaves from a tree poppy, *Bocconia frutescens* L. (Fig. 4.163b) and capsules of a *Ferdinandusa* sp. (Fig. 4.161c) with a heavy mealybug infestation, grafted together on the same plant.

Fig. 4.163 Papaveraceae: *Bocconia frutescens* + Rubiaceae: *Ferdinandusa* sp.?: (**a**) Fol. 6r; (**b**) *B. frutescens* foliage. (Source: JMK, CC BY-SA 3.0); (**c**) *F. panamensis* fruit. (Source: Dick Culbert, CC BY 2.0)

Distribution *B. frutescens*: Native from Mexico to South America. *Ferdinandusa* spp. are native from Central America to South America; for example, *F. panamensis* Standl. & L. O. Williams is native from Mexico to Colombia.

Names *B. frutescens*: *gordolobo*, *llorasangre*, *tabaquillo palo amarillo*, *palo de toro*, *pan cimarrón* (Spanish); plume poppy, tree poppy, tree celadine (English). *B. frutescens* also has Nahuatl names of *cococ patli/cococpatli*, *iztecauhticmixitl*, and *uitzocuitlapilxiuitl* (Díaz 1976; Estrada Lugo 1989); it may be the same as *cococ xihuitl* (stinging herb) in fol. 30v in the *Codex Cruz-Badianus* (Farfán and Elferink 2010; Miranda and Valdés 1964; Ortiz de Montellano 1975, 1990; Reko 1947). *F. panamensis*: unknown.

Uses *B. frutescens*: Standley (1920–1926) states: "The plant contains an alkaloid, protopine. The yellow or orange juice is very bitter and acrid and has a disagreeable odor. It is used for treating ulcers, skin eruptions, chilblains, bronchitis, and chronic ophthalmia, and to remove warts, and is said to have vermifuge and purgative properties. The leaves, too, are sometimes heated and applied as a poultice to wounds. In Colombia an infusion of the roots is valued as a remedy for jaundice and dropsy. In Jamaica, it is stated, the leaves are rubbed on house floors to keep away insects, and in Colombia the oil extracted from the seeds is used to destroy vermin on the head and skin." The medicinal value of *B. frutescens* is discussed by Breedlove and Laughlin (1993) and Chávez (2006). *F. panamensis*: unknown.

Polygonaceae + Asteraceae

Fol. 22r. Polygonaceae: *Oxyria digyna?* + Asteraceae: *Bartlettina* sp.? (Fig. 4.164)

Description This phytomorph (Fig. 4.164a) shows leaves with extreme cupping, which is usually a signal of infection, *such as the* infection of boxwood (*Buxus* spp.) with boxwood psyllid [*Cacopsylla* (*Psylla*) *buxi* (Linnaeus)]. However, mountain-sorrel, *Oxyria digyna* (L.) Hill (Fig. 4.164b), naturally exhibits degrees of cupping, and most of the inflorescence agrees with the Polygonaceae. The bluish inflorescence (Fig. 4.164c) cannot be identified beyond an Asteraceae, although it might be a *Bartlettina* sp. such as *Bartlettina sordida* (Less.) R. M. King & H. Rob. (*see* fol. 46v, Fig. 4.29).

Distribution *O. digyna*: circumboreal, south to New Mexico and Arizona.

Names *O. digyna*: mountain sorrel. *B. sordida:* blue mist flower, blue torch (English).

Uses *O. digyna*: The culinary value by the Alaska Native and Eskimo is discussed by Moerman (1998). *B. sordida*: ornamental.

Fig. 4.164 Polygonaceae: *Oxyria digyna?* + Asteraceae: *Bartlettina* sp.?: (**a**) fol. 22r; (**b**) *Oxyria digyna* (Source: Paul Slichter); (**c**) *Bartlettina sordida.* (Source: Raulbot, CC BY-SA 3.0)

Rubiaceae + Boraginaceae?

Fol. 15v. Rubiaceae: *Carapichea ipecacuanha* + Boraginaceae: *Tournefortia volubilis*? (Fig. 4.165)

Description This phytomorph (Fig. 4.165a) has the appearance of a "hybrid" phytomorph prepared from dried specimens, i.e., *Carapichea ipecacuanha* (Brot.) L. Andersson [*Cephaelis ipecacuanha* (Brot.) Stokes] (Fig. 4.165b) and an inflorescence of *Tournefortia volubilis* L. (Fig. 4.165c), "grafted" together on the same plant, united by a "cupule." The inflorescence bears similarities to the Boraginaceae, but the low foliage and tuberous roots bear more affinity to the Rubiaceae, and a cupule is unknown in extant plants, having gone extinct with the pteridosperms.

Distribution *C. ipacacuanha*: Native from Brazil to Nicaragua but widely cultivated. *T. volubilis*: Native from Texas to Mexico.

Names *C. ipecacuanha*: twining soldierbush (English). *T. volubilis*: *xulkin* (Yucatán, Maya); *comida de culebra*, Nicaragua; and *pringamoza*, Puerto Rico (Spanish) (Standley 1920–1926). The latter species might be *tlachichinoapatlahoac* in Hernández et al. (1651:292) and Hernández (1942:368) and (Díaz 1976).

Uses *C. ipecacuanha*: Used medically as a source of the emetic ipecac. *T. volubilis*: Ayensu (1981) records "Decoction for sores on people and animals. Leaves parched, ground up fine and dusted on sores. Boil for bathing aching or painful muscles. Used for female troubles and to 'restore man's manhood.'"

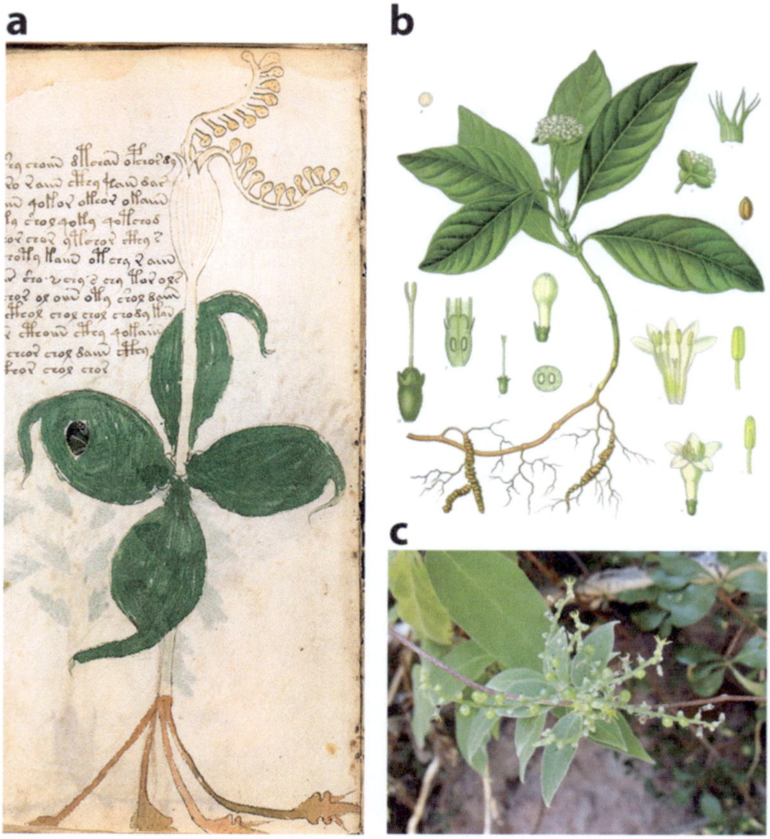

Fig. 4.165 Rubiaceae: *Carapichea ipecacuanha* + Boraginaceae: *Tournefortia volubilis*?: (**a**) fol. 15v; (**b**) *C. ipecacuanha* (Köhler et al. 1887); (**c**) *Tournefortia volubilis*. (Source: Sue Carnahan, SEINet Arizona-New Mexico Chapter)

Smilacaceae +?

Fol. 22v. Smilacaceae: *Smilax* sp. + ? (Fig. 4.166)

Description This phytomorph (Fig. 4.166a) shows a rhizome with epidermal prickles and jointed roots, very characteristic of briars, *Smilax* spp. (Fig. 4.166b), which have leaves with entire margins, never crenate, with typically cordate bases. The leaves have been apparently re-assembled from dried fragmented specimens, and the terminal inflorescence is too foliar to even guess a family.

Distribution *Smilax* is found in tropics and subtropics worldwide, including species in North America and South America.

Names Catbriars, greenbriers, smilaxes, sarsaparilla (English).

Uses Roots of numerous *Smilax* species are used for their medicinal value (Martinez 1969), and names and ethnobotanical uses are recorded by Standley (1920–1926) and Moerman (1998).

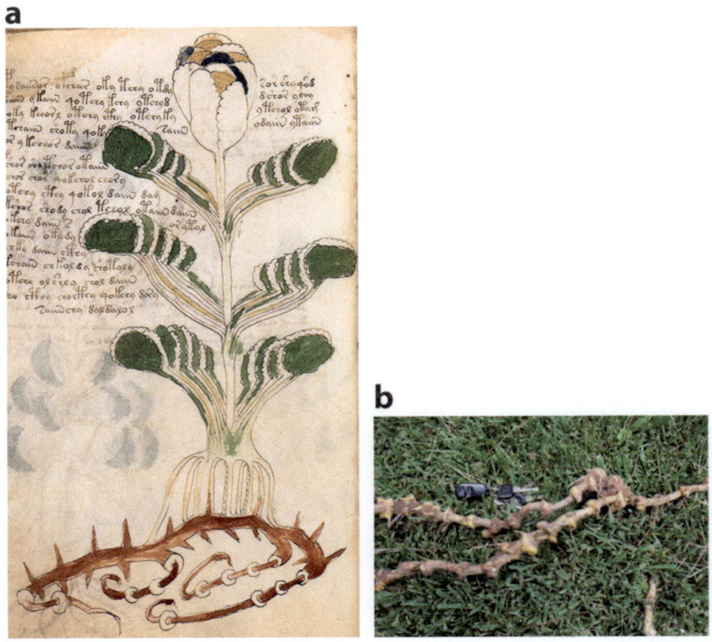

Fig. 4.166 Smilacaceae: *Smilax* sp.?: (**a**) fol. 22v; (**b**) *Smilax* sp. roots. (Source: http://www.walterreeves.com/landscaping/smilax-id-and-control)

Conclusion

With one exception, all the images identified in the *Voynich Codex* have been shown to be either plants indigenous to the New World or circumboreal (e.g., *Actaea rubra*). One European introduction has been indicated (*Tragopogon porrifolius*), likely a weedy contaminant in wheat introduced by the Spanish. This provides convincing evidence that the *Voynich Codex* is a post-Columbian manuscript. Although most of the plants are indigenous to Mexico or their range extends to Mexico, a number of plants are indigenous to North or South America (e.g., *Caulanthus heterophyllus*), sometimes out of the region of New Spain. This problem can be explained by the extensive trade routes of the Aztec merchants (*pochteca*). (See *Trade* in Chap. 2.) Most of the phytmorphs may have been illustrated from fresh specimens (e.g., *Viola bicolor*), but many seem to be drawn from dried samples (e.g., *Duranta erecta*), providing further evidence for purchase from the extensive Mexica markets. "Hybrid phytomorphs" seem to be unique to this herbal, perhaps to illustrate the use of two herbs within one preparation, or a mixture of dried samples from which the artist prepared the illustration.

The methods of plant illustration in the *Voynich Codex* are not those of the native Nahua of pre-Conquest New Spain, so what were their origins? Fray Motolinia, one of the twelve Franciscan priests who accompanied Cortés, remarked that the Nahua were extremely talented in copying Latin and Greek manuscripts, so much so that the original and copy were indistinguishable (Motolinia 1951). The Spanish friars routinely used European images of Biblical figures as inspiration for Nahua artists (*tlacuiloque*) (Camelo Arredondo et al. 1964; Morrill 2014). Furthermore, the library of the Colegio de Tlatelolco, the school for the Nahua elite created after the Conquest by Fray Zumarraga, the bishop of New Spain, had an eclectic collection of books including 51 from Lyon, 51 from Paris, 35 from Venice, 22 from Salamanca, 20 from Antwerp, 19 from Basle, 13 from Mexico, 11 from Cologne, 19 from Aleala de Henares, and 22 from Santiago de Tlatelolco (Gravier 2011).

The illustration style of plant illustrations in the Herbal section of the *Voynich Codex*, such as individual naturalistic illustrations interspersed with text, is common in ancient Dioscoridean illustrated herbals, such as the *Juliana Anica Codex* of 512 (Blunt and Raphael 1979; Collins 2000; Janick and Hummer 2012); they were continuously recopied for a millennium. Naturalism was reintroduced in the herbals of the *Secreta Saliterniana*, an Arab-influenced school based in Salerno, which also contain many of the iconographic features of the *Voynich Codex*. These Salerno-inspired herbals range in time from the *Tractatus de herbis*, British Museum, Egerton 747 of the first third of the fourteenth century, to the many variations of the *Livre des Simples Médecines* of the end of the 14th to first half of the fifteenth century (Collins 2000). The Salerno school greatly impacted European herbals prior to the Conquest of New Spain. There may have been European herbals in the well-stocked library of the Tlatelolco *Colegio* that influenced the *Voynich Codex*, as well as other herbals executed by native *tlacuiloque*. However, caution must be exercised until antecedent herbals can be identified (Emmart 1940; Peterson 1988).

It is likely that species identifications will be expanded and refined in the future as more taxonomists with expertise in Mexican, Mesoamerican, and New World flora become involved. Furthermore, we anticipate that the identification of plants in the Pharmaceutical section will be increased, as many are labeled in Voynichese. Likewise, the two vegetative phytomorphs in the Herbal section (fol. 27r, right side; fol. 52v) might be identified some day from the text. Because many are sketchily drawn and sometimes are limited only to roots or leaves, usually without inflorescences, it unlikely that the identification can be made on the morphology alone.

Literature Cited

Abdel-Hady, N.M., G.T.M. Dawoud, A.A. El-Hela, and T.A. Morsy. 2011. Interrelation of antioxidant, anticancer and antilieshmania effects of some selected Egyptian plants and their phenolic constituents. *Journal of the Egyptian Society of Parasitology* 41: 785–800.

Abdullah, E., R. Ahmad Raus, and P. Jamal. 2011. Evaluation of antibacterial activity of flowering plants and optimization of process conditions for the extraction of antibacterial compounds from *Spathiphyllum cannifolium* leaves. *African Journal of Biotechnology* 10 (81): 18679–18689.

Achigan-Dako, E.G., M.W. Pasquini, F. Assogba-Komlan, S. N'danikou, H. Yédomonhan, A. Dansi, and B. Ambrose-Oji. 2010. *Traditional vegetables in Benin*. Cotonou: Institut National des Recherches Agricoles du Bénin, Imprimeries du CENAP.

Afzal, K., M. Uzair, B. Ahmad Chaudhary, A. Ahmad, S. Afzal, and M. Saadullah. 2015. Genus *Ruellia*: Pharmacological and phytochemical importance in ethnopharmacology. *Acta Poloniae Pharmaceutica – Drug Research* 72: 821–827.

Aguilar, R. C. 2010. Plantas medicinales del Municipio de Tlalnelhuayocan, Veracruz. Thesis. Universidad Veracruzana.

Al-Shehbaz, I.A. 2012. Brassicaceae. In *The Jepson manual. Vascular plants of California*, ed. B.G. Baldwin, D.H. Goldman, D.J. Keil, R. Patterson, T.J. Bosatti, and D.H. Wilken, 2nd ed., 512–577. Berkeley: University of California Press.

Alcántara Rojas, B. 2008. In Nepapan Xochitl: The power of flowers in the works of Sahagún. In *Colors between two worlds: The Florentine Codex of Bernardino de Sahagún*, ed. L.A. Waldman, 106–431. Florence: Villa I Tatti.

Alcorn, J.B. 1984. *Huastec Mayan ethnobotany*. Austin: University of Texas Press.

Alviz, A.A., R.D. Salas, and L.A. Franco. 2013. Efecto diurético agudo de los extractos etanólico y acuoso de *Ceratopteris pteridoides* (Hook) en ratas normales. *Biomédica* 33: 115–121.

André, E. 1883. Onoseris drakeana. *Revue Horticole* 55: 180–182.

Argueta Villamar, A., L. M. Cano Asseleih, and M. E. Rodarte. 2009. Atlas de las Plantas de la medicina tradicional Mexicana. Biblioteca Digital de la Medecina Tradicional Mexicana. http://www.medicinatradicionalmexicana.unam.mx/atlas.php. 24 Jan 2014.

Asprey, G. F., and P. Thornton. 1953, 1954. Medicinal plants of Jamaica. Parts I and II. *West Indian Medical Journal* 2(4):233–52, 3(1):17–41.

Austin, D.F., R. Felger, and T.R. Van Devender. 2005. Nomenclature of *Ipomoea arborescens* (Convolvulaceae) in Sonora, Mexico. *Sida* 21: 1283–1292.

Ayensu, E.S. 1981. *Medicinal plants of the West Indies*. Algonac: Reference Publications.

Barlow, M. 1986. The Voynich manuscript—by Voynich? *Cryptologia* 10: 210–216.

Batres, L.D.P., C.A.B. Alfaro, and J. Ghaemghami. 2012. Mesoamerica aesthetics: Horticultural plants in hair and skin care. *Chronica Horticulturae* 50 (2): 12–15.

Bayer, R. J., A. L. Bogle, and D. M. Cherniawsky. 2006. *Petasites*. In: *Flora of North America North of Mexico*. Vol. 20, Magnoliophyta: Asteridae, Part 7: Asteraceae, Part 2, ed. Flora of North America Editorial Committee, 635–7. New York: Oxford University Press.

Belotto, E.M., T. Souza-Leal, A.L. Paraluppi, A.L.F.F. Furtado, F.O.G. Gaspi, and C. Pedroso-De-Morares. 2014. Caracterização anatômica, histoquímica e de classes de metabólitos secundários de folhas de *Guzmania* lingulata (L.) Mez. *Revista Brasileira de Plantas Medicinais* 16: 127–134.

Berg, C.C. 2001. Moreae, Artocarpeae, and *Dorstenia* (Moraceae). *Flora Neotropica* 83: 1–346.

Berlin, E.A., and B. Berlin. 1996. *Medical ethnobiology of the Highland Maya of Chiapas, Mexico: The gastrointestinal diseases*. Princeton: Princeton University Press.

Bertol, E., V. Fineschi, S.B. Karch, F. Mari, and I. Riezzo. 2004. *Nymphaea* cults in ancient Egypt and the new world: A lesson in empirical pharmacology. *Journal of the Royal Society of Medicine* 97: 84–85.

Betts, E.M. 1985. *Thomas Jefferson's garden book*, 1766–1824. Philadelphia: American Philosophical Society.

Blackman, B.K., M. Scascitelli, N.C. Kane, H.H. Luton, D.A. Rasmussen, R.A. Bye, D. Lentz, and L.H. Rieseberg. 2011. Sunflower domestication alleles support single domestication center in eastern North America. *Proceedings of the National Academy of Sciences (USA)* 108: 14360–14365.

Blaschke, M., M. Zehl, B. Hartl, C. Strauss, J. Winkler, E. Urban, G. Krupitza, and B. Kopp. 2017. Isolation of eudesmanes from *Pluchea odorata* and evaluation of their effects on cancer cell growth and tumor invasiveness in vitro. *Phytochemistry* 141: 37–47.

Bleiler, E.F., ed. 1976. *Early floral engravings: All 110 plates from the 1612 "Florilegium" by Emanuel Sweerts*. New York: Dover Publications.

Blunt, W., and S. Raphael. 1979. *The illustrated herbal*. London: Thames and Hudson.

Borschenius, F., L.S. Suárez Suárez, and L.M. Prince. 2012. Molecular phylogeny and redefined generic limits of *Calathea* (Marantaceae). *Systematic Botany Bot.* 37 (3): 620–635.

Borsutzky, M., T. Passie, W. Paetzold, H.M. Emrich, and U. Schneide. 2002. Hawaiianisch Holzrose: (Psycho-)Pharmakologische Wirkungen der Samen der *Argyreia nervosa*. *Der Nervenarzt* 73: 892–896.

Bown, D. 1988. *Aroids, plants of the arum family*. Portland: Timber Press.

Breedlove, D. E., and R. M. Laughlin. 1993. The flowering of man: A Tzotzil botany of Zinacantán, 2 vols. *Smithsonian Contributions to Anthropology* No. 35.

Buskuhl, H., R.A. Freitas, F. Delle Monache, A. Barison, F.R. Campos, Y.E. Corilo, M.N. Eberlin, and M.W. Biavatti. 2009. A new polyacetylene from *Vernonia scorpioides* (Lam.) Pers. (Asteraceae) and its *in vitro* antitumoral activity. *Journal of the Brazilian Chemical Society* 20: 1327–1333.

Bye, R.A., and E. Linares. 2013. Códice de la Cruz-Badiano: Medicine préhispánica. Primera parte. *Arqueología Mexicana* 50: 7–91.

Bye, R.A., E. Linares, and D.L. Lentz. 2009. México: Centro de origen de la domesticación del girasol. *Tip Revista Especializada en Ciencias Quimico-Biológicas* 12: 5–12.

Calzada, F., T. Juárez, N. García-Hernández, M. Valdes, O. Ávila, L. Mulia, and C. Velázquez. 2017. Antiprotozoal, antibacterial and antidiarrheal properties from the flowers of *Chiranthodendron pentadactylon* and isolated flavonoids. *Pharmacognosy Magazine* 13: 240–244.

Camporese, A., M.J. Balick, R. Avigo, R.G. Esposito, N. Morsellino, F. De Simone, and A. Tubaro. 2003. Screening of anti-bacterial activity of medicinal plants from Belize (Central America). *Economic Botany* 87: 103–107.

Camelo Arredondo, R., J. Gurría Lacroix, and C. Reyes Valerio. 1964. *Juan Gerson, Tlacuilo de Tecamachalco*. México: Departamento de Monomuntos Coloniales, Instituto Nacional de Anthropolgia e Historia.

Castro, V., R. Murillo, C.A. Klaas, C. Meunier, G. Mora, H.L. Pahl, and I. Merfort. 2000. Inhibition of the transcription factor NF-kB by sesquiterpene lactones from *Podachaenium eminens*. *Planta Medica* 66: 591–595.

Chávez, M. C. 2006. Flora medicinal de San Bartolo Tutotepec, Hidalgo. Ph.D. Thesis, Universidad Autónoma del Estado de Hidalgo, Pacuca de Soto, Hidalgo.

Chizmar Fernández, C. 2009. *Plantas comestibles de Centroamérica*. Costa Rica: Instituto Nacional de Biodiversidad, INBio, Santo Domingo de Heredia.

Clausen, J., R.B. Channell, and U. Nur. 1964. *Viola rafinesquii,* the only Melanium violet native to North America. *Rhodora* 66: 32–46.

Clayton, M., L. Guerrini, and A. de Avila. 2009. *Flora: The Aztec herbal*. London: Royal College Enterprises.

Collins, M. 2000. *Medieval herbals: The illustrative traditions*. London: The British Library.

Compton, J.A., A. Culham, and S.L. Jury. 1998. Reclassification of *Actaea* to include *Cimicifuga* and *Souliea* (Ranunculaceae): Phylogeny inferred from morphology, nrDNA ITS, and cpDNA *trn*L-F sequence variation. *Taxon* 47: 593–634.

Conard, H.S. 1905. *The waterlilies: A monograph of the genus Nymphaea*. Washington, DC: Carnegie Institution.

Correll, D.S., B.G. Schubert, H.S. Gentry, and W.O. Hawley. 1955. The search for plant precursors of cortisone. *Economic Botany* 9: 307–375.

de la Cruz, M., and J. Badiano. 1991. *Libellus de medicinalibus indorum herbis*. México: Fondo de Cultura Económica.

de Ávila Blomberg, A. 2012. Yerba del coyote, veneno del perro: La evidencia lexica para identificar plantas en el Códice de la Cruz Badiano. *Acta Botanica Mexicana* 100: 489–526.

de Echegaray, E. 1889. *Diccionario general etimológico de la lengua Española*. Madrid: Tomo primero. Ricardo Álvarez.

Díaz, J.L. 1976. *Índice y sinonimia de las plantas medicinales de México*. México: Instituto Mexicano para el Estudio de las Plantas.

Díaz, J.L. 1977. Ethnopharmacology of sacred psychoactive plants used by Indians of Mexico. *Annual Review of Pharmacology and Toxicology* 17: 647–675.

Díaz-Castelazo, D., V. Rico-Gray, P.S. Oliveira, and M. Cuautle. 2004. Extrafloral nectary-mediated ant-plant interactions in the coastal vegetation of Verucruz, Mexico: Richness, occurrence, seasonality, and ant foraging patterns. *Ecoscience* 11: 472–481.

Dobkin de Rios, M. 1974. The influence of psychotropic flora and fauna on Maya religion. *Current Anthropology* 15: 147–164.

Doskotch, R.W., and C.D. Hufford. 1969. Damsin, the cytotoxic principle of *Ambrosia ambrosioides* (Cav.) Payne. *Journal of Pharmaceutical Sciences* 58: 186–188.

Dressler, R.L. 1953. The pre-Columbian cultivated plants of Mexico. *Botanical Museum Leaflets* 16: 115–172.

Dunmire, W.W. 2004. *Gardens of New Spain: How Mediterranean plants and foods changed America*. Austin: University of Texas Press.

Dunn, D.B., and W.E. Harmon. 1977. The *Lupinus montanus* complex of Mexico and Central America. *Annals of the Missouri Botanical Garden* 64: 340–365.

Ebihara, A., J.-Y. Dubuisson, K. Iwatsuki, S. Hennequin, and M. Ito. 2006. A taxonomic revision of Hymenophyllaceae. *Blumea* 51: 221–280.

Emboden, W.A. 1979. *Nymphaea ampla* and other narcotics in Maya ritual and shamanism. *Mexico* 1 (4): 50–52.

Emboden, W.A. 1981a. Transcultural use of narcotic water lilies in ancient Egyptian and Maya drug ritual. *Journal of Ethnopharmacology* 3: 39–83.

Emboden, W.A. 1981b. Mayan-Egyptian uses of water lilies (Nymphaeaceae) in shamanic ritual drug use. In *Folk medicine and herbal healing*, ed. G.G. Meyer, K. Blum, and J.G. Cull, 275–286. Springfield: Charles C. Thomas Publishers.

Emboden, W.A. 1982a. The mushroom and the water lily: Literary and pictorial evidence for *Nymphaea* as a ritual psychogen in Mesoamerica. *Journal of Ethnopharmacology* 5: 139–148.

Emboden, W.A. 1982b. The water lily and the Maya scribe: An ethnobotanical interpretation. *New Scholar* 8: 103–127.

Emboden, W.A. 1983. The ethnobotany of the Dresden Codex with especial reference to the narcotic *Nymphaea ampla*. *Botanical Museum Leaflets* 29 (2): 87.

Emmart, E.W. 1940. *The Badianus manuscript (Codex Barberini, Latin 241)*. Baltimore: John Hopkins Press.

Encarcion Dimayuga, R., R.F. Murillo, and M.L. Pantoja. 1987. Traditional medicine of Baja California Sur (Mexico). II. *Journal of Ethnopharmacology* 20: 209–222.

Estrada Lugo, E. 1989. El Códice Florentino: Su información etnobotánica. Colegio de México. Postgraduados, Institución de Enseñanza e Investigación en Ciencias Agricolas, Montecillo.

Farfán, J.A.F., and J.G.R. Elferink. 2010. *Ethnobotany and Aztec sexuality*. Muenchen: Lincom Europa.

Félix-Silva, J., R. Brandt Giordani, A. Antonio da Silva Jr., S. M. Zucolotto, and M. de Freitas Fernandes-Pedrosa. 2014. *Jatropha gossypiifolia* L. (Euphorbiaceae): A review of traditional uses, phytochemistry, pharmacology, and toxicology of this medicinal plant. Evidence-Based Complementary and Alternative Medicine https://doi.org/10.1155/2014/369204.

Fernald, M.L. 1938. Noteworthy plants of southeastern Virginia. *Rhodora* 40: 434–459.

García Bidó, R. 2014. *Voces de Bohio: Vacabulario de la cultura Taina*. 2nd ed. ISBN: 978-1502964281.

Gates, W. 2000. *An Aztec herbal: The classic codex of 1552*. Mineola: Dover Publications, [first publ. Maya Society 1939, Baltimore].

Goldberg, A. 1967. The genus Melochia L. (Sterculiaceae). In *Contributions from the United States National Herbarium*, vol. 34, 191–363.
Gómez-Betancur, I., and D. Benjumea. 2014. Traditional use of the genus *Renealmia* and *Renealmia alpinia* (Rottb.) Maas (Zingiberaceae): A review in the treatment of snakebites. *Asian Pacific Journal of Tropical Medicine* 7: Suppl. 1):SS74–Suppl. 1):SS82.
González Stuart, A. 2004. Plants used in Mexican traditional medicine. http://www.herbal-safety.utep.edu/wp-content/uploads/2016/09/Plants-Used-in-Mexican-Traditional-Medicine-July-04.pdf. Accessed 31 Dec 2012.
Graham, J.G., A. Pudicho Torres, M. Pinedo Ceciliio, and J. Schunke Vigo. 2004. Interesting use of *Calathea standleyi* Macbride (Marantaceae). *Heliconia Society International Bulletin* 11 (1): 1–2.
Gravier, M.G. 2011. Sahagún's codex and book design in the indigenous context. In *Colors between two worlds. Florentine Codex of Bernardino de Sahagún*, ed. G. Wolf and J. Connors, 156–197. Milan: Officina Libraria.
Guerra, F. 1952. Indice etimologico. In *Libellus de medicinalibus Indorum herbis. El manuscrito pictórico mexicano-latino de Martín de la Cruz y Juan Badiano de 1552*, ed. M. de la Cruz, J. Badiano, and F. Guerra, 237–256. México: Estudio. Texto y Versión. Editorial Vargas Rea y El Diario Español.
Hancock, J.F. 2012. *Plant evolution and the origin of crop species*. Wallingford: CABI.
Harter, A., K.A. Gardner, D. Falush, D.L. Lentz, R.A. Bye, and L.H. Rieseberg. 2004. Origin of extant domesticated sunflowers in eastern North America. *Nature* 430: 201–205.
Heiser, C.B. 2008. How old is the sunflower in Mexico? *Proceedings of the Natlional Academy of Sciences (USA)* 105: E48.
Hernández, F. 1942. In *Historia de las Plantas de Nueva España*, ed. I. Ochoterena. Mexico: Imprenta University.
Hernández, F., F. Celsi, F. Colonna, B. Deversini, J. Faber, J. Greuter, V. Mascardi, N.A. Recchi, and J. Terentius. 1651. *Rerum medicarum Novae Hispaniae Thesarus, seu, Plantarum animalium mioneralium Mexicanorum historia*. Romae: Vitalis Mascardi.
Herrera, F. 2004. *Hippocrene concise dictionary.* Nahuatl-English, English-Nahuatl (Aztec). New York: Hippocrene Books.
Hitziger, M. 2016. Mayan phytotherapy in Guatemala: A transdisciplinary study for ethnographic documentation and local empowerment. Ph.D. thesis, ETH Zürich. https://doi.org/10.3929/ethz-a-010735971.
Hodel, D.R. 1992. *Chamaedorea palms*. Lawrence: Allen Press.
Hough, W. 1908. The pulque of Mexico. *Proceedings of the US National Museum* 33: 577–592.
Hutton, K. 2010. A comparative study of the plants used for medicinal purposes by the Creek and Seminole tribes. Ph.D. thesis, University of South Florida.
Ibarra-Manríquez, G., and G. Cornejo-Tenorio. 2013. *El género* Ficus *(Moraceae), silvestres y cultivados, de México.* Rapid Color Guide #394. Chicago: Field Museum.
Janick, J., and K.E. Hummer. 2012. The 1500th anniversary 512–2012 of the Juliana Anicia Codex: An illustrated Diorcoridean recension. *Chronica Horticulturae* 52 (3): 9–156.
Janick, J., and A.O. Tucker, eds. 2018. *Unraveling the Voynich Codex*. Cham: Springer Nature.
Janick, J., A. Whipkey, and J. Stolarczyk. 2013. Synteny of images in three illustrated Dioscoridean herbals: *Juliana Anicia Codex, Codex Neapolitanus,* and *Morgan 652. Notulae Botanicae Horti Agrobotanici Cluj-Napoca* 41 (2): 1–7.
Johnson, D.V. 1998. *Tropical palms. Non-wood forest products 10*. Rome: Food and Agriculture Organization of the United Nations.
Josabad Alonso-Castro, A., J. Jose Maldonado-Miranda, A. Zarate-Martinez, M.D.R. Jacobo-Salcedo, C. Fernández-Galicia, L. Alejandro Figueroa-Zuñiga, N. Abel Rios-Reyes, M. Angel de León-Rubio, N. Andrés Medellín-Castillo, A. Reyes-Munguia, R. Méndez-Martínez, and C. Carranza-Alvarez. 2012. Medicinal plants used in the Huasteca Potosina, México. *Journal of Ethnopharmacology* 143: 292–298.
Kabiru, A., and L.Y. Por. 2013. *Elephantopus species:* Traditional uses, pharmacological actions and chemical composition. *Advances in Life Science and Technology* 15: 6–13.
Karttunen, F. 1983. *An analytical dictionary of Nahuatl*. Norman: University of Oklahoma Press.

Kennedy, D. 2010. *Oaxaca al gusto: An infinite gastronomy*. Austin: University of Texas Press.

Khachitpongpanit, S., S. Singhatong, T. Sastraruji, and C. Jaikang. 2016. Phytochemical study of *Ruellia tuberosa* chloroform extract: Antioxidant and anticholinesterase activities. *Der Pharmacia Lettre* 8 (6): 238–244.

Killip, E.P. 1938. The American species of Passifloraceae. Chicago: Field Museum Press. *Fieldiana* 19: 1–331.

Kim, H., Q. Wang, C.F. Shoemaker, F. Zhong, G.E. Bartley, and W.H. Yokoyama. 2015. Polysaccharide gel coating of the leaves of *Brasenia schreberi* lowers plasma cholesterol in hamsters. *Journal of Traditional and Complementary Medicine* 5: 56–61.

Köhler, F.E., G. Pabst, W. Mülller, and K. Guinther. 1887. *Köhler's Medizinal-Pflanzen in naturgetreuen Abbildungen mit kurz erläuterndem Texte*. Germany: Gera-Untermhaus.

Konrath, E.L., B.M. Neves, C. dos S. Passos, P.S. Lunardi, M.G. Ortega, J.L. Cabrera, C.A. Gonçalves, and A.T. Henriques. 2012. *Huperzia quadrifariata* and *Huperzia reflexa* alkaloids inhibit acetylcholinesterase activity in vivo in mice brain. *Phytomedicine* 19: 1321–1324.

Král, M. 2014. Of dahlia myths and Aztec mythology: The dahlia in history. http://www.pugetsounddahlias.org/DahliaMyths.pdf. Accessed 28 Aug 2017.

Kuete, V., F.W. Fokou, O. Karaosmanoğlu, V.P. Beng, and H. Sivas. 2017. Cytotoxicity of the methanol extracts of *Elephantopus mollis, Kalanchoe crenata* and 4 other Cameroonian medicinal plants towards human carcinoma cells. *BMC Complementary and Alternative Medicine* 17 (1): 280.

Lee, K.-H., T. Ibuka, H. Furukawa, M. Kozuka, R.-Y. Wu, H.H. Hall, and H.-C. Huang. 1980. Antitumor agents XXXVIII: Isolation and structural elucidation of novel germacranolides and triterpenes from *Elephantopus mollis*. *Journal of Pharmaceutical Sciences* 68: 1050–1058.

Leighton, A. 1970. *Early American gardens "for meate or medicine"*. Boston: Houghton Mifflin Co.

Lentz, D.L., M.D. Pohl, J.L. Alvarado, S. Tarighat, and R. Bye. 2008a. Sunflower (*Helianthus annuus* L.) as a pre-Columbian domesticate in Mexico. *Proceedings of the National Academy of Sciences USA*. 105 (17): 6232–6237.

Lentz, D.L., M.D. Pohl, and R. Bye. 2008b. Reply to Rieseberg and Burke, Heiser, Brown, and Smith: Molecular, linguistic, and archaeological evidence for domesticated sunflower in pre-Columbian Mesoamerica. *Proceedings of the National Academy of Sciences USA* 105 (30): E49–E50.

Linares, E., and R. Bye. 2013. Códice de la Cruz-Badiano: Medicine préhispánica. Segunda parte. *Arqueología Mexicana* 51: 7–93.

Lindsay, R. 1994. Aztec fruit reappears in the mountains of Mexico. New Scientist 142(1922). https://www.newscientist.com/article/mg14219220-800-aztec-fruit-reappears-in-the-mountains-of-mexico/. Accessed 8 May 2018.

Ma, X., C. Tan, D. Zhou, D.R. Gang, and P. Xiao. 2007. Huperzine A from *Huperzia* species--an ethnopharmacolgical review. *Journal of Ethnopharmacology* 113 (1): 15–34.

Mali, P.Y., and S.S. Panchal. 2013. A review on phyto-pharmacological potentials of *Euphorbia thymifolia* L. *Ancient Science of Life* 32: 165–172.

Marshall, K.D., and S.L. Thornton. 2018. Worse than the disease? The rash of *Lomatium dissectum*. *Kansas Journal of Medicine* 11 (2): 1–6.

Martínez, A.L., A. Madariaga-Mazón, I. Rivero-Cruz, R. Bye, and R. Mata. 2017. Antidiabetic and antihyperalgesic effects of a decoction and compounds from *Acourtia thurberi*. *Planta Medica* 82: 534–544.

Martinez, M. 1969. *Las plantas medicinales de Mexico*. Fifth ed. Mexico: Ediciones Botas.

Mathias, M.E., and L. Constance. 1941. A synopsis of North American species of *Eryngium*. *American Midland Naturalist* 24: 361–87M.

McDonald, J.A., and B. Stross. 2012. Water Lily and the Cosmic Serpent: Equivalent conduits of the Maya spirit realm. *Journal of Ethnobiology* 32: 74–107.

McMahon, B. 1806. *The American gardener's calendar*. Philadelphia: B. Graves.

McPherson, G. 1981. Studies in *Ipomoea* (Convolvulaceae) I. The Arborescens group. *Annals of the Missouri Botanical Garden* 68: 527–545.

Mears, J. 1980. The evolution of the pseudoguaianolides of *Parthenium* L. (Asteraceae, Ambrosiinae). *Proceedings of the Academy of Natural Sciences of Philadelphia* 132: 156–172.

Meira, M., E.P. da Silva, J.M. David, and J.P. David. 2012. Review of the genus *Ipomoea:* Traditional uses, chemistry and biological activities. *Revista Brasileira de Farmacognosia* 22: 682–713.

Meyer, F.C. 1951. *Valeriana* in North America and the West Indies (Valerianaceae). *Annals of the Missouri Botanical Garden* 38: 377–503.

Mickel, J.T., and A.R. Smith. 2004. *The pteridophytes of Mexico*. New York: New York Botanical Garden.

Miller, M.D. 1970. Isolation and identification of isolysergic acid amide and isolysergic acid amide as the principal ergoline alkaloids in *Argyreia nervosa*, a tropical wood rose. *Journal of the Association of Official Analytical Chemists* 53: 123–128.

Millspaugh, C.F. 1892. *American medicinal plants*. Philadelphia: Yorston & Co.

Miranda, F., and J. Valdés. 1964. Comentarios botánicos. In *Libellus de medicinalibus indorum herbis manuscrito Azteca de 1552 segun traduccion Latina de Juan Badiano version Espanola con estudios y comentarious por diversos autores*, ed. M. de la Cruz, 243–284. México: Instituto Mexicano del Seguro Social.

Moerman, D.E. 1998. *Native American ethnobotany*. Portland: Timber Press.

Moerman, D.E. 2009. *Native American Medicinal Plants: An Ethnobotanical Dictionary*. Portland: Timber Press.

Moody, M.L., and L.H. Rieseberg. 2012. Sorting through the chaff, nDNA gene trees for phylogenetic inference and hybrid identification of annual sunflowers (*Helianthus* sect. *Helianthus*). *Molecular Phylogenetics and Evolution* 64: 145–155.

Moore, M. 1990. *Los remedios: Traditional herbal remedies of the Southwest*. Santa Fe, NM: Red Crane Books.

Morales, S.M., and H.O. Cruz. 2007. *Manual de Medicina Veterinaria Tradicional en la Huasteca Tamaulipeca, México*. México: Instituto Tamaulipeco para la Cultura y Las Artes.

Morrill, P.C. 2014. *The Casa del Deán: New World imagery in a sixteenth century Mexican mural cycle*. Austin: University of Texas Press.

Morton, J. K. 2005. Silene. In *Flora of North America*. Vol. 5, Magnoliophyta: Caryophyllidae, part 2, Flora of North America Editorial Committee (ed.), 168–216. New York: Oxford University Press.

Motolinia, T. de. 1951. *Motolinia's history of the Indians of New Spain*. Transl. F.B. Steck. Washington, DC: Academy of American Franciscan History.

Nana, F.W., A. Hilou, J.F. Millogo, and O.G. Nacoulma. 2012. Phytochemical composition, antioxidant and xanthine oxidase inhibitory activities of *Amaranthus cruentus* L. and *Amaranthus hybridus* L. extracts. *Pharmaceuticals* 5: 613–628.

Navarrete, A., D. Niño, B. Reyes, C. Sixtos, E. Aguirre, and E. Estrada. 1990. On the hypocholesteremic effect of *Eryngium heterophyllum*. *Fitoterapia* 61: 183–184.

Nieland, L., and W. Finley. 2009. *Lone star wildflowers*, 164–165. Lubbock: Texas Tech University Press.

Nesom, G.L. 1983. The evolutionary history if *Blumea viscosa* (Asteraceae) and a first report from North America. *Sida* 10: 30–32.

Ngadjui, B.T., and B.M. Abegaz. 2003. The chemistry and pharmacology of the genus *Dorstenia* (Moraceae). *Studies in Natural Products Chemistry* 29: 761–805.

Ocaranza, F. 2011. Historia de la medicina en México. 2nd ed. Cauhtémoc, México: Cien de Mexico.

Odonne, G., G. Herbette, V. Eparvier, G. Bourdy, R. Rojas, M. Sauvain, and D. Stien. 2010. Antileishmanial sesquiterpene lactones from *Pseudelephantopus spicatus*, a traditional remedy from the Chayahuita Amerindians (Peru). Part III. *Journal of Ethnopharmacology* 137: 875–879.

Olajuyigbe, O., and A. Ashafa. 2014. Chemical composition and antibacterial activity of essential oil of *Cosmos bipinnatus* Cav. leaves from South Africa. *Iranian Journal of Pharmaceutical Research* 13: 1417–1423.

O'Neill, H. 1944. Botanical observations on the Voynich MS. *Speculum* 19: 126.

Ooi, K.L., T.S. Tengku Muhammad, L.Y. Lam, and S.F. Sulaiman. 2014. Cytotoxic and apoptotic effects of ethyl acetate extract of *Elephantopus mollis* Kunth. In human liver carcinoma HepG2Cells through caspase-3 activation. *Integrative Cancer Therapies* 13 (3): NP1–NP9.

Ortiz de Montellano, B. 1975. Empirical Aztec medicine. *Science* 188: 215–220.

Ortiz de Montellano, B. 1990. *Aztec medicine, health, and nutrition*. New Brunswick: Rutgers University Press.

Pagno, T., L.Z. Blind, M.W. Biavatti, and M.R.O. Kreuger. 2006. Cytotoxic activity of the dichloromethane fraction from *Vernonia scorpioides* (Lam.) Pers. (Asteraceae) against Ehrlich's tumor cells in mice. *Brazilian Journal of Medical and Biological Research* 39: 1483–1491.

Paniagua-Ibáñez, M., A. López-Caamal, P. Mussali-Galante, E. Sánchez-Salinas, Ma.L. Ortiz-Hernández, R. Ramírez-Rodríguez, and E. Tovar-Sánchez. 2015. Morphological variation of *Cosmos bipinnatus* (Asteraceae) and its relation to abiotic variables in Central Mexico. *Revista Chilena de Historia Natural* 88: 14.

Pardeshi, M.H., A.A. Deshmukh, and K.A. Gajare. 2017. *Ruellia tuberosa Linn.* Acts as an anti-fertility agent that reduces sperm count, motility and viability in male Swiss albino mice (*Mus-musculus*). *International Journal of Current Pharmaceutical Research* 9 (1): 105–109.

Park, M.S., and P.E. Elvander. 2012. Saxifragaceae. In *The Jepson manual. Vascular plants of California*, ed. B.G. Baldwin, D.H. Goldman, D.J. Keil, R. Patterson, T.J. Bosatti, and D.H. Wilken, 2nd ed., 1234–1244. Berkeley: University of California Press.

Pereda-Miranda, R., D. Rosas-Ramírez, and J. Castañeda-Gómez. 2010. Resin glycosides from the morning glory family. *Fortschritte der Chemie Organischer Naturstoffe* 92: 77–153.

Peterson, J.F. 1988. The Florentine Codex imagery and the colonial tlacuilo. In *The work of Bernardino de Sahagun, pioneer ethnographer of sixteenth-century Aztec Mexico*, ed. J. Klor de Alva, H.B. Nicholson, and E. Quiñones Keber, 273–293. Austin: University of Texas Press.

Polllard, A.J. 1986. Variation in *Cnidoscolus texanus* in relation to herbivory. *Oecologia* 70: 411–413.

Pollo, L.A.E., C.F. Bosi, A.S. Leite, C. Rigotto, J. Kratz, C.M.O. Simões, D.E.P. Fonseca, D. Coimbra, G. Caramori, A. Nepel, F.R. Campos, A. Barison, and M.W. Biavatti. 2013. Polyacetylenes from the leaves of *Vernonia scorpioides* (Asteraceae) and their antiproliferative and antiherpetic activities. *Phytochemistry* 95: 375–383.

Prather, L.A. 1999. Systematics of *Cobaea* (Polemoniaceae). *Systematic Botany Monographs American Society of Plant Taxonomists* 57: 1–81.

Ragasa, C., A.B. Alimboyoguen, and C.-C. Shen. 2009. Antimicrobial triterpenoids from *Elephantopus mollis*. *NRCP Research Journal* 10 (1): 33–38.

Rajalakshmi, P., V. Sumathi, and M. Pugalenth. 2016. Antioxidant activity of *Erigeron karvinskianus* DC. and *Ageratina adenophora* (Spreng.) King (leaves). *International Journal of Food Science and Nutrition* 1 (5): 64–68.

Ramírez, J., and G.V. Alcocer. 1902. *Sinonimia vulgar y científica de las plantas mexicanas*. Mexico: Oficina Tipográfica de la Secretaría de Fomento.

Rands, R.L. 1953. The water lily in Maya art: A complex of alleged Asiatic origin. *Anthropological Papers* No. 34. *Bureau of American Ethnology Bulletin* 151: 75–153.

Rauha, L.K., C.D.S. Horinouchi, A.M.V. Loddi, E.F. Pietrovski, R. Neris, F. Souza-Fonseca-Guimarães, D.F. Buchi, M.W. Biavatti, M.F. Otuki, and D.A. Cabrini. 2011. Effectiveness of *Vernonia scorpioides* ethanolic extract against skin inflammatory processes. *Journal of Ethnopharmacology* 138: 390–397.

Rausa, Ahmad, R. Norhayati Abdullahb, N. Fakhriah Ismaila, and M. Shahbuddina. 2015. Extraction and evaluation of anti inflammatory activity of *Spathiphylllum cannifolium*. *Jurnal Teknologi (Sciences & Engineering)* 77 (24): 89–93.

Redonda-Martinez, R. 2017. Diversidad y distribución de la tribu Vernonieae (Asteraceae) en México. *Acta Botanica Mexicana* (119): 115–138.

Reko, B.P. 1947. Nombres bótanicos del manuscrito Badiano. *Boletín de la Sociedad Botánica de México* 5: 23–43.

Richards, J.H., M. Dow, and T. Troxler. 2010. Modeling *Nymphoides* architecture: A morphological analysis of *Nymphoides aquatica* (Menyanthaceae). *American Journal of Botany* 97: 1761–1771.

Rico-Gray, V. 1993. Use of plant-derived food resources by ants in the dry tropical lowlands of coastal Verucruz, Mexico. *Biotropica* 25: 301–315.

Rieseberg, L., and J.M. Burke. 2008. Molecular evidence and the origin of the domesticated sunflower. *Proceedings of the National Academy of Sciences USA* 105 (30): E46.

Roberts, E. and A. Hill. 2009. Opuntia chaffeyi. Tephrocactus Study Group 15(2):28–30. http://tephro.com/index.php/the-journal/article-index-by-title.html?id=21. Accessed 5 Oct 2018.

Ross-Ibarra, J., and A. Molina-Cruz. 2002. The ethnobotany of chaya (*Cnidoscolus aconitifolius* ssp. *aconitifolius* Breckon): A nutritious Maya vegetable. *Economic Botany* 56: 350–365.

Roys, R.L. 1976. *The ethno-botany of the Maya. Suppl. S. Cosminsky*. Philadelphia: Institute for the Study of Human Issues.

Ruiz-López, M.A., P.M. Garcia-López, R. Rodríguez, and J.F. Zamora Natera. 2010. Mexican wild lupines as a source of quinolizidine alkaloids of economic potential. *Polibotánica* 29: 159–164.

Russell, N.H. 1965. Violets (*Viola*) of Central and Eastern United States: An introductory survey. *Sida* 2: 1–113.

Safford, W.E. 1912. Sacred flowers of the Aztecs. *Volta Review* 14: 87–93.

Sahagún, B. de. 1963. Florentine Codex general history of the things of New Spain. Book 11 – Earthly things. Transl. C. E. Dibble and A. J. A. Anderson. Salt Lake City: University of Utah Press.

Sandoval, E., R.A. Bye, G. Rios, and M.I. Aguilar. 2005. Microscopic analysis and histochemical observations of the medicinal root of *Iostephane heterophylla*. *Boletin de la Sociedad Botánica de México* 77: 65–73.

Seelinger, M., R. Popescu, P. Seephonkai, J. Singhuber, B. Giessrigl, C. Unger, S. Bauer, K.-H. Wagner, M. Fritzer-Szekeres, T. Szekeres, R. Diaz, F.M. Tut, R. Frisch, B. Feistel, B. Kopp, and G. Krupitza. 2012. Fractionation of an extract of *Pluchea odorata* separates a property indicative for the induction of cell plasticity from one that inhibits a neoplastic phenotype. *Evidence-Based Complementary and Alternative Medicine*. https://doi.org/10.1155/2012/701927.

Sharma, P., and R. Vijayvergia. 2013. In vitro α-amylase inhibitory activity and GC-MS analysis of *Petrea volubilis*. *International Journal of Science and Research* 4 (4): 190–194.

Shinners, L.H. 1961. *Viola rafinesquii*: Nomenclature and native status. *Rhodora* 63: 327–335.

Siméon, R. 2010. *Diccionario de la lengua Nahuatl o Mexicana*. México: Siblo Veintiuno.

Song, H., Z. Zheng, J. Wu, Jia Lai, Q. Chu, and X. Zheng. 2016. White pitaya (*Hylocereus undatus*) juice attenuates insulin resistance and hepatic steatosis in diet-induced obese mice. *PLoS One* 11 (2): e0149670.

Sousa, S.M., and V.E. Rudd. 1993. Revision del Genero *Styphnolobium* (Leguminosae: Papilionoideae: Sophoreae). *Annals of the Missouri Botanical Garden* 80: 270–283.

Stockdale, M.K., and S.E. Bryenton. 1978. Indian plant foods of the Florida Panhandle. *The Florida Anthropologist* 31 (3): 109–116.

Standley, P.C. 1914. A revision of the genus Cobaea. *Contributions from the United States National Herbarium* 17: 448–458.

Standley, P.C. 1920–1926. Trees and shrubs of Mexico. *Contributions from the United States National Herbarium* 23: 1–1721.

Standley, P. C and L. O. Williams. 1973. Labiatae, mint family. In *Flora of Guatemala*, ed. P. C. Standley and L. O. Williams. Part IX, No. 3. *Fieldiana* 24:237–317.

Subba, B., and R.C. Kandel. 2012. Chemical composition and bioactivity of essential oil of *Ageratina adenophora* from Bhaktapur District of Nepal. *Journal of Nepal Chemical Society* 30: 78–86.

Swan, D.C. 2010. The North American lotus (*Nelumbo lutea* Willd.); Sacred food of the Osage People. *Ethnobotany Research & Applications* 8: 249–253.

Taiz, L., and S.L. Taiz. 2011. The biological section of the Voynich manuscript: A textbook of medieval plant physiology. *Chronica Horticulturae* 51 (2): 9–23.

The Plant List. 2013. Version 1.1. Published on the Internet; http://www.theplantlist.org/.

Therrell, L., D. Cole, V. Claassen, C. Ryan, and M. A. Davies. 2006. Plant selection, collection, and propagation techniques. Wilderness and backcountry site restoration guide. USDA Forest Service. https://www.fs.fed.us/t-d/pubs/htmlpubs/htm06232815/page11.htm#310. Accessed 5 June 2018.

Tilley, D., L. St, D. John, N. Shaw Ogle, and J. Cane. 2010. *Plant guide for fernleaf biscuitroot (Lomatium dissectum). USDA-Natural Resources Conservation Service*. Aberdeen: Idaho Plant Materials Center.

Tucker, A.O., and J. Janick. 2016. Identification of phytomorphs in the. *Voynich Codex Horticultural Reviews* 44: 1–64.

Tucker, A.O., and J. Janick. 2018. Phytomorph and geomorph identification, chapter 4. In *Unravelling the Voynich Codex*, ed. J. Janick and A.O. Tucker. Cham: Springer Nature.

Tucker, A.O., and R.H. Talbert. 2013. A preliminary analysis of the botany, zoology, and mineralogy of the Voynich manuscript. *HerbalGram* 100: 70–85.

USDA, ARS. 2015. National Genetics Resources Program. Germplasm Resources Information Network – GRIN [Online Database]. National Germplasm Resources Laboratory, Beltsville, Maryland. http://www.ars-grin.gov/cgi-bin/npgs/html/taxgenform.pl. 4 Oct 2018.

Valdés Gutiérrez, J., H. Flores Olivera, and H. Ochoterena-Booth. 1992. La botánica en el Codice de la Cruz. In *Estudios actuales sobre el libellus de medicinalibus Indorum herbis*, ed. J. Kumate, M.E. Pineda, C. Viesca, J. Sanfilippo, I. de la Peña Páez, J. Valdéz Gutiérrez, H. Flores Olivares, H. Ochoterena-Booth, and Z. Lozoya, 129–180. México: Secretaria de Salud.

Vázquez García, M., G. Lino von Poser, M. Apel, R. Cerros Tlatilpa, A. Mendoza-Ruiz, M. Luisa Villarreal, A. Teresinha Henriques, and A. Cardoso Taketa. 2017. Anticholinesterase activity and identification of huperzine A in three Mexican lycopods: *Huperzia cuernavacensis*, *Huperzia dichotoma* and *Huperzia linifolia* (Lycopodiaceae). *Pakistan Journal of Pharmaceutical Sciences* 30 (1. Suppl): 235–239.

Velázquez, C., F. Calzada, B. Esquival, E. Barbosa, and S. Calzada. 2009. Antisecretory activity from the flowers of *Chiranthodendron pentadactylon* and its flavonoids on intestinal fluid accumulation induced by *Vibrio cholerae* toxin in rats. *Journal of Ethnopharmacology* 126: 455–458.

Villaseñor, J.L. 2016. Checklist of the native vascular plants of Mexico. *Revista Mexicana de Biodiversidad* 87: 559–902.

Vogel, S. 2015. Vertebrate pollination in Compositae: Floral syndromes and field observations. *Stapfia* 103: 5–26.

Walshe-Roussel, B. 2014. An ethnobiological investigation of Q'eqchi' Maya and Cree of Eeyou Istchee immunomodulatory therapies. Ph.D. thesis, University of Ottawa, Canada.

Watson, T.J. 1977. The taxonomy of *Xylorhiza* (Asteraceae, Astereae). *Brittonia* 29: 199–216.

Whitley, G.R. 1985. The medicinal and nutritional and properties of *Dahlia* spp. *Journal of Ethnopharmacology* 14: 75–82.

Wiersema, J.H., A.R. Novelo, and J.R. Bonilla-Barbosa. 2008. Taxonomy and typification of *Nymphaea ampla* (Salisb.) DC. sensu lato (*Nymphaeaceae*). *Taxon* 57 (3): 967–974.

Williams, D.E. 1993. *Lycianthes moziniana* (Solanaceae): An underutilized Mexican food plant with new crop potential. *Economic Botany* 47: 387–400.

Winther, J.L., and W.E. Friedman. 2008. Arbuscular mycorrhizal associations in *Lycopodiaceae*. *New Phytologist* 177: 790–801.

Witters, W.L. 1975. Extraction and identification of clavine and lysergic acid alkaloids from morning glories. *Ohio Journal of Science* 75 (4): 198–201.

Wunderlin, R. P., B. F. Hansen, A. R. Franck, and F. B. Essig. 2018. Atlas of Florida plants. https://cdn.plantatlas.org/img/specimens/USF/118489.jpg.

Zepeda, G.C., and L.O. White. 2008. Herbolaria y pintura mural: Plantas medicinales en los murals del Convento del Divino Salvador de Malinalco. *Estado de México. Polibotanica* 25: 173–199.

Chapter 5
Aztec Plants in 16th Century New World and European Herbals

Illustrated 16th Century Herbals of Aztec Plant Resources

Extensive book burning by the Inquisition in New Spain in the sixteenth century (*see* Fig. 2.3) destroyed much contemporary information about Aztec plants. However, detailed images and text of Aztec plants survive from a number of contemporary resources (Table 5.1):

- An Aztec herbal entitled *Little Book of the Medicinal Herbs of the Indians* (*Libellus de Medicinalibus Idorum Herbis*), known as the *Badianus Manuscript* or, more properly, *Codex Cruz-Badianus* of 1552.
- The 13 volumes of Bernardino de Rivera, known as Sahagún, entitled the *General History of the Things of New Spain*, especially book XI, entitled *Earthy Things* (Sahagún 1963), also known as the *Florentine Codex*, written between 1545 and 1590.
- Botanical works of Francisco Hernández, naturalist and protomedico, based on a scientific expedition to New Spain between 1570 and 1577 but published later.
- The *Voynich Codex*, dated between 1565 and 1572 (Janick and Tucker 2018).
- The Spanish book *Historia medicinal de las cosas que se traen de nuestras Indias Occidentales* (1569, 1571) of Nicolás Monardes, translated as *Joyful newes of the the newe found worlde* by John Frampton (1577), consisting of discoveries and reports of travelers and explorers in New Spain regarding plants, animals, and minerals.
- *Herball* of John Gerarde (1597, 2nd ed. 1633).

A. O. Tucker, J. Janick, *Flora of the Voynich Codex*,
https://doi.org/10.1007/978-3-030-19377-5_5

Table 5.1 New world species listed in herbals of New Spain and Europe

Herbal	Year written	Total phytomorphs	New world plants illustrated
New Spain			
Codex Cruz-Bandianus[a]	1522	184	All
Florentine Codex[b]	1545–1590,	*ca.* 385	All
Hernández [c]	1570–1577	683	All
Voynich Codex	1565–1572 est.	359	169/170[f]
Europe			
Monardes[d]	1569, 1574, 1577, 1580	10	est. 6[g]
Gerarde[e]	1597, 1633	2766	>27[h]

[a]*Libellus de Medicinalibus Idorum Herbis,* 1522

[b]*Florentine Codex. General history of the things of New Spain*, 1951–1982

[c]*Rerum medicarum Novae Hispaniae Thesaurus, seu, Plantarum animalium mineralium Mexicanorum historia,* 1651

[d]*Joyfull newes of the newe founde worlde,* 1577

[e]*Herball,* amended edition of 1633 with sixteenth century woodcuts

[f]One species, *Tragopogon porrifolius*, a European species, is probably a late introduction to Mexico as a contaminant of imported wheat or rye

[g]Tobacco (*Nicotiana tobacco*), sassafras (*Smilax sp.*), carlo sancto *(unknown),* guacatone, (*Chiranthodendron sp.*), flower of mechoacan (*Ipomoea jalapa?),* fruit of balsam (species unknown, could be *Abies balsamea)*

[h]Balsam tree (*Abies balsamea*), cacao (*Theobroma cacao*), capsicum peppers (*Capsicum annuum*), datura (*Datura metel*), Drakes root (*Dorstenia contrajerva*), goldenrod (*Soldigao canadensis*), guaicum (*Guaiacum officinale),* kidney bean (*Phaseolus sp.*), maize (*Zea mays*), marvel-of-the-world (*Mirabilis jalapa*), mechoacan (*Ipomoea jalapa*), *pumpkins and squash* (*Cucurbita sp.*), *milkweed* (*Ascepias syriaca*), *nightshades* (*Datura* spp.), papaya (*Carica papaya*), passionflower (*Passiflora edulis),* pineapple (*Ananas comosus*), potato (*Solanum tuberosum*), sassafras (*Sassafras albidum*), smilax (*Smilax ornata),* sunflower (*Helianthus annuus*), sweetpotato (*Ipomoea batatas*), tobacco (*Nicotiana tabacum*), tomato (*Solanum lycopersicum*), Virginia spiderwort (*Tradescantia* sp.), yucca *(Yucca sp.)*

Codex Cruz-Badianus

This seminal 1552 herbal was written and illustrated at the famous *Colegio Imperial de Santa Cruz* at Taltelolco (Emmart 1940). This institution was formally established in 1536 at the site of an earlier school for Aztec elites (*calmecac*) to train sons of the Aztec nobility for the clergy. It was strongly opposed by the Dominicans, who questioned its orthodoxy and the ability of the natives to learn Christian doctrine, concluding that they should not be baptized and therefore could not enter the priesthood. By 1555, the First Council of Mexico decreed as part of the doctrine of blood purity (*limpieza de sangre*) that indigenous people, along with mestizos, Moors, and mulattos, were banned from serving as priests (Kirk and Rivett 2014:293).

The *Colegio* (under the deanship of Sahagún) became the intellectual center of New Spain. Many of the indigenous students, such as Antonio Valeriano, Martin Jacovita, and Pedro de San Buenaventura, had extraordinary talents and great intellectual ability (Emmart 1940; SilverMoon 2007). The authors of the renowned herbal known now as the *Codex Cruz-Badianus* were two indigenous members of the *Colegio*: Martin (*Martinus* in Latin) de la Cruz and Juan Badiano (*Juannes Badianus* in Latin), whose Spanish names were conferred upon their baptism. The author, Martin de la Cruz, was the *Colegio*'s indigenous doctor, who gave instruction in medicine. The illustrator is unknown. Juan Badiano, a Latin teacher at the *Colegio* and a former student, translated the Cruz text into Latin. He was born and reared in the village of Xochimilco, the site of the famous "floating" gardens or *chinampas* (SilverMoon 2007).

The herbal was sent to Spain as a gift soon after its completion in 1552 and was not mentioned by Sahagún or Francisco Hernández. At least four copies were made. The original was once owned by Cardinal Francesco Barberini, and it ended up in the Vatican Library until 1990, when John Paul II returned it to Mexico. It now resides in the National Institute of Anthropology and History in Mexico City. This herbal contains colored illustrations and descriptions in Latin of 185 plants.

Florentine Codex

Bernardino de Rivera, a Franciscan friar, missionary priest, and ethnographer of the Aztecs, was born in Sahagún, Spain (ca. 1499) and is generally known simply as Sahagún (Fig. 5.1a). He arrived in New Spain in 1529, learned Nahuatl, and spent more than 50 years studying the culture and history of the Aztecs; he also authored a dictionary of Nahuatl and Spanish. Sahagún was one of the founders of the *Colegio de Santa Cruz* at Tlatelolco in 1536 to train sons of the Aztec nobility to the priesthood. He served as dean in 1560, but retired to the main convent of San Francisco in Mexico in 1563. He died in 1590.

Sahagún made the *Colegio de Santa Cruz* a center for research, translation, and literary production, and it became a focus for the intellectual elite of New Spain. In the 1540s, Fray Toribio de Benavente (1482–1568), who was one of the 12 apostles of New Spain and was known as *Motolinía* (the poor one), charged Sahagún with the task of collecting information on the Aztecs, resulting in the seminal work translated as the *General History of the Things of New Spain*, written in Nahuatl and Spanish. Indigenous students at the *Colegio de Santa Cruz* served as translators and mediators. This celebrated work represents a collaborative effort by Sahagún and indigenous students and faculty to memorialize the preconquest culture of the Aztecs (SilverMoon 2007).

The work was translated into Nahuatl in 1569, and the final copy was completed in 1585. Three revisions exist. As the best-preserved manscript resides in Florence, it is usually referred to as the *Florentine Codex*, a premier example of ethnology and a key resource for understanding Aztec culture. A recent English translation, published in 12 volumes by the School of America, includes an insert of 965 illustrations from the historian Francesco del Paso y Tonoso, of which about 385 are plants.

Fig. 5.1 Herbalist portraits

The Botanical Works of Francisco Hernández de Toledo (1514–1587)

Francisco Hernández (1514–1587), known as Pliny of the Indies, was a medical doctor, naturalist, botanist, poet, and historian—truly a Renaissance man (Fig. 5.1b). Born in La Pueble de Montalban, near Toledo, Spain, he served as court physician to King Philip II in 1567. He is generally considered to have been of Jewish heritage, but the evidence is circumstantial; nothing is known of his early life (Varey et al. 2000a:33). In 1570 the king appointed him "General protomedico of our Indies, islands and mainland of the Ocean Sea" and he embarked on a seminal scientific expedition to New Spain to study medicinal plants. Accompanied by his son Juan, Hernández traveled for seven years, collecting and classifying specimens, interviewing indigenous people through interpreters, and conducting medical studies (Varey 2000a). He was assisted by three indigenous painters (baptized Pedro Vázquez and Antón and Baltasar Elías). He visited at least one of the Aztec botanical gardens. Hernández's 2500-page manuscript covered 3000 plants. He returned to Spain in 1577.

Parts of Francisco Hernández's extensive descriptions of his findings were published in a translation in Mexico in 1615 entitled, *Plantas y Animales de la Nueva Espana, y sus virtudes por Francisco Hernández, y de Latin en Romance por Fr. Francisco Ximenez* (Varey 2000b). Over 1200 plants are included. A text of the Roman edition of Hernández's work, entitled *Rerum Medicarum Novae Hispaniae Thesaurus (*1651) is available online, with 683 phytomorphs. The original mansucript was destroyed by fire in the Escorial in 1765.

Voynich Codex (1565–1572)

There are 359 plant images (phytomorphs) in the *Voynich Codex*, 131 in the Herbal section and 228 in the Pharmaceutical Section (Janick and Tucker 2018). Chapter 4 of this book identifies 129 images of the Herbal section (98.5%) and 41 images (18.0%) of the Pharmaceutical section.

Nicolás Bautista Monardes (1493–1585)

Nicolás Monardes, a physician and botanist, spent his entire life in Spain, where he authored various books on medical matters (Fig. 5.1c). His most famous work, *Historia medicinal de las cosas que se traen de nuestras Indias Occidentales* was published in three parts (1565, 1571, and 1574.) and translated to English in 1577 by John Frampton as *Joyful newes of the the newe found worlde*. Although Monardes never visited the New World, he had an extensive correspondence with various travelers, missionaries, and explorers, who provided information regarding medical uses of New World plants, animals, and minerals. He included only 10 woodcuts of plants, the most notable being tobac(c)o and sassafras; none occurs in the *Voynich Codex,* but an illustration of an armadillo is duplicated. The work was widely translated and may have been the impetus for King Philip II of Spain to send Francisco Hernández to explore the medicinal plants of New Spain.

John Gerard(e) (1545–1612)

John Gerard (Latinized to Gerarde), a barber surgeon in London (Fig. 5.1d), is the author of the best-known botanical work in Renaissance England (Arber 1938, Janick et al. 2012). His famous *Herball* of 1597 (1800 illustrations) is based on an incomplete translation of the *Stripium Historiae Pemptades sex* (1583) of the Flemish botanist Rembert Dodoens (1517–1585), translated from Latin to English by Robert Priest, a London physician. Gerard finished the task and reconfigured it in the arrangement of Mathias de L'Obel and Petrus Pena's *Nova stirpium adveraria* of 1570.

The graceful prose of Gerard contains interjected comments, many based on plant performance in his gardens. The woodcuts used by Gerard were from the collection of Nicolaus Bassaeus of Frankfurt; these illustrations (about 1800, 16 original) were mostly recycled from the *Eicones Plantarum seu stirpium* or *Kräuterbuch* of the German botanist Jacobus Theodorus Tabernaemontanus, published in 1590. Gerard's 1597 herball is famous for the first description and image of potato (*Solanum tuberosum*). The amended edition of 1633, edited by the London apothecary Thomas Johnson, included new material, incorporating many more New World crops, with 2766 woodcuts obtained from the publisher, Christophe Plantin (1513–1588).

Plant Families, Genera, and Species of Mexico and the New World

Diversity of Mexican Flora

There is great diversity of native vascular plants in Mexico. An inventory by Jose Luis Villaseñor (2016) lists 297 families, 2854 genera, and 23,314 species (1039 species of ferns and lycophytes, 149 gymnosperms, and 22,126 angiosperms). The floristic richness of Mexico is considered the fourth greatest in the world, and the number of endemic species is second only to South Africa. Our objective is to compare the phytomorphs in the *Voynich Codex* with illustrations in four other sources:

the *Codex Cruz Badianus* herbal of 1522 (Emmart 1940); the illustrations in the *Florentine Codex* of Sahagún, completed in 1585 (trans. by Dibble and Anderson 1963); illustrations of Francisco Hernández conpleted between 1570 and 1573 and collected in the *Rerum Medicarum Novae Hispaniae Thesaurus (*1651) and *Historia de las Plantas de Nueva España* (1942–1945); and the amended edition of the famous 1597 English *Herball* of John Gerard, enlarged and revised by Thomas Johnson in 1633.

Aztec Plant Names

Appendix A lists 1607 Aztec plant names from various sources. In this list, multiple identifications were made of the same phytomorph because of their lack of critical botanical characteristics. Furthermore, the species identifications were made by non-botanists, by unknown methods. These identifications thus remain hypotheses unless confirmed subsequently by later sources, and thus are not credited. Thirdly, some of the previously identified splecies are not necessarily native to Mexico, but native to surrounding areas of North and Central America and the West Indies; that is, they could have been grown at one of the botanic gardens or obtained at one of the large markets.

This list also includes a few words of non-Nahuatl that seem to have been used in Central Mexico. A few introduced, non-native species, such as ginger, *Zingiber officinale* Roscoe, or the weedy sow thistle, *Sonchus oleraceus* L., could have been plausibly cited by the sixteenth century sources. Species that are actually native to geographically distant areas and not introduced, and some other preposterous identifications, have been ignored: for example, *Aloe variegata* L., correctly *Gonialoe variegata* (L.) Boatwr. & J.C. Manning of southern Africa, was identified in the *Florentine Codex* by Estrada Lugo (1989). Species that are taxonomically incorrect have also been discarded or modified; for example, Estrada Lugo (1989) claimed *Bourreria tenella* Humb., Bonpl. & Kunth and *Portulaca rubra* L. in the *Florentine Codex*, but these species were apparently never published (or at least not listed in any modern databases). The citation of illegitimate names, such as *Salvia polystachya* Ortega by Estrada Lugo (1989) in the *Florentine Codex*, are a conundrum. Some names, such as *Saponaria americana* by Estrada Lugo (1989) in the *Florentine Codex*, appear to be entirely fictitious. Some Nahuatl names simply cannot be identified to species. Emmart (1940) created plate numbers for folio numbers of phytomorphs in the *Codex Cruz-Badianus*, but both are cited. There have been many post–sixteenth century commentaries, which can be found in the Literature Cited for Appendix A.

Plants cited as being used by the Nahua, arranged by family with updated nomenclature, are summarized in Appendix B. (Some species of bacteria and algae have no Nahuatl name recorded). Of the 149 plant familes listed in Appendix B, 85 are

Table 5.2 The 149 families of plants identified as used by the Nahua[a]

The 86 families of plants identified from sixteenth century sources:
Adiantaceae, Adoxaceae, Aizoaceae, Alismataceae, Anacardiaceae, Alstroemeriaceae, Altingiaceae, Amaryllidacea, Anacardiaceae, Annonaceae, Aristolochiaceae, Aspleniaceae, Basellaceae, Begoniaceae, Berberidaceae, Betulaceae, Bignoniaceae, Bixaceae, Burseraceae, Calophyllaceae, Cannabaceae, Cannaceae, Caprifoliaceae, Caricaceae, Chenopodiaceae, Chrysobalanaceae, Combretaceae, Commelinaceae, Corariaceae, Cornaceae, Cupressaceae, Cyatheaceae, Cyperaceae, Dilleniaceae, Ebenaceae, Ephedraceae, Equisetacreae, Fagaceae, Garryaceae, Hydrangeaceae, Iridaceae, Juncaceae, Lauraceae, Loasaceae, Loganiaceae, Loranthaceae, Lythraceae, Magnoliaceae, Malpighiaceae, Maracelianceae, Marciliaceae, Melanthiaceae, Melastomataceae, Meliaceae, Menispermaceae, Molluginaceae, Myricaceae, Myrtaceae, Nyctaginaceae, Orchidaceae, Orobanchaceae, Oxalidaceae, Pentaphylacaceae, Phrymaceae, Phytolaccaceae, Picramniaceae, Piperaceae, Plantaginaceae, Plumbaginaceae, Polygalaceae, Polypodiaceae, Portulacaceae, Primulaceae, Rhamnaceae, Rutaceae, Sapindaceae, Sapotaceae, Selaginellaceae, Stegnospermataceae, Tropaeolaceae, Typhaceae, Vitaceae, Vitariaceae, Xanthorrhoeaceae, Zamiaceae, Zygophyllaceae
The 14 unique families from *the Voynich Codex*:
Anemiaceae, Araliaceae, Cabombaceae, Grossulariaceae, Hymenophyllaceae, Nyctaginaceae, Ophioglossaceae, Osmundaceae, Penthoraceae, Pteridaceae, Saxifragaceae, Schizaeaceae, Taxodiaceae, Violaceae
The 49 families in both groups:
Acanthaceae, Amaranthaceae, Apiaceae, Apocynaceae, Araceae, Arecaceae, Asparagaceae/Agavaceae, Asteraceae, Boraginaceae, Brassicaceae, Bromeliaceae, Cactaceae, Campanulaceae, Caryophyllaceae, Convolvulaceae, Crassulaceae, Cucurbitaceae, Dioscoreaceae, Ericaceae, Euphorbiaceae, Fabaceae, Geraniacea, Gesneriaceae, Huperziaceae/Lycopodiaceae, Lamiaceae, Liliaceae, Malvaceae, Marantaceae, Menyathaceae, Moraceae, Nelumbonaceae, Nymphiaceae, Onagraceae, Papaveraceae, Passifloraceae, Pinaceae, Poaceae, Polemoniaceae, Polygonaceae, Ranunculaceae, Rosaceae, Rubiaceae, Salicaceae, Smilacaceae, Solanaceae, Urticaceae, Valerianaceae, Verbenaceae, Zingiberaceae

[a]134 (89.9%) were identified from sixteenth century sources; 64 (43%) were identified in the *Voynich Codex*

unique to the other sixteenth century sources, 14 are unique to the *Voynich Codex*, and 49 are found in both groups (Table 5.2). This represents 50.2% of the 297 familes in Mexico reported by Villaseñor (2016). Of the 1122 species listed in Appendix B, 923 unique to sixteenth century sources, 175 are unique to the *Voynich Codex,* and 24 are common to both. This would indicate that Nahua people were aware of about 1122 species, representing 4.8% of the 23,313 species in Mexico, attesting to the great botanical knowledge of the Aztecs.

Voynich Codex Phytomorphs

In the *Voynich Codex*, 170 of 359 phytomorphs are identified in Chap. 4, including 129 of 131 in the Herbal section and 41 of 229 in the Pharmaceutical section.These consist of 136 genera (4.7% of the total in Mexico) and 64 families, 22.8% of the

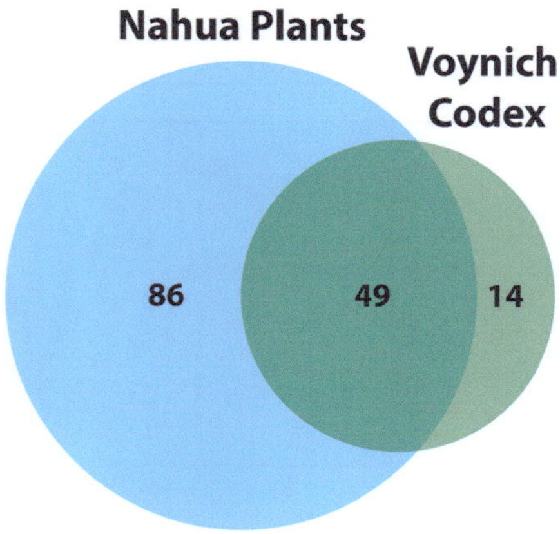

Fig. 5.2 Venn diagram of the number of plant families of vascular plants from various sixteenth century sources (149 total), identified in Table 5.2. There were 134 families from various sixteenth century sources and 64 from the *Voynich Codex*, with 49 in common. A caveat for these figures must be inserted here: The amorphous phytomorphs in sixteenth century illustrations often have more than one identification from botanists

total in Mexico. These 64 families represent 43.0% of the 149 families listed in Table 5.2, summarized in a Venn diagram (Fig. 5.2). The actual numbers may be higher, because 12 pages of the Herbal section are missing. As noted in Chap. 4, of 170 phytomorphs identified, only one (*Tragopogon porrifolius*) is an Old World species, probably a late introduction to Mexico.

There were overlaps of 17 phytomorphs representing 15 families in the *Voynich Codex* with the four herbal sources investigated (Fig. 5.3). Two phytomorphs show overlaps in all five herbals (*Opuntia* sp., *Capsicum* sp); three overlap in four herbals (*Helianthus* sp., *Bocconia* sp., *Smilax* sp.), four overlap in three herbals (*Agave* sp., *Manihot* sp., *Taxodium* sp., *Passiflora* sp.), and eight overlap in two herbals (*Chiranthodendron* sp., *Dahlia* sp., *Dorstenia* sp, *Iostephane* sp., *Ipomoea* sp., *Philodendron sp.*, *Salix* sp., *Urtica*/*Urera* sp.). Species overlap was also found for armadillo (*Dasypus* sp.), New World placental mammals (Fig. 5.4).

The duplication of phytomorphs and families in the *Voynich Codex* with other Mexican herbals in the sixteenth century as well as Gerard's *Herball* is evidence that the *Voynich Codex* must be included among contemprorary herbals devoted to New World plants. Although ignored, the *Voynich Codex* needs to be considered as one of the primary sources of early information on Mexican plants, taking its place as one of the great resources of New Word botany.

Voynich Codex (1568–1572)	Cruz-Badianus (1552)[a]	Sahagún (1585)[b]	Hernández (1651)[c] (1942)[d]	Gerarde (1597, 1633)[e]
***Opuntia* sp., Cactaceae**				
				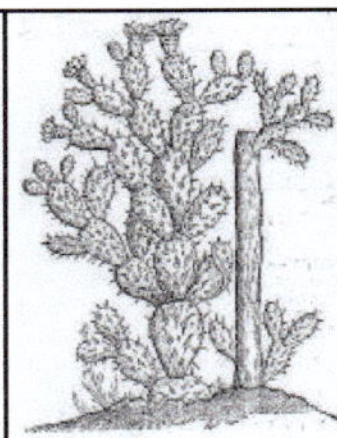
folio 100r#8	folio 49v *tlatoc nochtli* sown or planted nopal	plate 433 *nopalli*	p. 78 Cap. 45[c] *nocheznopalli/ nopalnocheztli*	p. 1512 Ficus indica The Indian fig tree
***Capsicum* sp., Solanaceae**				

folio 101r#3 *Capsicum annnuum*	Folio 46r *C. annnuum*	plate 572 *tlalamatl*	p. 135 Cap. 3[c] chilli	p. 364 Capsicum longioribus siliquis, Long codded Ginny Pepper

Fig. 5.3 Examples of Phytomorph Overlaps in New World and European Sources. (For identification of Nahuatl names, see Appendix A. [a]Illustrations from Emmart (1940); [b]Illustration numbers from Sahagún (1963); [c]Page and Cap. numbers from Hernández (1651); [d]Page, Cap., and Fig. numbers from Hernández 1942; [e]Illustrations and names from Gerarde (1633))

Voynich Codex (1568–1572)	Cruz-Badianus (1552)[a]	Sahagún (1585)[b]	Hernández (1651)[c] (1942)[d]	Gerarde (1597, 1633)[e]
***Helianthus* sp., Asteraceae**				

folio 93r		plate 757 *chimalacatl*	p. 228, Cap. 15[c] *chimalatl peruina*	p. 751 Flos Solis maior, The greater Sunfloure (Dodoens, 1568)
***Bocconia* sp., Papaveraceae**				

folio 89v(2) (foliage only)	folio 13v *cochizxihuitl* sleep herb	plate 536 *cococ xiuitl* sleep herb	p. 158, Cap. 31[c] *cococ xihuitl* sleep herb	
***Smilax* sp., Smilacaceae**				
				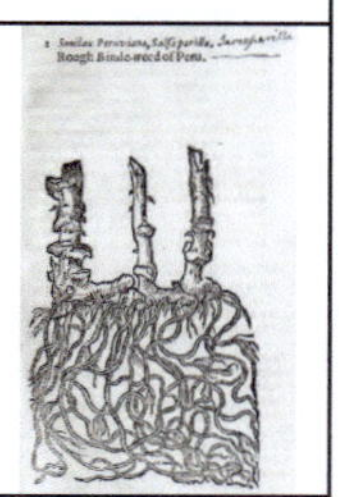
folio 22v (roots only)	folio 8v *tequammaytl* man eater's hand		p. 288, Cap. 41[c] *mecapatli paratla mecapatli o zarzaparrilla*	p. 859 Smilax peruviana, rough bindeweed of Peru

Fig. 5.3 (continued)

Voynich Codex (1568–1572)	Cruz-Badianus (1552)[a]	Sahagún (1585)[b]	Hernández (1651)[c] (1942)[d]	Gerarde (1597, 1633)[e]
***Agave* sp., Asparagaceae/Agavaceae**				

folio 100r#4		plate 596 *metl*	p. 270 Cap. 12[c] *metl*	
***Manihot* sp., Euphorbiaceae**				

folio 93v *M. rubricaulis*		plate 440 *quauhcamotli* *M. esculenta*	p. 116 Cap 21[c] *ytzticpatli* *M. rubricaulis*	
***Taxodium* sp., Taxodiaceae**				

folio 100r #15 *Taxodium* sp., cf. *T. mucronatum* (*T. huegelii, T. mexicanum*)?		plate 379 *aueuetl* *Taxodium mucronatum*	p. 92 Cap. 66[c] *aueuetl &* *ahoehoetl* *ahoehoetl/aueuetl/ahuehuetl* *T. mucronatum*	

Fig. 5.3 (continued)

Voynich Codex (1568–1572)	Cruz-Badianus (1552)[a]	Sahagún (1585)[b]	Hernández (1651)[c] (1942)[d]	Gerarde (1597, 1633)[e]
***Passiflora* sp., Passifloraceae**				

folio 23v *Passiflora* sect. *Decaloba*			p. 301 Cap. 58 [c] *coanenepilli* or *contrayerua coanenepilli / cohuanenepolli / cuanenepilli*	p. 1592 Clematis trifloia, flos Passionis, the Maracoc or Passion Floure
***Chiranthodendron* sp., Malvaceae**				
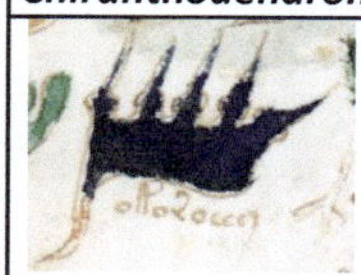				
folio 102r#11 *Chiranthodendron pentadactylon*			p. 383 Cap. n/a [c] *macpalxochi quahuitl macpalxochi quahuitl*	
***Dahlia* sp., Asteraceae**				
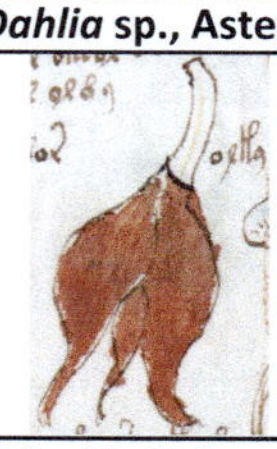				
Folio 99v#18 *Dahlia* sp.			Vol 1, p.28, Cap.27. Fig 8 [d] *Acocotli quauhnahuacense Dahlia vaiabilis Dahlia sp.* aff. *D. coccinea*	

Fig. 5.3 (continued)

Voynich Codex (1568–1572)	Cruz-Badianus (1552)[a]	Sahagún (1585)[b]	Hernández (1651)[c] (1942)[d]	Gerarde (1597, 1633)[e]
***Dorstenia* sp., Moraceae**				
				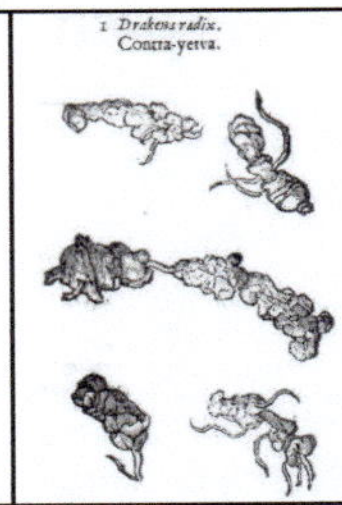
folio 36v *Dorstenia contrajerva*				p. 1621 Drakena radix Contra-yerva
***Iostephane heterophyla*, Asteraceae**				

folio 24v *I. heterophylla*			Vol 1, p.116, Cap.132, Fig. 5.34[d] *Acuitzehuariracua*	
***Ipomoea* sp., Convolvulaceae**				
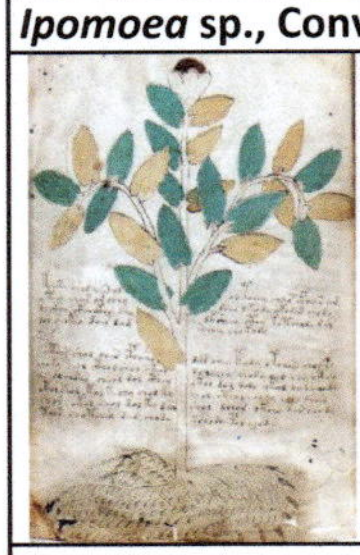				
folio 1v *I. arborescens*	folio 9r *xiuhhamolli* herbaceous soap. *I. murucoides*			

Fig. 5.3 (continued)

Voynich Codex (1568–1572)	Cruz-Badianus (1552)[a]	Sahagún (1585)[b]	Hernández (1651)[c] (1942)[d]	Gerarde (1597, 1633)[e]
***Philodendron* sp., Araceae**				

folio 100r #2 *P. mexicanum*		plate 709 *uacalxochitl* *P. tripartitum*		
***Salix* sp., Salicaceae**				

folio 100r#11 Salix sp.	plate 106 *quetzala[h]uexot* *l* */quetzalhuexotl/* *quetzalauexotl/* *quetzalāhuexōtl* precious willow water *S. lasiolepis*			
***Urtica/Urera* sp., Urticaceae**				

folio 25r *Urtica* *chamaedryoides*	plate 26 *atzitzicaztli* water nettle *Urera* *caracasana*			

Fig. 5.3 (continued)

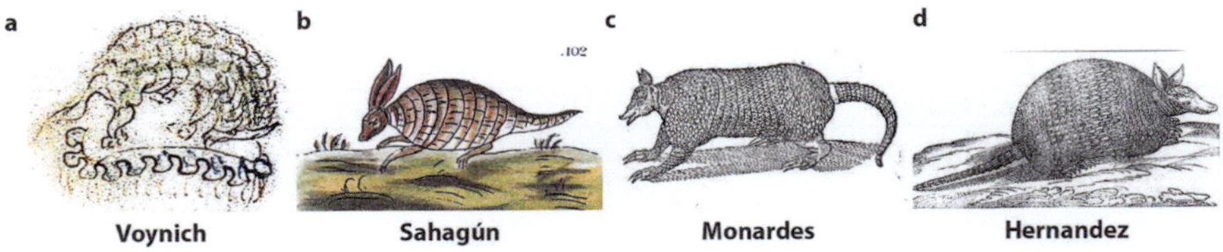

Fig. 5.4 Armadillo (*Dasypus* sp.) illustrations: (**a**) Voynich Codex, folio 80v; (**b**) Sahagún 1963, plate 201; (**c**) Monardes 1577; (**d**) Hernández 1651 (Varey 2000b, Fig. 59)

Conclusions

The information presented in this volume, *The Voynich Flora: An Exploration of Aztec Plants,* can be summarized as follows:

- The Aztecs had a tremendous knowledge of plants based on their vast botanical gardens, the number of plants named, plant expeditions, and knowledge of their medical uses. Although fifteenth century plant books in Mexico have been lost, a sixteenth century herbal, *Codex Cruz Badianus of* 1552 (written by indigenous Nahuas), plus information obtained from the books of Sahagún, Hernández, Monardes, and Gerard provides information about plants known to the Aztecs in Colonial New Spain.
- Identification has been made of all but two of the 131 phytomorphs of the Herbal section and of 41 of the 229 plants in the Pharmaceutical section in the *Voynich Codex*. Each phytomorph has been described and information provided on distribution, names, and medicinal uses. Only one European species has been described, providing hard evidence that the *Voynich Codex* is a manuscript of sixteenth century Mexico in Colonial New Spain.
- The congruence of families, genera, and species with other sixteenth century works on Aztec plants confirms that the *Voynich Codex* must be included as part of the oeuvre of sixteenth century Aztec botany.

Literature Cited

Arber, A. 1938. *Herbals: Their origin and evolution, a chapter in the history of botany 1470–1670*. Darien: Hafner Publishing Co.

Emmart, E.W. 1940. *The Badianus manuscript (Codex Barberini, Latin 141) Vatican library. A Aztec herbal of 1552*. Baltimore: Johns Hopkins Press.

Estrada Lugo, E. 1989. *El Códice Florentino: Su información etnobotánica*. México: olegio de Postgraduados, Institución de Enseñanza e Investigación en Ciencias Agricolas, Montecillo.

Gerarde, J. 1597. *The herball or general historie of plants*. London: John Norton.

Gerarde, J, and T. Johnson. 1633. *The herball or general historie of plants ....very much revised and enlarged by Thomas Johnson*. London: Adam Islip, Joice Norton, and Richard Whitakers (Republished 1975. Mineola, MN: Dover).

Hernández, F. 1942. *Historia de las Plantas de Nueva España*. Ed. I. Univ., Mexico: Ochoterena. Imprenta.

Hernández, F., F. Celsi, F. Colonna, B. Deversini, J. Faber, J. Greuter, V. Mascardi, N.A. Recchi, and J. Terentius. 1651. *Rerum medicarum Novae Hispaniae Thesaurus, seu, Plantarum animalium mioneralium Mexicanorum historia*. Romae: Vitalis Mascardi.

Janick, J., H.S. Paris, and M.C. Daunay. 2012. The cucurbits and nightshades of renaissance England: John Gerard and William Shakespeare. *Horticultural Reviews* 40: 215–257.

Janick, J., and A.O. Tucker, eds. 2018. *Unraveling the Voynich Codex: Mexican*. Cham: Springer Nature.

Kirk, S., and S. Rivett. 2014. *Religious transformations in the early modern Americas*. Philadelphia: University of Pennsylvania Press.

Monardes, N. 1493–1588. *Historia Medicinal de las Cosas que se traen de Nuestras Ind Occidentales*, Seville, Spain, 1565.

Monardes, N. 1577. Joyful newes out of the newe founde worlde. In *wherein is declared the rare and singular vertues of diverse…Herabes…Englished by John Frampton Marchaunt*. London: W. Norton.

Reko, B.P. 1947. Nombres bótanicos del manuscrito Badiano. *Boletín de la Sociedad Botánica de México* 5: 23–43.

Sahagún, B. de. 1951–1982. *Florentine Codex. General history of the things of New Spain* [1540–1585] 12 vol. Trans. A. J. O. Anderson and C. E. Dibble. Salt Lake City: University of Utah Press.

Sahagún, B. de. 1963. *Florentine Codex. General history of the things of New Spain.* Book 11 – *Earthy things.* Transl. C. E. Dibble and A. J. O. Anderson. Salt Lake City: University of Utah Press.

SilverMoon. 2007. The imperial college of Tlatelolco and the emergence of a new Nahua intellectual elite in New Spain (1500–1760). PhD Thesis, Duke University, Durham.

Varey, S., R. Chabrán, and D.B. Wiener, eds. 2000a. *Searching for the secrets of nature: The life and works of Dr. Francisco Hernández*. Stanford: Stanford University Press.

Varey, S., R. Chabrán, and D.B. Wiener, eds. 2000b. *The Mexican treasury: The writings of Dr. Francisco Hernández*. Stanford: Stanford University Press.

Villaseñor, J.L. 2016. Checklist of the native vascular plants of Mexico. *Revista Mexicana de Biodiversidad* 87: 559–902.

Appendix A: Names of Plants Used by the Nahua, 16th Century to the Present

N.B. Not all plant names have been identified (yet), and those that lack identifications are indicated by "=?" only. Also, the same Nahuatl name may refer to more than one species from different sources.

aauaton/haauaton=FC565=*Gaultheria myrsinoides* (*Pernettya ciliata*) (Estrada Lugo 1989)

aacxoyatic/aacxoiatic/aacxoiatic/haacxoyatic=FC603=*Ipomoea capillacea* or *Polygala verticillata* (Estrada Lugo 1989)

acabacahuitztli=*Cynodon dactylon* (*Capriola dactylon*) (Díaz 1976)

acacapacquilitl (aquatic edible herb that makes a slapping noise)=pl. 79 Badianus (C-B fol. 44r)=FC=*Jaegeria bellidiflora* (Díaz 1976; Pico and Nuez 2000a; Estrada Lugo 1989; Miranda and Valdés 1964; Reko 1947) or *Bidens laevis* (*B. chrysanthemoides*) (Díaz 1976; Miranda and Valdés 1964) or *Jaegeria pedunculata* (Linares and Bye 2013; Valdés Gutiérrez et al. 1992) or *Aganippea dentata* (Estrada Lugo 1989; Guerra 1952)

acacapaxihuitl=*Rorippa palustris* (Díaz 1976)

açacatzontli/azacatzontli (aquatic grass hair/head)=pl. 79 Badianus (C-B fol. 44r)=*Tillandsia chaetophylla* (Reko 1947) or Bromeliaceae (Miranda and Valdés 1964) or *Phragmites australis* (*P. communis*) (Guerra 1952) or Cyperaceae? or Juncaginaceae?

acacoyotl=p. 262 Hernandez Rerum/p. 112,113 Hernandez Historia=*Polianthes* sp. or *Lithospermum* sp.? (Ramírez and Alcocer 1902) or *Piper sanctum* (Díaz 1976)

acahual=*Bidens pilosa* or *Heterotheca inuloides* (Díaz 1976)

acahuale=*Bidens aurea* or *Simsia amplexicaulis* (Díaz 1976)

acahualli=*Helianthus annuus* (Dressler 1953; Ramírez and Alcocer 1902)

acahualillo=*Bidens laevis* (Díaz 1976)

acahuatl=*Heterotheca inuloides* (Díaz 1976)

acaltzauhtli=*Galeoglossum tubulosum* (*Cranichis tubulosa*) (Díaz 1976; Ossenbach 2009)

A. O. Tucker, J. Janick, *Flora of the Voynich Codex*,
https://doi.org/10.1007/978-3-030-19377-5

acamallotetl (*acamallotl* stone)=pl. 30, 48, 79, 113 Badianus (C-B fol. 18v, 28v, 44r, 61r)=*Pouteria campechiana* (*Lucuma salicifolia*) (Miranda and Valdés 1964) or *Pachira aquatica* (Guerra 1952)
acametl=FC=p. 273 Hernandez Rerum/p. 1053 Hernandez Historia=*Agave atrovirens* (Estrada Lugo 1989)
acaquilitl (edible herb similar to the reed)=FC=*Bidens laevis* (Díaz 1976; Estrada Lugo 1987 in Pico and Nuez 2000a)
acaquitl mecatl=*Ipomoea nil* (Díaz 1976)
acatl/ācatl (reed, primary name)=pl. 79 Badianus (C-B fol. 44r)=*Phragmites australis* (*P. communis*) (de Ávila Blomberg 2012; Díaz 1976; Estrada Lugo 1989; Guerra 1952; Miranda and Valdés 1964; Reko 1947) or *Quercus laurina* or *Q. polymorpha* or *Q. rugosa* or *Q. peduncularis* (Díaz 1976)
acatzacuhtli=*Galeoglossum tubulosum* (*Cranichis tubulosa*)?
acaxaxan=p. 263 Hernandez Rerum/p. 92 Hernandez Historia=*Tradescantia* sp.?
acaxaxan=p. 263 Hernandez Rerum/p. 94 Hernandez Historia=*Commelina* sp.?
acaxilotic=FC=*Plantago mexicana* (Díaz 1976 ; Ortiz de Montellano 1975, 1990)
acatzan aycxitl=p. 340 Hernandez Rerum/p. 104 Hernandez Historia=*Pellaea ternifolia* (Díaz 1976)
acaxochitl=FC720a=p. 347 Hernandez Rerum/p. 95,96 Hernandez Historia=*Lobelia cardinalis* (*L. fulgens*) or *L. laxiflora* (Díaz 1976; Estrada Lugo 1989)
acaxochitl chichiltic=*Lobelia cardinalis* (*L. fulgens*) (Díaz 1976) or *L. laxiflora* (Estrada Lugo 1989)
acecintle=*Acer negundo* (Farfán and Elferink 2010)
achachayotli=*Urvillea ulmacea* (Díaz 1976)
achichiantic=*Salvia ballotifolia* (Díaz 1976)
achichiantic yohalense=*Salvia ballotiflora*? (Farfán and Elferink 2010)
achichilacachtic=*Ludwigia adscendens* (*Jussiaea repens*) (Díaz 1976)
achilli/achilto/ā-chīlli (water pepper)=pl. 65 Badianus (C-B fol. 37r)=FC649=*Persicaria hydropiper* (*Polygonum hydropi*per) (de Ávila Blomberg 2012; Guerra 1952; Ortiz de Montellano 1990; Reko 1947) or *Persicaria punctata* (*Polygonum acre*) (Díaz 1976; Estrada Lugo 1989) or *Persicaria hydropiperoides* (*Polygonum hydropi*peroides) (Díaz 1976; Linares and Bye 2013; Miranda and Valdés 1964; Valdés Gutiérrez et al. 1992) or *Euphorbia hypericifolia* (Estrada Lugo 1989)
achiotl/achiotetl=p. 74 Hernandez Rerum=FC=*Bixa orellana* (Díaz 1976; Dressler 1953; Estrada Lugo 1989; Ocaranza 2011; Ortiz de Montellano 1990)
achochoquilitl/axoxoquilitl (edible herb with a nice taste)=FC=*Bidens pilosa* (Estrada Lugo 1987 in Pico and Nuez 2000a; Estrada Lugo 1989) or *B. laevis* (*B. chrysanthemoides*) (Díaz 1976; Estrada Lugo 1989) or *B. aurea* (*B. tetragona*) (Díaz 1976; Estrada Lugo 1989)
acimpatli=*Haplophyton cimicidum* (Díaz 1976)
acocochtli=*Arracacia atropurpurea* (Díaz 1976)
acocoquilitl=*Arracacia* sp. (Díaz 1976)
acocote=*Arracacia atropurpurea* or *Ipomoea wolcottiana* (Díaz 1976)
acocotle=*Bidens pil*osa (Díaz 1976)

acocotli/acocohtli/cocohtli (water-cane/water-pipe)=FC=*Arracacia atropurpurea*? (Díaz 1976; Estrada Lugo 1989; Guerra 1952) or *Dahlia* sp. aff. *D. coccinea*? (de Ávila Blomberg 2012; Reko 1947; Sorensen 1970; Zepeda and White 2008) or *Lagenaria siceraria* (*L. vulgaris*) (Díaz 1976; Estrada Lugo 1989) or *D. pinnata* (*D. variabilis*) (Estrada Lugo 1989)

acocotli quauhnahuacensi=p. 31 Hernandez Rerum/p. 28 Hernandez Historia=*Dahlia* sp. aff. *D. coccinea*?

acocotli xalacocatli (hollow aquatic cane)=*Arracacia atropurpurea* (Pico and Nuez 2000b)

acocotlil tepoztlanensi=p. 31 Hernandez Rerum=*Dahlia* sp.?

acocoh[xihui]tl/acōcohtli (primary name, water gullet plant)=pl. 94 Badianus (C-B fol. 51v)=*Arracacia atropurpurea* (Linares and Bye 2013; Guerra 1952; Miranda and Valdés 1964; Valdés Gutiérrez et al. 1992) or *Dahlia* sp. (Guerra 1952)

acocoxiuitl=FC=*Arracacia atropurpurea* (Díaz 1976; Estrada Lugo 1989) or *A. trifida* (Estrada Lugo 1989) or *Dahlia coccinea* (Estrada Lugo 1989)

acocoxochitic=*Drymaria ovata* (Díaz 1976)

acocoxochitl =FC=*Dahlia coccinea* or *D. pinnata,* or *D. lehmannii* (Díaz 1976 ; Dressler 1953; Estrada Lugo 1989) or *Lobelia cardinalis* or *L. laxiflora* (Estrada Lugo 1989)

acotoli ligustici facie=p. 32 Hernandez Rerum=*Dahlia* sp.?

actimpatli=*Haplophyton cimicidum* (Díaz 1976)

acuyo/acueyo (quelite of the bubbled)=*Piper auritum* or *P. sanctum* (Díaz 1976 ; Pico and Nuez 2000b)

acuilotl=*Philadelphus coulteri* (Díaz 1976)

acuitlacpalli=FC=*Sagittaria macrophylla* (Estrada Lugo 1989)

acuitzehuariracua=p. 254 Hernandez Rerum/p. 116 Hernandez Historia=*Iostephane heterophylla* (Díaz 1976)

acxoyatic=p. 134 Hernandez Rerum/p. 37 Hernandez Historia=*Polygala verticillata* or *Ipomoea capillacea* (Díaz 1976)

acxoyatl=p. 348 Hernandez Rerum/p. 40 Hernandez Historia=*Asclepias linaria* or *Abies religiosa* (Díaz 1976)

acxoyatl (primary name)=pl. 95 Badianus (C-B fol. 52r)=FC=*Pseudotsuga* sp. (de Ávila Blomberg 2012; Guerra 1952; Reko 1947) or *Abies religiosa* (Estrada Lugo 1989; Linares and Bye 2013; Miranda and Valdés 1964; Valdés Gutiérrez et al. 1992)

ahacaxilotic=FC=*Plantago mexicana* (Estrada Lugo 1989)

ahaipuih=*Plumeria rubra* (*P. acutifolia*) (Díaz 1976)

ahate=*Annona squamosa* (Díaz 1976)

ahatede panucho/ahate depannuco/ahate penucino=p. 348, 454 Hernandez Rerum/p. 74,75 Hernandez Historia=*Annona squamosal*?

ahoacaquahuitl=*Litsea parvifolia* or *Persea americana* (*P. gratissima*) (Díaz 1976)

ahoacatlmontes=*Litsea parvifolia* (Díaz 1976)

ahoapatlahuac=*Quercus elliptica* (*Q. lanceolata*) (Díaz 1976)

ahoaquahuitl=*Quercus laurina* (*Q. berbineris*) or *Q. polymorpha* or. *Q. peduncularis* (*Q. tomentosa*) or *Q. rugosa* (Díaz 1976)

ahoatan/ahoaton/ahuaton=p. 211 Hernandez Rerum/p. 47 Hernandez Historia=*Gaultheria myrsinoides* (*Pernettya ciliata*) (Díaz 1976)
ahoatepatli=*Achimenes erecta* (*A. coccinea*) (Díaz 1976)
ahoatetzmolin=*Quercus elliptica* (*Q. lanceolata*) (Díaz 1976)
ahoatl=*Quercus acutifolia* (Díaz 1976)
ahoaxaquahuitl=*Persea americana* (*P. gratissima*) (Díaz 1976)
ahoaxatl=*Persea americana* (*P. gratissima*) (Díaz 1976)
ahoaxocotl=*Byrsonima crassifolia* (Díaz 1976)
ahuacaquahuitl/ahoacaquahuitl/ahuacatl=FC=p. 89 Hernandez Rerum=*Persea americana* (Farfán and Elferink 2010; Ortiz de Montellano 1975)
ahhuachcho/ tonatiuh yxiuh ahhauachcho (the sun's dewy herb)=pl. 42, 51 Badianus (C-B fol. 23v, 30r)=*Malvaviscus* sp. (Guerra 1952)
ahhuiyac tlatlanquaye (fragrant possessor of many knees)=pl. 83, 88 Badianus (C-B fol. 46r, 48v)=Lamiaceae? or *Peperomia* sp. (Guerra 1952; Miranda and Valdés 1964; Reko 1947) or *Capsicum* sp.
ahoehoetl/aueuetl/ahuehuetl=p. 92 Hernandez Rerum/p. 148 Hernandez Historia=*Taxodium mucronatum* (*T. huegelii*) (Díaz 1976; Dressler 1953; Estrada Lugo 1989; Farfán and Elferink 2010)
ahquiztli/haquiztli/ā-quīztli (water comes out)=pl. 74, 75 Badianus (C-B fol. 41v, 42r)=FC=*Amaranthus* sp. (Valdés Gutiérrez et al. 1992) or *Amaranthus hybridus* (Linares and Bye 2013) or *Rhus* sp. (de Ávila Blomberg 2012) or *Toxicodendron radicans* (Díaz 1976; Guerra 1952; Miranda and Valdés 1964) or *Anacardium* sp. (Estrada Lugo 1989) or *Toxicodendron* sp. (Guerra 1952)
ahuacatl/ahuatate=*Persea americana* (Díaz 1976; Dressler 1953; Farfán and Elferink 2010)
ahuachpatli=*Persea americana* (*P. gratissima*) (Díaz 1976)
ahuahuaxochitl=*Heliocereus speciosus* (Díaz 1976)
ahuapatli/ahuapahtli/ahoapatli=p. 133 Hernandez Rerum/p. 49 Hernandez Historia=*Galinsoga parviflora* (Díaz 1976) or *Calea zacatechichi* (Díaz 1976)
ahuatl tepiton (insignificant little oak)=pl. 86 Badianus (C-B fol. 47v)=*Rhamnus serrata* (Díaz 1976; Guerra 1952; Linares and Bye 2013; Reko 1947; Valdés Gutiérrez et al. 1992) or *Quercus* sp. (Miranda and Valdés 1964)
ahuaxocotl=*Malpighia glabra* (Díaz 1976)
ahuehuecuahuitl=*Taxodium mucronatum* (*T. huegelii*) (Díaz 1976)
ahuehuete/ahuehuetl=*Taxodium mucronatum* (*T. huegelii*) (Díaz 1976)
ahuexocaquilitl (edible herb found in the *ahuexotl* [*Salix* sp.])=FC=*Bidens aurea* (Estrada Lugo 1987 in Pico and Nuez 2000a)
ahuhuetl=FC=*Bombax ceiba* (Estrada Lugo 1989)
ahuichichi/ahuichichic=*Cayaponia racemosa* (Díaz 1976)
ahuiyac tlatancuaye=*Capsicum annuum* (Díaz 1976)
ahuiyacxiuitl (fragrant herb)=pl. 66 Badianus (C-B fol. 37v)=*Lippia graveolens* (Guerra 1952)
aile=*Alnus acuminata* subsp. *arguta* (*A. arguta*) or *Alnus jorullensis* (*A. firmifolia*) (Díaz 1976)
ailes/ailites=*Agastache mexicana* (*Cedronella mexicana*) (Díaz 1976)

aiotli/anozo yecayotlki=FC=*Cucurbita maxima* or *Cucurbita* sp. (Estrada Lugo 1989)

aitztolin=FC582=*Rhynchospora corymbosa* (*R. aurea*) (Estrada Lugo 1989) or *Cyperus* sp.?

aja=*Brosimum alicastrum* (Díaz 1976)

aji=FC=*Capsicum annuum* (Estrada Lugo 1989)

ajocopaque=*Gaultheria acuminata* (*G. ovata*) (Díaz 1976)

ajopatla=*Gaultheria acuminata* (*G. nitida*) (Díaz 1976)

akitz=*Cascabela gaumeri* (*Thevetia gaumeri*) or *Cascabela thevetia* (*Thevetia nerifolia, T. peruviana*) or *Cascabela thevetioides* (*Thevetia yccotli*) (Díaz 1976)

akolqelite=*Microsechium palmatum* (*M. helleri*) (Díaz 1976)

alahoacapatli leptophyllo/leptofilo=p. 298 Hernandez Rerum/p. 12 Hernandez Historia=*Polemonium grandiflorum* (Díaz 1976)

algodon=FC=*Gossypium hirsutum* (*G. mexicanum, G. palmeri*) (Estrada Lugo 1989)

amacapuli/amacapulin=FC=*Morus celtidifolia* (Díaz 1976; Estrada Lugo 1989; Standley 1920–1926:204)

amacoztic=p. 82 Hernandez Rerum/p. 251 Hernandez Historia=*Ficus nymphaeifolia* or *F. petiolaris* (Díaz 1976)

amacuahuitl=*Cordia boissieri* (Díaz 1976)

amamalacotl/anamalacotl/amalacotl (swirl of water)=FC647=*Hydrocotyle ranunculo*ides (Díaz 1976; Estrada Lugo 1989; Pico and Nuez 2000b) or *Bryophyllum pinnatum* (*B. calycinum*) (Díaz 1976) or *Bryophyllum* sp. or *Solidago paniculata* (Estrada Lugo 1989)

amamaxtla/amamaxtla purgante (aquatic belt)=FC=*Rumex mexicanus* (Díaz 1976; Estrada Lugo 1987 in Pico and Nuez 2000a; Ortiz de Montellano 1990) or *R. patientia* (Díaz 1976)

amapa=*Cordia alliodora* or *Tabebuia heterophylla* (*T. pentaphylla*) (Díaz 1976)

amaquahuitl=*Cordia boissieir* or *C. tinifolia* (Díaz 1976) or *Ficus* sp. (Altamirano 1896)

amaquauitl=FC=*Ficus maxima* (*F. radula*) (Standley 1920–1926:209) or *F. cotinifolia* or *F. obtusifolia* (*F. involuta*) or *Cordia tinifolia* (Estrada Lugo 1989)

amate=*Ficus aurea* (*F. telcolutensis*) or *F. congesta* (*F. fasciculata*) or *F. insipida* (*F. glabrata*) or *F. obtusifolia* (*F. involuta*) (Díaz 1976)

amate amarillo=*Ficus nymphaeifolia* or *F. petiolaris* (*F. jaliscana*) (Díaz 1976)

amate blanco=*Ficus insipida* (*F. glabrata*) (Díaz 1976)

amate de madrono=*Ficus pertusa* (*F. arbutifolia*) (Díaz 1976)

amate de vara=*Ehretia tinifolia* (Díaz 1976)

amate prieto=*Ficus aurea* (*F. tecolutensis*) or *F. maxima* (*F. guadalajarana*) (Díaz 1976)

amatillo=*Ficus pertusa* (*F. padifolia*) (Díaz 1976)

amatlaxihoitl/amatlaxioitl=*Rivina humilis* (Díaz 1976)

amatzacuhtli/amatzauhtle=*Prosthechea pastoris* (*Encyclia pastoris, Epidendrum pastoris*) or *Laelia speciosa* (Berdan 2007; Díaz 1976) or *Encyclia concolor* (Berdan 2007)

amatzauhtli=p. 349 Hernandez Rerum/p. 255 Hernandez Historia=*Prosthechea pastoris* (*Encyclia pastoris, Epidendrum pastoris*)?
amaxtlatl=FC=*Rumex mexicanus* or *Rumex* sp. (Estrada Lugo 1989)
amazauhtli=p. 349 Hernandez Rerum=Orchidaceae?
amazquitl/amezquite=*Ficus pertusa* (*F. arbutifolia, F. complicata, F. padifolia*) (Díaz 1976)
amocoztic=*Ficus nymphaeifolia* (Díaz 1976)
amohuitli=*Hamelia patens* (*H. erecta*) (Díaz 1976)
a-mo-kia=*Panaeolus sphinctrinus* (Díaz 1976)
amole/amolli=FC460b=*Agave filifera* or *A. parryi* or *Agave lophantha* (*A. univittata, A. heteracantha*) or *Manfreda guttata* (*Agave guttata*) or *Phytolacca octandra* or *Polianthes tuberosa* or *Prochnyanthes mexicana* (*P. viridescens*) or *Sapindus marginatus* or *Stegnosperma halimifolium* (Díaz 1976 ; Estrada Lugo 1989) or *Zephyranthes carinata* (Díaz 1976) or *Sapindus saponaria* (Díaz 1976 ; Estrada Lugo 1989)
amolquelite=*Microsechium palmatum* (*M. helleri*) or *Phytolacca icosandra* (Díaz 1976)
amolzochitl=*Polianthes geminiflora* (*Bravoa geminiflora*) (Díaz 1976)
amoxtle/amomotli/amoxtli=pl.61, 90 Badianus (C-B fol. 34r, 49v)=*Nostoc commune* (Godínez et al. 2001) or bryophyte (Guerra 1952)
amozotl (aquatic edible herb that sticks to the clothes non-sharp)=p. 349 Hernandez Rerum/p. 362 Hernandez Historia=FC=*Urtica dioica* (Estrada Lugo 1987 in Pico and Nuez 2000a)
anacahuite=*Cordia boissieri* or *C. sebestena* or *C. tinifolia* (Díaz 1976)
anchinol/anchinoli=*Heimia salicifolia* (Díaz 1976)
a-neg-la-kya= *Datura wrightii* (*D. innoxia*) (Díaz 1976)
a-ni=*Panaeolus sphinctrinus* (Díaz 1976)
anil=*Indigofera suffruticosa* (*I. anil*) (de Batres et al. 2012; Díaz 1976)
anpnas=FC=*Annona muricata* or *A. reticulata* (Estrada Lugo 1989)
anonyma=p. 351 Hernandez Rerum/p. 482 Hernandez Historia=*Stevia* sp.?
aocoxochitl=p. 353 Hernandez Rerum/p. 346 Hernandez Historia=*Didymaea mexicana* (Díaz 1976)
apanxaloa/apancholoa=p. 353 Hernandez Rerum/p. 35 Hernandez Historia=*Cuphea aequipetala* (Díaz 1976; Farfán and Elferink 2010)
aparegua/aparequa=*Discocnide mexicanum* (*Urticastrum mexicanum*) (Díaz 1976)
aphatzi puntzumeti=*Calea integrifolia* (Díaz 1976)
apintli=*Manfreda guttata* (*Agave guttata*) (Altamirano 1896; Díaz 1976) or *Prochnyanthes mexicana* (*P. viridescens*) (Díaz 1976)
apipiltzin=*Psilocybe aztecorum* (Díaz 1976)
apitzalpatli=p. 285 Hernandez Rerum/p. 4 Hernandez Historia=*Malvastrum coromandelianum*?
apitzalpatli=p. 285 Hernandez Rerum/p. 6 Hernandez Historia=*Flourensia thurifera* (*Helianthus thurifer*) (Díaz 1976)
apitzalpatli=*Ricinus communis* (Comas et al. 1995)
apitzalatli tzontololotli=p. 199 Hernandez Rerum/p. 10 Hernandez Historia=*Perymenium rude* (Díaz 1976) or *Priva* sp.?

apompo=*Pachira aquatica* (*P. macrocarpa*) (Díaz 1976)
aposmi/aposi=*Erythrina flabelliformis* (Díaz 1976)
apoyomatli=p. 33 Hernandez Rerum/p. 392 Hernandez Historia=*Cyperus articulatus* (Díaz 1976; Farfán and Elferink 2010)
aquahuitl/ā-cuahuitl (water tree)=pl. 39, 81, 83 Badianus (C-B fol. 24r, 45r, 46r)=*Chilopsis linearis* (Guerra 1952; Miranda and Valdés 1964; Valdés Gutiérrez et al. 1992) or *Astianthus viminalis* (de Ávila Blomberg 2012; Díaz 1976; Reko 1947) or *Tebebuia rosea* (Linares and Bye 2013)
aqualachti=*Cayaponia racemosa* (Díaz 1976)
aquiche=*Guazuma ulmifolia* (*G. tomentosa*) (Díaz 1976)
aquillot=*Philadelphus coulteri* (Díaz 1976)
aquilotl/acuilotl=p. 107 Hernandez Rerum/p. 224 Hernandez Historia=*Philadelphus coulteri*?
aquiztli=p. 354 Hernandez Rerum/p. 118 Hernandez Historia=FC=*Paullinia fuscescens* (Díaz 1976; Estrada Lugo 1989) or *Anacardium* sp. or *Rhus* sp. (Reko 1947)
araroba=*Vataireopsis araroba* (*Andira araroba*) (Díaz 1976)
atatapalacatl/atlacuezona/atzatzamolli=FC=*Nymphaea mexicana* (Estrada Lugo 1989; Zepeda and White 2008) or *Nymphaea gracilis* (*Castalia gracilis*) (Estrada Lugo 1989)
atecioatl=p. 445 Hernandez Rerum/p. 68 Hernandez Historia=*Chromolaena collina* (*Eupatorium collinum*) (Díaz 1976)
atecomatl=*Crescentia alata* (Díaz 1976)
atecuicixtli/atecuiyxtli/atecuiztle=*Rhynchosia phaseoloides* (*Dolicholus phaseoloides*) (Díaz 1976)
atehuapatli=p. 264 Hernandez Rerum/p. 66 Hernandez Historia=*Pilea* sp. or *Achimenes erecta* (*A. coccinea*) (Díaz 1976)
ateje=*Cordia bullata* var. *globosa* (Díaz 1976)
atematzalquilitl=*Lupinus elegans* (Díaz 1976)
atempahtli=*Haplophyton cimicicum* (Díaz 1976)
atenxihuitl=p. 355 Hernandez Rerum/p. 70 Hernandez Historia=*Ranunculus petiolaris* (*R. geoides*) (Díaz 1976)
atepocapatli=p. 33-34 Hernandez Rerum/p. 154 Hernandez Historia=*Lasianthaea aurea* (*Zexmenia aurea*) (Díaz 1976)
ates=*Annona squamosa* (Díaz 1976)
atka-t=*Psilocybe cordispora* or *P. hoogshagenii* (Díaz 1976)
atlacotl=FC648=*Aster* sp. (Estrada Lugo 1989)
atlancane/atlanchane/atlanchana=p. 264 Hernandez Rerum/p. 80 Hernandez Historia=*Lythrum maritimum* or *Cuphea lan*ceolata (Ocaranza 2011) or *Cuphea aequipetala or C. lanceolata* or *L. maritimum* (Díaz 1976)
atlatlancuata=*Iresine calea* (Díaz 1976)
atlatzonpili/atlatzonpillin/atlatzompili=p. 117 Hernandez Rerum/p. 128 Hernandez Historia=FC698=*Malvaviscus* aff. *arboreus* (Zepeda and White 2008) or *M. drummondii* (Díaz 1976; Estrada Lugo 1989) or *Hedeoma costata* var. *costata* (*H. quinquenervata*) (Díaz 1976)

atehuapatli=*Achimenes erecta* (*A. coccinea*) (Farfán and Elferink 2010)
atlepatli=FC454=*Ranunculus hydrocharoides* (*R. stolonifer*) (Díaz 1976; Estrada Lugo 1989) or *R. dichotomus* (Estrada Lugo 1989)
atlinam/atlinan=p. 286 Hernandez Rerum/p. 83 Hernandez Historia=*Rumex pulcher* (Díaz 1976)
atochietl/ātōchietl (rabbit's tobacco of damp places)=pl. 24 Badianus (C-B fol. 15v)=*Poliomintha* sp.? or *Cunila lythrifolia* (Bye and Linares 2013; Valdés Gutiérrez et al. 1992) or *Clinopodium* sp. (de Ávila Blomberg 2012) or *Lippia alba* (*L. geminata*) (Guerra 1952; Reko 1947) or *Hedeoma acinoides* or *H. piperita* (Díaz 1976) or *Hedeoma* sp. (Miranda and Valdés 1964)
atonahuitzpatli=*Peperomia umbilicata* (Díaz 1976; Farfán and Elferink 2010) or *Bouvardia erecta* or *Tripogandra disgrega* (*Tradescantia disgrega*) (Díaz 1976)
atoyaxocotl/ā-tōyā-c-xocotl (river, apple or fruit)=pl. 10, 81 Badianus (C-B fol. 8v, 45r)=pl. 104 Badianus (C-B fol.56v)=FC418=*Spondias purpurea* (de Ávila Blomberg 2012; Díaz 1976; Estrada Lugo 1989; Guerra 1952; Miranda and Valdés 1964; Reko 1947)
atuto=*Vitex mollis* or *V. pyramidata* (Díaz 1976)
atzacuhtli=*Cranicis speciosa*?
atzapotl/atztzapotl=FC414=*Pouteria campechiana* (*Lucuma salicifolia*) (Díaz 1976; Estrada Lugo 1989) or *Casimiroa greggii* (*Sargentia greggii*) or *Sideroxylon* sp. (Estrada Lugo 1989)
atzatzamolli=FC447c,483=*Nymphaea gracilis* (Díaz 1976)
atzautli/atzauhtli=*Cranichis speciosa* (Díaz 1976; Ossenbach 2009)
atzauhtli acuatico=*Cranichis tubularis* (Díaz 1976)
atzcalxochitl=*Sprekelia formosissima* or *Zephyranthes carinata* (Díaz 1976)
atzitzicaztli/atzizicaztli/ā-tzītzicāztli (water nettle)=pl. 26 Badianus (C-B fol. 16v)=p. 356 Hernandez Rerum/p. 384 Hernandez Historia=*Urtica chamaedryoides* (*U. chichiaztli*) or *U. urens* or *Urera caracasana* (Bye and Linares 2013; de Ávila Blomberg 2012; Díaz 1976; Farfán and Elferink 2010; Guerra 1952; Miranda and Valdés 1964; Reko 1947; Valdés Gutiérrez et al. 1992)
atzomiatl/atzoyatl=FC=*Mirabilis jalapa* (Estrada Lugo 1989; Ortiz de Montellano 1990) or *Castilleja tenuiflora* (*C. canescens*) (Estrada Lugo 1989)
atzoyatl=p. 170 Hernandez Rerum/p. 195,196 Hernandez Historia=*Mirabilis jalapa* or *Castilleja tenuiflora* (*C. can*escens) (Díaz 1976)
auacatl/ahuacatl/auacaquauitl=FC=*Persea americ*ana (Díaz 1976; Estrada Lugo 1989; Ortiz de Montellano 1975, 1990)
auaquauitl/ahuaquauitl=FC38=*Quercus* sp. (Ortiz de Montellano 1990; Standley 1920–1926:173)
auatetzmoili=FC382=*Quercus* sp. (Estrada Lugo 1989; Standley 1920–1926:173)
auatl/āhuatl (oak)=pl. 70 Badianus (C-B fol. 39v)=*Quercus* sp. (de Ávila Blomberg 2012; Guerra 1952; Miranda and Valdés 1964)
aueuetl=FC379=*Taxodium mucronatum* (*T. huegelii*) (Díaz 1976)
avitle=*Justicia spigera* (*Jacobinia mohintli*) (Díaz 1976)
axicalli=*Lagenaria siceraria* (*L. vulgaris*) (Díaz 1976)

axin/axquauitl=FC=*Spondias mombin* (*S. lutea*) or *S. purpurea* or *Spondias* sp. (Estrada Lugo 1989)
axixcozahuizpatli=p. 153 Hernandez Rerum/p. 475 Hernandez Historia=?
axixpatli=p. 355 Hernandez Rerum/p. 14 Hernandez Historia=*Dalea lutea* (*Parosela plumosa*) or *Persicaria punctata* (*Polygonum acre*) or *Persicaria hydropiper* (*Polygonum hydropiper*) or *Verbena carolina* or *V. litoralis* (Díaz 1976)
axixpatli coztic=*Rumex mexicanus* (Díaz 1976)
axixpatli texaxahuacense=p. 112 Hernandez Rerum/p. 18 Hernandez Historia=*Vitis tiliifolia* (*V. caribaea*)?
axixquahuitl=*Buddleja americana* (Díaz 1976)
axixtlacotl=*Baccharis salicina* (*B. glutinosa*) or *Salvia longispicata* (Díaz 1976)
axochiatl=*Lonicera pilosa* or *Oenothera laciniata* (Díaz 1976)
axocopaque=*Gaultheria acuminata* or *G. procumbens* (Díaz 1976)
axocotl (water sour fruit)=pl. 70, 80, 104 Badianus (C-B fol. 39v, 44v, 56v)=*Spondias purpurea* (Díaz 1976; Guerra 1952; Miranda and Valdés 1964; Reko 1947)
axoxoco/axocopac/axocopaconi (sour aquatic edible herb)=FC464b, 623, 624=*Rumex mexicanus* (Estrada Lugo 1987 in Pico and Nuez 2000a; Estrada Lugo 1989) or *Gaultheria acuminata* (Díaz 1976; Estrada Lugo 1989) or *R. pulcher* subsp. *woodsii* (*R. brevipes*) (Estrada Lugo 1989)
axoloquahuitl=*Rhus terebinthifolia* (Díaz 1976)
axquauitl/axin=FC=*Jatropha curcas* (Estrada Lugo 1989; Ortiz de Montellano 1990)
ayacachquahuitl=p. 357 Hernandez Rerum/p. 285 Hernandez Historia=*Jatropha peltata* (*J. platyphylla*) or *Crescentia alata* (Díaz 1976)
ayacahuite/ayauhquahuitl/ayauhquauitl/aiauhquauitl/āyauh-cuahuitl (fog tree)=pl. 62, 70, 80, 91 Badianus (C-B fol. 35v, 39v, 44v, 50r)=FC376=*Pinus ayacahuite* (de Ávila Blomberg 2012; Díaz 1976; Estrada Lugo 1989; Guerra 1952; Linares and Bye 2013; Miranda and Valdés 1964; Reko 1947; Valdés Gutiérrez et al. 1992)
ayal/ayale=*Crescentia alata* (Díaz 1976)
ayanqueni=*Helicteres baruensis* (Díaz 1976)
ayauhtli (Ayauh, water goddess)=pl. 46 Badianus (C-B fol. 27v)=*Pinus ayacahuite* (Miranda and Valdés 1964) or *Cucurbita* sp. (Guerra 1952)
ayauhton/ayauhtona/ayauhtonan=FC480b=*Cuphea jorullensis* (Díaz 1976) or *Porophyllum coloratum* or *Porophyllum tagetoides* (Pico and Nuez 2000b)
ayauhtona/ayauhtonan[yxiuh] (our-mother-of-the-fog's herb)=pl. 99 Badianus (C-B fol. 54r)=FC=*Ipomopsis pinnata* (Linares and Bye 2013; Valdés Gutiérrez et al. 1992) or *Cuphea* sp. (Guerra 1952; Miranda and Valdés 1964) or *Cuphea jorullensis* (Estrada Lugo 1989) or *Porophyllum coloratum* (*P. seemannii*) (Estrada Lugo 1989)
ayauhxochitl=*Verbena officinalis* (Díaz 1976)
ayecocimatl/ayecohtli/ayacote/ayecotl/ayocotli/aiecotli (primary name)=pl. 50 Badianus (C-B fol. 29v)=*Phaseolus coccineus* (*P. multiflorus*) (Bye and Linares 2013; de Ávila Blomberg 2012; Díaz 1976; Estrada Lugo 1989; Guerra 1952;

Miranda and Valdés 1964; Reko 1947; Valdés Gutiérrez et al. 1992; Zepeda and White 2008)
aylin (water ylin, tree)=pl. 70 Badianus (C-B fol. 39v)=*Alnus* sp.? (Guerra 1952; Miranda and Valdés 1964)
ayohuactli=*Jatropha curcas* (Ortiz de Montellano 1990)
ayohuitztic=*Solanum fontanesianum* (Díaz 1976)
ayohuitztli=*Celosia argentea* (*C. cristata*) or *Solanum angustifolium* (*S. cornutum*) (Díaz 1976)
ayohuiztle=*Solanum angustifolium* (*S. cornutum*) (Díaz 1976)
ayonelhuatl/ayoh-nelhuatl (squash root)=pl. 109 Badianus (C-V fol. 59r)=*Cucurbita* sp. (de Ávila Blomberg 2012; Guerra 1952; Miranda and Valdés 1964; Reko 1947; Valdés Gutiérrez et al. 1992) or *Cucurbita pepo* (Linares and Bye 2013)
ayoquilitl/ayoquiltic=*Manihot foetida* (Díaz 1976)
ayotectli=*Cayaponia laciniosa* (*Bryonia variegata*) or *Cayaponia racemosa* or *Manihot foetida* (Díaz 1976)
ayotic=*Euphorbia seguieriana* (*E. campestris*) (Díaz 1976)
ayotli=*Cucurbita* sp. (Dressler 1953) or *C. pepo* (Díaz 1976)
ayoxochquilitl=FC464c=*Cucurbita* sp. (Estrada Lugo 1989)
ayozotic=*Cucurbita foetidissima* (Díaz 1976)
aytzmitl (aquatic purslane)=FC=*Ludwigia peploides* (Estrada Lugo 1987 in Pico and Nuez 2000a)
azazal=*Cayaponia racemosa* (Díaz 1976)
azcalxochitl=FC707=*Sprekelia formosissima* (*Amaryllis formosissima*) (Estrada Lugo 1989)
azcapan yxhuatlahcolpahtli/āzcanpan ixhua tlahzolpahtli (rubbish-dump remedy that comes up around anthills)=pl. 20, 76 Badianus (C-B fol. 13v, 42v)=*Datura stramo*nium (Bye and Linares 2013; de Ávila Blomberg 2012; Díaz 1976; Guerra 1952; Reko 1947; Miranda and Valdés 1964; Valdés Gutiérrez et al. 1992) or *D. innoxia* (*D. meteloides*) (Díaz 1976; Miranda and Valdés 1964)
azcatzontecomatl=FC=*Acalypha monostachya* (*A. hederacea*) or *Polygala alba* (Estrada Lugo 1989)
azote=*Ipomoea murucoides* (Díaz 1976)
azpan=FC=*Epilobium mexicanum* or *Oenothera elata* subsp. *hookeri* (*O. hookeri*) (Estrada Lugo 1989)
azumiatl=*Sambucus canadensis* (*S. mexicana*) (Díaz 1976)
bledos=FC=*Amaranthus hybridus* or *A. palmeri* or *Amaranthus cruentus* (*A. paniculatus*) or *A. spinosus* (Estrada Lugo 1989)
çacacili=FC532=*Lithospermum* sp. (Estrada Lugo 1989)
çaçacichit/zacachichic=p. 198 Hernandez Rerum/p. 812 Hernandez Historia=*Conyza filaginoides*
cacahoanantzin=*Licania arborea* (Díaz 1976)
cacahoaquahuitl/quauhpatlathli=p. 80 Hernandez Rerum/p. 991 Hernandez Historia=*Theobroma bicolor* (Bye and Linares 1990) or *T. cacao* (Díaz 1976; Ortiz de Montellano 1990)
cacahoatl=*Arachis hypogaea* (Díaz 1976)

cacahuananche=*Licania arborea* or *Gliricidia sepium* (Díaz 1976)
cacahuate=*Arachis hypogaea* (Díaz 1976)
cacahuaxochitl=p. 226 Hernandez Rerum/p. 1063 Hernandez Historia=*Quararibea funebris* (Alcántara Rojas 2008; Reko 1947; Wasson 1980)
cacalacaxochitl=*Plumeria rubra* (Díaz 1976)
cacaloxochitl/cācālō-xōchitl (crow flower)=pl. 70, 81, 97 Badianus (C-B fol. 39v, 45r, 53r)=FC692=*Plumeria rubra* (*P. acutifolia, P. bicolor*) (Alcántara Rojas 2008; Comas et al. 1995; de Ávila Blomberg 2012; Díaz 1976; Estrada Lugo 1989; Guerra 1952; Linares and Bye 2013; Miranda and Valdés 1964; Ortiz de Montellano 1990; Reko 1947; Valdés Gutiérrez et al. 1992; Zepeda and White 2008) or *Plumeria alba* (Díaz 1976) or *Crotalaria cajanifolia* (*C. eriocarpa*) (Estrada Lugo 1989) or *Tigridia pavonia* (Estrada Lugo 1989)
cacamaxochitl/cacauaxochitl=FC684=*Quararibea funebris* (*Lexarza funebris*) (Estrada Lugo 1989)
cacamilxochitl=p. 358 Hernandez Rerum/p. 1061 Hernandez Historia=Ranunculaceae?
cacamotic=FC=*Ipomoea purga* (Díaz 1976; Ortiz de Montellano 1990) or *I. jalapa* (Díaz 1976) or *I. batatas* or *I. mexicana* or *I. schaffneri* or *Merremia quinquefolia (I. quinquefolia*) or *Turbina corymbosa* (*Ipomoea sidaefolia*) (Estrada Lugo 1989)
cacamotic tlanoquiloni=p. 299 Hernandez Rerum/p. 528 Hernandez Historia=*Ipomoea jalapa*?
çacanochopalli=FC=*Opuntia megarhiza*?
çacanoualli/cacanoualli/zaxanoualli=FC=*Andropogon* sp. (Estrada Lugo 1989)
cacaoatl=*Theobroma cacao*? (Ocaranza 2011)
çaçapipilol xochitl/zacapillolxochitl=p. 120 Hernandez Rerum/p. 824 Hernandez Historia=*Lonicera pilosa*?
çacatetztli=FC637=*Panicum* sp.?
cacatlaxcalli=FC=*Cuscuta americana* or *C. odontolepis* or *C. tinctoria* (Estrada Lugo 1989)
cacatlaolxochitl/zacatlaolxochitl=p. 359 Hernandez Rerum/p. 818 Hernandez Historia=FC=*Cassytha filiformis* (Estrada Lugo 1989)
cacaoaquauitl/cacahoatl=FC=*Theobroma cacao* (Estrada Lugo 1989; Standley 1920–1926:805; Zepeda and White 2008) or *T. angustifolium* or *T. bicolor* (Estrada Lugo 1989)
cacapatli/zacapatli=p. 359 Hernandez Rerum/p. 814 Hernandez Historia=*Menyanthes trifoliata*?
çacateztli/cacateteztli=FC447a, 484=*Panicum* sp. (Estrada Lugo 1989)
cacaua quahuitl/cacahoaquahuitl (cacao)=p. 79 Hernandez Rerum/p. 909 Hernandez Historia=*Theobroma cacao* (Bye and Linares 1990)
cacauaxochitl/cacahuaxōchitl (cacao flower)=pl. 65, 97, 98 Badianus (C-B fol. 37r, 53r, 53v)=*Quararibea funebris* (de Ávila Blomberg 2012; Díaz 1976; Linares and Bye 2013; Miranda and Valdés 1964; Schultes 1957; Valdés Gutiérrez et al. 1992)
çacauitzpatli/zacahuitzpatli=p. 198 Hernandez Rerum/p. 820 Hernandez Historia=?

cachruiaqua/cahriuaca=p. 359 Hernandez Rerum/p. 1013 Hernandez Historia=?
cacomitl=FC=*Tigridia pavonia* or *Tigridia sp.* (Estrada Lugo 1989)
cacxomitl/cacomite=*Tigridia pavonia* (Dressler 1953; Zepeda and White 2008)
calabazas=FC=*Cucurbita foetidissima* or *C. argyrosperma* (*C. mixta*) or *C. moschata* or *C. pepo* (Estrada Lugo 1989)
calancapacle/calancapatle=*Packera bellidifolia* (*Senecio vulneraria*) (Díaz 1976)
caltoli/caltolli=FC639a=*Cyperus* sp. (Estrada Lugo 1989) or *Schoenoplectus* sp.?
camopahtli=*Echeandia flavescens* (*Anthericum leptophyllum*) (Díaz 1976)
camopaltic=*Stevia jorullensis* (*S. clinopodia*) (Díaz 1976)
camatotoncapatli=p. 360 Hernandez Rerum/p. 1020 Hernandez Historia=?
camolxochitl/chamolxochitl=p. 361 Hernandez Rerum/p. 926 Hernandez Historia=*Caesalpinia pulcherrima*
camopaltic=p. 360 Hernandez Rerum/p. 845 Hernandez Historia=*Stevia jorullensis* (*S. clinopodia*)?
camotli=FC441=*Ipomoea batatas* (*Convolvulus batatas*) (Dressler 1953; Estrada Lugo 1989; Farfán and Elferink 2010) or *Dioscorea* sp. (Farfán and Elferink 2010)
canas=FC=*Bambusa* sp. or *Saccharum officinarum* (Estrada Lugo 1989)
capollaxipehualli (barkless cherry tree)=pl. 78 Badianus (C-B fol. 43v)=*Prunus serotina* (*P. serotina* subsp. *capuli*) (Guerra 1952; Reko 1947) or *Chrysobalanus icaco* (Farfán and Elferink 2010)
capulxihuitl/capulin/capoli/capolin (capulin plant, cherry tree)=pl. 53 Badianus (C-B fol. 31r)=FC425=p. 238 Sahagún=p. 95 Hernandez Rerum/p. 902 Hernandez Historia=*Prunus serotina* ((*P. capollin*, *P. serotina* subsp. *capuli*) (Comas et al. 1995; Estrada Lugo 1989; Guerra 1952; Ocaranza 2011; Ortiz de Montellano 1990) or *Conostegia* sp.? *Rhamnus* sp.? (Reko 1947)
catzotl=*Pachyrhizus erosus* (*P. angulatus*) (Díaz 1976)
caxtlatlapan=FC674=*Ipomoea jalapa* or *Ipomoea* sp. or *Merremia tuberosa* (*Ipomoea tuberosa*) (Estrada Lugo 1989)
cayolizcan/zayolizcan/çayolitizcan=p. 66 Hernandez Rerum/p. 848 Hernandez Historia=FC=*Buddleja americana* (Díaz 1976; Ortiz de Montellano 1990)
ceçentlahcol (half by half/half apiece)=pl. 67 Badianus (C-B fol. 38r)=*Echeveria* sp. (Linares and Bye 2013; Valdés Gutiérrez et al. 1992) or *Vernonia* sp. (Guerra 1952; Miranda and Valdés 1964; Reko 1947)
celicpatli/cecicpatlil pahuatlanense/cececpatli/cecepatli=p. 363 Hernandez Rerum/p. 859 Hernandez Historia=*Herpestis* sp.? or *Iresine calea* (Farfán and Elferink 2010)
cempoalxochitl/cempualçuchil=p. 154 Hernandez Rerum/p. 645,646 Hernandez Historia=FC677b=*Tagetes erecta* or *T. patula* (Dressler 1953; Estrada Lugo 1989) or *T. erecta* (Comas et al. 1995; Díaz 1976; Farfán and Elferink 2010; Ocaranza 2011; Ortiz de Montellano 1990; Zepeda and White 2008) or *Tagetes tenuifolia* (*T. peduncularis*) (Estrada Lugo 1989)
cenanam de quahuana hoac=p. 361 Hernandez Rerum=*Lantana camara*?
cenanan tetzcoquensi/cenanan tetzcoquense=p. 185 Hernandez Rerum/p. 740 Hernandez Historia=*Asclepias lanuginosa*?

centli/cintli=*Zea mays* (Dressler 1953; Estrada Lugo 1989)
centzonxochitl/cen-tzon-xōchitl (400 flowers)=pl. 74, 75, 79, 80, 108 Badianus (C-B fol. 41v, 42r, 44r, 44v, 58v)=*Galium* sp. (de Ávila Blomberg 2012; Miranda and Valdés 1964) or *Galium trifidum* (Linares and Bye 2013; Valdés Gutiérrez et al. 1992) or *Phlox* sp.? (Guerra 1952)
ceoalchichiltic=*Vitis tiliifolia* (Díaz 1976)
chalcaloxochil/chacalxochitl/chacamalxochitl/chachalxochitl=*Caesalpinia pulcherrima* (Díaz 1976)
chachalacametl=p. 361 Hernandez Rerum/p. 1057 Hernandez Historia=*Citharexylum affine* (Díaz 1976)
chalcuitlatl=*Alternanthera sessilis* (*A. repens*) or *Lythrum alatum* (Díaz 1976)
chamolxochitl=*Caesalpinia pulcherrima* or *Conyza filaginoides* (Díaz 1976)
chapolxiuitl/chapolxihuitl=FC525=*Euphorbia bracteata* (*Pedilanthus pavonis*) (Díaz 1976; Estrada Lugo 1989) or *Euphorbia tithymaloides* (*P. tithymaloides*) (Estrada Lugo 1989)
chapolycxitl=*Dodonaea viscosa* (Díaz 1976)
charapetl=p. 363 Hernandez Rerum=?
chapulizctli/chapuliztle=*Dodonaea viscosa* (Díaz 1976)
chatalhuich=p. 70 Hernandez Rerum=*Senna septemtrionalis* (*Cassia laevigata*) (de Batres et al. 2012)
chayotectli/chayotestle=*Sechium edule* (Díaz 1976)
chayotli=*Sechium edule* (Dressler 1953)
chia/chian=pl. 59 Badianus (C-B fol. 34r)=FC=*Hyptis* sp. (de Ávila Blomberg 2012) or *Salvia* sp. (Miranda and Valdés 1964) or *Salvia hispanica* (Dressler 1953; Estrada Lugo 1989; Guerra 1952; Ocaranza 2011) or *Mesosphaerum suaveolens* (*Hyptis suaveolens*) or *S. polystachya* (Estrada Lugo 1989)
chiantzotzoltic/chiantzozolli/chiantzotzotl=FC=p. 234 Hernandez Rerum/p. 208 Hernandez Historia=*Salvia hispanica* (Díaz 1976; Dressler 1953; Estrada Lugo 1989; Reko 1947)
chiantzozolli=p. 234 Hernandez Rerum/p. 209 Hernandez Historia=*Nama pringlei*?
chicallotl/chicalotl=p. 215 Hernandez Rerum/p. 804 Henandez Historia=FC=*Argemone grandiflora* or *A. mexicana* or *A. ochroleuca* (Estrada Lugo 1989; Farfán and Elferink 2010; Ocaranza 2011)
chichiantic=p. 227 Hernandez Rerum/p. 218 Hernandez Historia=*Lantana hirsuta*?
chichicahoaztic/chichicahoazton=p. 143 Hernandez Rerum/p. 352 Hernandez Historia=*Eryngium carlinae*? (Díaz 1976; Farfán and Elferink 2010)
chichicallotl=*Argemone mexicana* (Díaz 1976)
chichicamole/chichicamule=*Cucurbita radicans* or *Microsechium palmatum* (*M. guatemalense, M. helleri*) (Díaz 1976)
chichicamotic=*Ipomoea hederifolia* (Díaz 1976)
chichicaquilitl/ocoquilitl (bitter aquatic edible herb)=p. 364 Hernandez Rerum/p. 356 Hernandez Historia=FC472c=*Sonchus oleraceus* (Estrada Lugo 1989; Farfán and Elferink 2010; Ocaranza 2011) or *Mimulus glabratus* (Díaz 1976;

Estrada Lugo 1987 in Pico and Nuez 2000a) or *Launaea intybacea* (*Lactuca intybacea*) (Díaz 1976; Estrada Lugo 1989) or *Mimulus luteus* (Estrada Lugo 1989)

chichicaste=*Mikania scandens* (Díaz 1976)

chichicastli/chichicastle/tsitsikaxtli=*Urtica chamaedryoides* (*U. chichicaztli*) or *U. urens* or *Urera caracasana* (Díaz 1976)

chichicayotl=*Cayaponia laciniosa* (*Bryonia variegata*)? (Ocaranza 2011)

chichicayotli=*Cayaponia racemosa* or *Cucurbita foetidissima* (Díaz 1976)

chichicaztle=*Lemna minor* or *Urera caracasana* or *Wigandia urens* (*W. caracasana*) (Díaz 1976)

chichic patli=FC613=*Bulnesia arborea* (*Guaiacum arboreum*) (Díaz 1976; Estrada Lugo 1989) or *Cephalanthus occidentalis* or *Hintonia latiflora* (*Coutarea latiflora*) or *Mandevilla tubiflora* (*Echites tubiflorus*) (Díaz 1976)

chichicquauitl/chichicquahuitl=FC=*Hintonia latiflora* (Díaz 1976) or *Cornus florida* var. *urbiniana* (*C. urbiniana*) (Estrada Lugo 1989) or *Garrya laurifolia* (Díaz 1976; Estrada Lugo 1989) or *G. laurifolia* (*G. oblonga*) or *G. laurifolia* Benth. subsp. *racemosa* (*G. racemosa*) (Estrada Lugo 1989) or or *Bulnesia arborea* (*Guaiacum arboreum*) (Estrada Lugo 1989) or *Garrya ovata* Benth.?

chichictexcalamatl/chichīc texcalāmatl/texcalamatlchichic (bitter paper-tree of the boulders)=pl. 78 Badianus (C-B fol. 43v)=*Ficus petiolaris* (de Ávila Blomberg 2012; Farfán and Elferink 2010; Guerra 1952; Reko 1947) or *F. cotinifolia* (Linares and Bye 2013) or *Ficus* sp. (Valdés Gutiérrez et al. 1992)

chichictzompontonic=*Pseudognaphalium conoideum* (*Gnaphalium conoideum*) (Díaz 1976)

chichicuahuitl=*Garrya laurifolia* or *G. ovata* or *Penstemon campanulatus* (Díaz 1976)

chichicxihuitl (bitter plant or herb)=pl. 55 Badianus (C-B fol. 32r)=*Arracacia* sp. (Reko 1947) or *Calea zacatechichi* (Díaz 1976; Miranda and Valdés 1964) or *Dalea* sp. or *Arracacia* sp. (Guerra 1952)

chichientic=FC535=*Salvia* sp. (Estrada Lugo 1989)

chichihualxochitl=FC694=*Carica papaya* (Ortiz de Montellano 1975, 1990)

chichillotl=*Argemone mexicana* (Díaz 1976)

chichilquiltic=FC543=*Amaranthus* sp. (Estrada Lugo 1989)

chichiltic cihuapatli=p. 367 Hernandez Rerum/p. 887 Hernandez Historia=?

chichilticpatli=p. 366 Hernandez Rerum/p. 995 Hernandez Historia=?

chichiltic tepetlauhxochitl=p. 368 Hernandez Rerum=*Laelia autumnalis* or *Stanhopea tigrina* (Ossenbach 2009)

chichiltictepetzacuxochitl=*Laelia autumnalis* or *Laelia speciosa* (Berdan 2007)

chichimecapatli=p. 138 Hernandez Rerum/p. 924 Hernandez Historia=*Euphorbia radians* (Díaz 1976)

chichimecatl/chichimecapatli=*Macroptilium gibbosifolium* (*Phaseolus heterophyllus*) (Díaz 1976)

chichipeloti=*Zea mays* (Rivera Morales 1941)

chichioalxochitl/chichiualxochitl=p. 365 Hernandez Rerum/p. 564 Hernandez Historia=*Carica papaya* (Díaz 1976)

chichioaquahuitl/chichiualquauitl=FC=*Jatropha cardiophylla* (Díaz 1976; Estrada Lugo 1989)
chichipiltic=FC=*Asclepias tuberosa* or *Astragalus crassicarpus* (Estrada Lugo 1989)
chichiquahuitl=*Cornus florida* var. *urbiniana* (*C. urbiniana*) or *Garrya laurifolia* (Díaz 1976)
chichiquilitl=*Solanum americanum* (*S. nigrum*) (Díaz 1976)
chichiual memeya/chichioalmemeya=p. 172 Hernandez Rerum/p. 964 Hernandez Historia=?
chicomacatl (seven reeds)=pl. 80 Badianus (C-B fol. 44v)=*Struthanthus interruptus* (*S. venetus*)? (Guerra 1952; Reko 1947) or *Psittacanthus calyculatus* (Díaz 1976; Miranda and Valdés 1964) or *Tillandsia prodigiosa* (Linares and Bye 2013; Valdés Gutiérrez et al. 1992) or *Ficus* sp.?
chicozapotl=*Manilkara zapota* (*Achras sapota*) or *Casimiroa edulis* (Díaz 1976)
chictecitl=FC=*Capsicum annuum* var. *glabriusculum* or *C. frutescens*?
chietla=*Ficus nymphaeifolia* (Díaz 1976)
chilamacatl=FC=*Helianthus annuus* (Ortiz de Montellano 1975)
chile=FC=*Capsicum annuum* (Estrada Lugo 1989)
chiikan=*Mucuna pruriens* (Díaz 1976)
chilacalote=*Bocconia arborea* (Díaz 1976)
chilacastle=*Lemna minor* (Díaz 1976)
chilacayote=*Melothria pendula* (*M. guadalupensis*) (Díaz 1976)
chilcuague=*Erigeron affinis* or *Heliopsis longipes* (Díaz 1976)
chilacole=*Erythrina coralloides* (Díaz 1976)
chilicote=*Erythrina flabelliformis* (Díaz 1976)
chillapatli=p. 102 Hernandez Rerum/p. 425 Hernandez Historia=*Tilia* sp.
chillazotl=*Argemone mexicana* (Díaz 1976; Ortiz de Montellano 1990)
chilli=p. 135, 136, 137 Hernandez Rerum/p. 430, 431, 433, 434 Hernandez Historia=*Capsicum annuum* (Farfán and Elferink 2010; Ocaranza 2011)
chilli Indie Orientalis=p. 169 Hernandez Rerum/p. 1071 Hernandez Historia=*Zingiber officinale*?
chilmecatl=*Erigeron affinis* or *Heliopsis longipes* (Díaz 1976)
chilochuchi=*Pseudobombax ellipticum* (*Bombax ellipticum*) (Díaz 1976)
chilpan=p. 202 Hernandez Rerum/p. 831 Hernandez Historia=FC678e=*Penstemon campanulatus* (Díaz 1976; Estrada Lugo 1989)
chilpanoxochitl=*Lobelia laxiflora* (Díaz 1976)
chilpantlacotl=*Penstemon barbatus* (Díaz 1976)
chilpantlaçolli/chilpantlazolli=p. 140 Hernandez Rerum/p. 738 Hernandez Historia=*Lobelia laxiflora* (Díaz 1976)
chilpanton=FC542=*Lobelia laxiflora* (Díaz 1976; Estrada Lugo 1989; Farfán and Elferink 2010; Ortiz de Montellano 1990)
chilpanxochitl=p. 210 Hernandez Rerum/p. 829 Hernandez Historia=*Lobelia laxiflora* (Díaz 1976) or *Penstemon barbatus*?
chilpatli=p. 139 Hernandez Rerum/p. 423 Hernandez Historia=*Croton cortesianus* (Díaz 1976)

chiltepin/chiltecpin/chiltecpitl=FC=*Capsicum frutescens* (*C. minimum*) or *C. baccatum* (*C. microcarpum*) (Estrada Lugo 1989) or *C. annuum* var. *glabriusculum*?

chimalacatl/chimalacaxochitl=FC757b=p. 101, 102 Hernandez Historia=*Helianthus annuus* (Díaz 1976; Dressler 1953; Estrada Lugo 1989; Lentz et al. 2008a, b; Ortiz de Montellano 1990) or *Verbesina crocata* (Díaz 1976; Estrada Lugo 1989)

chimalatl peruina=p. 228 Hernandez Rerum=*Helianthus annuus* (Díaz 1976; Dressler 1953; Zepeda and White 2008)

chimalxochitl=*Helianthus annuus* (Zepeda and White 2008)

china michuacanensi/china mechoacanense=p. 213 Hernandez Rerum/p. 769 Hernandez Historia=*Smilax* sp.?

chinchayote=*Sechium edule* (Díaz 1976)

chintul=*Sprekelia formosissima* (Díaz 1976)

chipaoacyztic=*Iostephane heterophylla* (Díaz 1976)

chiople=*Critonia hemipteropoda* (*Eupatorium hemipteropodum*) (Díaz 1976)

chipa(h)uacxihuitl (clean/light colored herb)=pl. 8 Badianus (C-B fol. 7v)=*Dioscorea* sp.? *Smilax* sp.? (Guerra 1952; Reko 1947) or *Dioscorea galeottiana* (Bye and Linares 2013; Miranda and Valdés 1964; Valdés Gutiérrez et al. 1992) or *D. convolvulacea* (Miranda and Valdés 1964)

chipequa=p. 203 Hernandez Rerum/p. 898 Hernandez Historia=?

chipilli=FC=*Crotalaria* sp. or *Crotalaria pumila* (Estrada Lugo 1989)

chiyauaxihuitl (greasy herb)=pl. 100 Badianus (C-B fol. 54v)=*Mirabilis* sp. (Linares and Bye 2013; Valdés Gutiérrez et al. 1992) or *Tagetes* sp.? (Reko 1947) or *Tagetes* sp. (Guerra 1952)

chochoyatl=*Roldana ehrenbergiana* (*Senecio canicidus*) (Díaz 1976)

chocolin=*Erythrina coralloides* or *E. mexicana* (Díaz 1976)

cholagogue=*Colubrina triflora* (*C. guatemalensis*) (Díaz 1976)

chucunguariqua=p. 368 Hernandez Rerum/p. 501 Hernandez Historia=*Dyschoriste* sp.?

chupire/chupireni=*Euphorbia calyculata* (Farfán and Elferink 2010)

cicimatic/cimatl=p. 265 Hernandez Rerum/p. 203 Hernandez Historia=FC608=*Canavalia villosa* (Estrada Lugo 1989) or *Ramirezella strobilophora* (Ochoterena-Booth 2000)

cihoapaltli=*Montanoa tomentosa* (Díaz 1976)

cihoapatli tepoztemica/cihuapatli tepoztlanico=p. 370 Hernandez Rerum/p. 892 Hernandez Historia=Lamiaceae?

cihoapatli xantetelco/cihuapatli xantelecense=p. 370 Hernandez Rerum/p. 893 Hernandez Historia=*Pluchea salicifolia* (*P. adnata*)?

ci(h)uapahtli/cihnapahtli/cihuapatli/chihuā-pahtli (women's remedy)=pl. 106 Badianus (C-B fol. 57v)=FC=*Montanoa tomentosa* (*Eriocoma floribunda*) (de Ávila Blomberg 2012; del Pozo 1966; Estrada Lugo 1989; Farfán and Elferink 2010; Guerra 1952; Linares and Bye 2013; Miranda and Valdés 1964; Ortiz de Montellano 1975; Valdés Gutiérrez et al. 1992)

cihuacoanenepilli=*Passiflora suberosa* (Díaz 1976)

cihuapastle=*Chromolaena odorata* (*Eupatorum odoratum*) (Díaz 1976) *cihuapatli*=*Alomia alata* (Farfán and Elferink 2010) or *Bouvardia bouvardioides* (Farfán and Elferink 2010) or *Chromolaena odorata* (*Eupatorium odratum*) (Díaz 1976) or *Hamelia calycosa* (Farfán and Elferink 2010) or *Montanoa tomentosa* (Díaz 1976; Estrada Lugo 1989) or *Montanoa floribunda* or *Pluchea odorata* or *Tibouchina longifolia* (*T. bougaeana*) or *Vernonanthura deppeana* (*Vernonia deppeana*) (Farfán and Elferink 2010)

cihuapatli yacapichtlensi/cihuapatli yacapictlense=p. 157 Hernandez Rerum/p. 889 Hernandez Historia=*Baccharis* sp. or *Stevia* sp.?

cimapatli=p. 371 Hernandez Rerum/p. 981 Hernandez Historia=?

cimatl (primary name)=pl. 89 Badianus (C-B fol. 49r)=FC443, 460a=p. 265 Hernandez Rerum/p. 202 Hernandez Historia=*Phaseolus coccineus* (*P. multiflorus*) (Díaz 1976; Estrada Lugo 1989; Guerra 1952; Linares and Bye 2013; Miranda and Valdés 1964; Reko 1947) or *Desmodium amplifolium* (Díaz 1976; Estrada Lugo 1989) or *D. orbiculare* or *D. parviflorum* (Díaz 1976) or *Phaseolus* sp. (Valdés Gutiérrez et al. 1992) or *Canavalia villosa* or *D. scoparium* or *Macroptilium atropurpureum* (*Phaseolus atropurpureus*) or *Cochliasanthus caracalla* (*Phaseolus caracalla*) or *Phaseolus* sp. (Estrada Lugo 1989)

cimatl chico=*Desmodium orbiculare* (Díaz 1976)

cimiapactli=*Montanoa floribunda* (Díaz 1976)

cintul=*Sprekelia formosissima* or *Anticlea frigida* (*Stenanthium frigidum*, *Veratrum frigidum*) (Díaz 1976)

cioapatli=*Erigeron glacialis* (*E. scaposus*) (Díaz 1976)

cirial=*Crescentia alata* (Díaz 1976)

cirian=*Crescentia alata* or *C. cujete* (Díaz 1976)

citzicastli/tzitzicastli/atzitsicaztli=*Urera caracasana* or *Urtica chichiaztli* or *Gronovia scandens* (Comas et al. 1995)

ciuapatli=FC598=*Montanoa tomentosa* (Díaz 1976; Ortiz de Montellano 1990)

ciuipatli/cihuapatli=p. 187 Hernandez Rerum/p. 880 Hernandez Historia= Lamiaceae?

ciuipatli/cihuapatli=p. 186 Hernandez Rerum/p. 882 Hernandez Historia=?

ciuipatli/cihuapatli=p. 186 Hernandez Rerum/p. 884 Hernandez Historia=?

coacamachalli=*Pithecellobium dulce* (Díaz 1976)

coaçihuizpatli/coacihuitzpatli=p. 188 Hernandez Rerum/p. 1068 Hernandez Historia=*Chromolaena odorata* (*Eupatorium odoratum*) or *Mollugo verticillata* (Díaz 1976)

coaciuizpatli/cohacihuizpatli=p. 301 Hernandez Rerum/p. 953 Hernandez Historia=*Tetranema roseum* (*T. mexicanum*)?

coahuilxochitl/coahoitzxochitl=p. 371 Hernandez Rerum/p. 607 Hernandez Historia=*Lantana camara* (Díaz 1976)

coalxoxouhqui=FC=*Ipomoea jalapa* or *I. rosea* or *Turbina corymbosa* (*Ipomoea sidaefolia*) (Estrada Lugo 1989)

coamecaxochitl=*Disocactus flagelliformis* (*Aporocactus flagelliformis*) (Díaz 1976)

coamichin=*Ficus congesta* (*F. fasciculata*) or *Ficus pertusa* (*F. arbutifolia, F. complicata, F. padifolia*) (Díaz 1976)

coanenepilli/cohuanenepolli/cuanenepilli=FC522=p. 300–302 Hernandez Rerum=*Passiflora jorullensis* (Díaz 1976; Estrada Lugo 1989; Ortiz de Montellano 1975, 1990) or *P.* aff. *sicyoides* (Zepeda and White 2008) or *P. suberosa* (Díaz 1976; Estrada Lugo 1989)
coanepilli de codorniz=*Physalis nicandroides* (Díaz 1976)
coapatli=p. 28 Hernandez Rerum/p. 591 Hernandez Historia=*Commelina tuberosa*? (Ocaranza 2011) or *Campyloneurum phyllitidis* or *Euphorbia prostrata* or *Lippia umbellata* or *Acourtia cordata* (*Perezia hebeclada*) or *Niphidium crassifolium* (*Pleopeltis crassifolia, Polypodium crassifolium*) (Díaz 1976)
coapatli asphodelino=p. 300 Hernandez Rerum/p. 597 Hernandez Historia=*Tradescantia crassifolia* (Díaz 1976)
coapatlis de tepequaquilco=p. 29 Hernandez Rerum=Asteraceae?
coapatli quauhnacensi/coapatli quauhnahuacense=p. 36 Hernandez Rerum/p. 593 Hernandez Historia=Asteraceae
coapatlis teoitztle/copatli teoiszla=p. 29 Hernandez Rerum/p. 595 Hernandez Historia=?
coaquiltic=*Rhodosciadium tuberosum* (Deanea tuberosa) (Díaz 1976)
coate=*Eysenhardtia polystachya* (Díaz 1976)
coatl=*Caesalpinia bonduc* (*C. bonducella*) (Díaz 1976)
coaztli=*Eysenhardtia polystachya* (Díaz 1976)
coatl xoxouhqui=FC449=*Turbina corymbosa* (*Rivea corymbosa*) (Díaz 1976; Estrada Lugo 1989) or *Datura ceratocaula* or *D. meteloides* (Estrada Lugo 1989)
coatlycedal=*Adiantum capillus-veneris* (Díaz 1976)
coatzontecomatl=*Matelea reticulata* (*Gonolobus reticulatus*) or *Stanhopea tigrina* (Díaz 1976)
coatzontecoxochitl/coatzontecomaxochitl/coatzontecomaxochitl=p. 266 Hernandez Rerum/p. 379 Hernandez Historia=FC719=*Stanhopea oculata* (Safford 1912; Zepeda and White 2008) or *Stanhopea hernandezii* or *Stanhopea tigrina* (Berdan 2007; Díaz 1976; Estrada Lugo 1989; Ossenbach 2009)
coaxihuitl/cuaxihuitl=FC=p.267 Hernandez Rerum/p. 600 Hernandez Historia=*Lythrum alatum* (Díaz 1976; Estrada Lugo 1989) or *L. vulneraria* or *Turbina corymbosa* (*Ipomoea burmanni*) (Estrada Lugo 1989)
coçatli coztic/cozatlicoztic=p. 156 Hernandez Rerum/p. 651 Hernandes Historia=*Tagetes patula*?
cochitzapotl=FC=*Casimiroa edulis* (Díaz 1976; Estrada Lugo 1989)
cochiztzapotl (sleep zapote/white zapote)=FC412=*Casimiroa edulis* (del Pozo 1966) (Ortiz de Montellano 1975, 1990; Zepeda and White 2008) or *Couepia polyandra* (Díaz 1976; Estrada Lugo 1989)
cochizxihuitl (sleep herb)=pl. 20 Badianus (C-B fol. 13v)=*Russelia polyedra* (Valdés Gutiérrez et al. 1992) or *Bocconia integrifolia* (Díaz 1976; Guerra 1952; Reko 1947; Viesca and Aranda 1996) or *Mentzelia hispida* (Bye and Linares 2013) or *Bocconia* sp. (Miranda and Valdés 1964)
cococaquilitl (hot aquatic quelite)=p. 171 Hernandez Rerum/p. 621 Hernandez Historia=*Agrimonia* sp. or *Adenophyllum glandulosum* (*A. coccineum*) (Díaz

1976; Pico and Nuez 2000b) or *Vigethia mexicana* (*Wyethia mexicana*) (Díaz 1976)

cococaton=Stellaria nemorum (Díaz 1976)

cococcuahuitl=Croton gossypiifolius (*C. sanguifluius*)? (Comas et al. 1995) or *Pterocarpus*? Or *Jatropha dioica* (*J. spatulata*)?

cocochiatl=Mimosa sensitiva (Díaz 1976)

cocochihuitl=Bocconia arborea (Díaz 1976)

cococ patli/cococpatli=FC=*Bocconia frutescens* (Díaz 1976; Estrada Lugo 1989) or *Tecoma stans* (Estrada Lugo 1989)

cocoxihuitl=FC=*Bocconia arborea* (Estrada Lugo 1989) or *B. frutescens* or *Momordica charantia* (Díaz 1976)

cocoxihuitl ocimono=Phyla scaberrima (*Lippia dulcis*) (Díaz 1976)

cocomecaxihuitl=Smilax rotundifolia? (Ocaranza 2011)

cocotemecatl/cococtemecatl=p. 141 Hernandez Rerum/p. 622 Hernandez Historia=Asteraceae

cococ xihuitl (stinging herb)=pl. 52 Badianus (C-B fol. 30v)=p. 158 Hernandez Rerum/p. 609 Hernandez Historia=p. 102 Sahagún=FC536=*Bocconia frutescens* (Farfán and Elferink 2010; Miranda and Valdés 1964; Ortiz de Montellano 1975, 1990; Reko 1947) or *B. arborea* (Miranda and Valdés 1964)

cocolmeca=Smilax cordifolia or *S. spinosa* (*S.mexicana*) or *S. pseudochina* (Díaz 1976)

cocolmecan=Smilax spinosa (*S. mexicana*) (Díaz 1976)

cocopi=FC=*Zea mexicana* (*Euchlaena mexicaqna*) (Díaz 1976; Estrada Lugo 1989)

cocoxihuitl=Bocconia arborea (Díaz 1976; Farfán and Elferink 2010) or *B. frutescens* (Díaz 1976)

cocoxihuitl expurgatoria/cococxihuitl purgante=p. 142 Hernandez Rerum/p. 615 Hernandez Historia=*Croton* sp.?

cococxihuitl teuhcaltzincensi/cococxihuitl/ teuhcaltzincene/coco-c xihutl (stinging herb)=pl. 52 Badianus (C-B fol. 30v)=p. 142 Hernandez Rerum/p. 612 Hernandez Historia=FC=*Bocconia* sp.? (de Ávila Blomberg 2012) or *B. frutescens* (Estrada Lugo 1989; Guerra 1952) or *B. integrifolia* (Estrada Lugo 1989) or Verbenaceae

cocoxochitl (cane flower/hollow-stem flower)=p. 371 Hernandez Rerum=*Dahlia* sp. (Sorensen 1970)

çoçoyatic=FC519=*Anticlea frigida* (*Stenanthium frigidum*)? or *Zygadenus* sp.? or *Schoenocaulon* sp.?

cocoyol=Acrocomia aculeata (*A. mexicana*) (Díaz 1976)

cocoztamatl/cocoztamal=p. 118 Hernandez Rerum/p. 641 Hernandez Historia=?

cocoztic=Thalictrum hernandezii (Díaz 1976; Estrada Lugo 1989)

cocoz xochipatli/cocozxuchipatli=p. 188 Hernandez Rerum/p. 989 Hernandez Historia=Asteraceae?

cocuite=Piscidia piscipula (Díaz 1976)

coen/coentic=p. 252 Hernandez Rerum/p. 745 Hernandez Historia=*Pachyrhizus erosus* (P. palmatilobus) (Díaz 1976)

coentli=Cosmos diversifolius (Díaz 1976)

cohacihuizpatli=Tetranema roseum (*T. mexicanum*) (Díaz 1976)

cohayelli=Eryngium carlinae (Díaz 1976)
cohuatli/coatli=pl. 34 Badianus (C-B. fol. 20v)=FC501=*Eysenhardtia polystachya* (Díaz 1976; Estrada Lugo 1989; Guerra 1952; Miranda and Valdés 1964; Ocaranza 2011; Ortiz de Montellano 1990; Reko 1947) or *Jatropha dioica* (*J. spathulata*) (Estrada Lugo 1989)
coilotapalo=Cecropia obtusifolia (*C. mexicana*) (Díaz 1976)
collotemecatl/colotemecatl (trailing stone cord)=pl. 77 Badianus (C-B fol. 43r)=p. 373 Hernandez Rerum/p. 755 Hernandez Historia=*Ipomoea* sp.? (Reko 1947) or *Operculina* sp. (Valdés Gutiérrez et al. 1992; Miranda and Valdés 1964)
coloc=Psidium guajava (Díaz 1976)
colomecatl (scorpion/curved vine)=pl. 77 Badianus (C-B fol. 43r)=*Ipomoea* sp. (Guerra 1952; Linares and Bye 2013) or *Operculina* sp.?
colorin=Erythrina americana or *E. coralloides* or *E. mexicana* (Díaz 1976)
colorin de peces=Piscidia piscipula (Díaz 1976)
colotzitzicaztli/cōlō-tzītzicāztli (scorpion/curved nettle)=pl. 85 Badianus (C-B fol. 47r)=p. 373 Hernandez Rerum/p. 386 Hernandez Historia=FC=*Cnidoscolus* sp. (de Ávila Blomberg 2012; Linares and Bye 2013; Miranda and Valdés 1964; Valdés Gutiérrez et al. 1992) or *Mentzelia* sp. (Guerra 1952; Reko 1947) or *Cnidoscolus urens* (*Jatropha urens*) (Díaz 1976; Estrada Lugo 1989) or *Jatropa* sp. (Comas et al. 1995; Estrada Lugo 1989) or *Gronovia scandens* or *Mentzelia lobata* or *Urera* sp. (Estrada Lugo 1989)
coltotzin (little coltotl)=pl. 113 Badianus (C-B fol. 61r)=*Dalea exserta* (*D. lagopus*) (Díaz 1976; Reko 1947) or *D. leporina* (Linares and Bye 2013) or *Dalea* sp. (Miranda and Valdés 1964; Valdés Gutiérrez et al. 1992) or *Marina nutans* (*Dalea nutans*) (Guerra 1952)
comalquilitl=Artemisia ludoviciana subsp. *redolens* (*A. redolens*) (Díaz 1976)
copal=FC=*Bursera* sp. (Estrada Lugo 1989) or *Bursera bipinnata* (*Elaphrium bipinnatum*) or *Heliocarpus americanus* (Estrada Lugo 1989)
copalche=Croton niveus (Díaz 1976)
copalchi=Hintonia latiflora (*Coutarea latiflora*) or *Croton guatemalensis* or *Croton niveus* (Díaz 1976)
copalchi blanco=Croton niveus (Díaz 1976)
copalchi de jojutla=Hintonia latiflora (*Coutarea latiflora*) (Díaz 1976)
copalchi de Michoacan=Exostema caribaeum (Díaz 1976)
copalchi delgado=Croton niveus (Díaz 1976)
copalchichic=Croton niveus (Díaz 1976)
copalcocote/copalcoxocotl=Cyrtocarpa procera (Díaz 1976)
copalcuajiote=Bursera multijuga (Díaz 1976)
copali=Cyrtocarpa procera (Díaz 1976)
copaliyac xiuhtontli/copal-ihyā-c xiuh-tōntli (insignificant little herb that stinks like copal)=pl. 52, 55 Badianus (C-B fol. 30v, 32r)=*Tagetes* sp. (de Ávila Blomberg 2012; Guerra 1952; Valdés Gutiérrez et al. 1992) or *Dyssodia papposa* (Bye and Linares 2013a)
copaljiote=Pseudosmodingium perniciosum (Díaz 1976)
copaljocote=Cyurtocarpa procera (Díaz 1976)

copalli=FC=*Bursera copallifera* (*B. jorullensis*) or *Bursera lancifolia* (*B. trijuga*) (Díaz 1976) or *Bursera* sp. (Estrada Lugo 1989) or *Bursera bipinnata* (*Elaphrium bipinnatum*) or *Heliocarpus americanus* (Estrada Lugo 1989)

copalli quahuitl/copalquahuitl=p. 45 Hernandez Rerum/p. 531 Hernandez Historia=FC=*Bursera glabrifolia* (*Elaphrium aloexylon*) or *Bursera copallifera* (*B. jorullensis*) (Díaz 1976) or *Bursera bipinnata* (*Elaphrium bipinnatum*) or *Heliocarpus americanus* (Estrada Lugo 1989)

copalquahuitl patlahoac=*Bursera ariensis* (*B. sessiliflora*) or *Bursera copallifera* (*B. jorullensis*) (Díaz 1976)

copalquahuitlpatlahoac (copal tree)=p. 10 Badianus (C-B. fol. 8v)=p. 46 Hernandez Rerum/p. 533 Hernandez Historia=FC=*Bursera copallifera* (*B. jorullensis, Elaphrium jorullense*) (Díaz 1976; Guerra 1952; Miranda and Valdés 1964; Reko 1947) or *Bursera* sp. (Estrada Lugo 1989)

copalli quauhxiotl/copalquauhxiotl=p. 47 Hernandez Rerum/p. 547 Hernandez Historia=*Bursera microphylla* (*Elaphrium microphyllum*) or *Bursera multijuga* (Díaz 1976)

copalquin=*Croton eluteria* (*C. cascarilla*) (Díaz 1976)

copalxihuitl=p. 103 Hernandez Rerum/p. 543 Hernandez Historia=*Salvia formosa* (Díaz 1976)

copalxiu=*Melampodium divaricatum* (Díaz 1976)

copalxocotl/copal-xocotl (copal tart-fruit)=pl. 104 Badianus (C-B fol. 56v)=p. 82 Hernandez Rerum/p. 541 Hernandez Historia=*Cyrtocarpa edulis* or *C. procera* (de Ávila Blomberg 2012; Díaz 1976; Guerra 1952; Linares and Bye 2013; Miranda and Valdés 1964; Valdés Gutiérrez et al. 1992)

copalyhyac=*Protium heptaphyllum* (Díaz 1976)

corpionchi=*Galphimia glauca* (Díaz 1976)

costiczapotl=*Pouteria campechiana* (Dressler 1953)

costomate/costomatl=*Physalis coztomatl* (Díaz 1976; Estrada Lugo 1989) or *Solanum lycopsericum* (Díaz 1976)

çotolin=FC=*Dioon edule*?

cotzotl=*Pachyrhizus erosus* (*P. angulatus*) (Díaz 1976)

couanenepilli/cohuanenepilli/cōānenepīlli (serpent tongue)=pl. 59, 80 Badianus (C-B fol. 34r, 44v)=*Dahlia coccinea* (Linares and Bye 2013; Valdés Gutiérrez et al. 1992) or *Passiflora jorullensis* (de Ávila Blomberg 2012; Guerra 1952; Reko 1947) or *Boerhavia coccinea* (*B. caribaea*) or *Potentilla candicans* (Miranda and Valdés 1964)

couaxocotl/cōāxocotl (snake tart-fruit)=pl. 68, 70 Badianus (C-B fol. 38v, 39v)=*Physalis* sp.? (de Ávila Blomberg 2012) or *Rauvolfia tetraphyllla* (*R. heterophylla*) (Valdés Gutiérrez et al. 1992) or *Crataegus* sp.? (Guerra 1952; Miranda and Valdés 1964; Reko 1947) or *Manilkara zapota* (*Calocarpum mammosum*) (Guerra 1952)

çoyatl=FC=*Brahea dulcis*?

coyol=*Acrocomia aculeata* (*A. mexicana*) (Díaz 1976)

coyolli=p. 71 Hernandez Rerum/p. 511 Hernandez Historia=?

coyoltomatl=FC486=*Cayratia japonica* (*Vitis mollis*) (Díaz 1976)

coyoluian=*Buddleja americana* or *B. cordata* (Díaz 1976)
coyolxochitl/coyoxochitl=p. 267,268 Hernandez Rerum/p. 504, 505, 506 Hernandez Historia=FC=*Bomarea edulis* (*B. hirtella*) (Dressler 1953; Estrada Lugo 1989) or *Tabebuia heterophylla* (*T. pentaphylla*) (Estrada Lugo 1989)
coyotlychcaum=*Asclepias verticillata* (Díaz 1976)
coyotomate=*Vitex mollis* or *V. pyramidata* (Díaz 1976)
coyotomatl/coyototomatl=FC=p. 204 Hernandez Rerum/p. 705 Hernandez Historia=*Physalis* sp. (Daunay et al. 2008; Estrada Lugo 1989; Farfán and Elferink 2010) or *Physalis nicandroides* (Díaz 1976) or *Vitex mollis* (Díaz 1976; Estrada Lugo 1989)
coyoxochitl/coyoxochitl yacapichtlense=p. 374 Hernandez Rerum/p. 796 Hernandez Historia=?
coyoxihuitl/coyoxiuitl (coyote herb)=pl. 63, 65, 79, 81, 88 Badianus (C-B fol. 36r, 37r, 44r, 45r, 48v)=*Loeselia mexicana* (Guerra 1952)
coyoxihuitl tlatztalehualtic/coyōxihuitl tlaztalēhualtic (pink coyote herb)=p. 61 Badianus (C-B fol. 35r)=*Salvia microphylla* (Linares and Bye 2013) or *Polanisia* sp. (de Ávila Blomberg 2012) or *Loeselia mexicana* (Guerra 1952; Reko 1947) or *Salvia* sp. (Miranda and Valdés 1964; Valdés Gutiérrez et al. 1992)
cozahuique=*Ficus congesta* (*F. fasciculata*) (Díaz 1976)
cozcanantzi (revered mother of the jewel)=pl. 67 Badianus (C-B fol. 38r)=*Ipomoea bracteata* (Guerra 1952; Linares and Bye 2013; Reko 1947) or *Quamoclit bracteata* (*Eriogonium bracteatum*) (Valdés Gutiérrez et al. 1992)
cozcaquauhxiuitl=pl. 94 Badianus (C-B. fol. 51v)=FC592=*Acourtia moschata* (*Perezia moschata*) (Díaz 1976; Estrada Lugo 1989)
coznochnopalli=FC431=*Opuntia pseudo-tuna* or *Opuntia* sp. (Estrada Lugo 1989)
cozomecatl=*Smilax congesta* (*S. mexicana*) or *S. pseudochina* (Díaz 1976)
cozticayotli=*Cucurbita moschata* (Díaz 1976)
cozticcoatzontecoxochitl=*Euchile citrina* (*Encyclia citrina*)
coztictepetzacuxochitl=*Govenia superba* or *G. liliacea*?
coztic mecapatli tilacensi/cozticmecapatli tilacense=p. 242 Hernandez Rerum/p. 834 Hernandez Historia=Fabaceae?
cozticoatzontecoxochitl=*Euchile citrina* (*Cattleya citrina, Prosthechea citrina*) (Díaz 1976; Ossenbach 2009)
cozticzacatzacuxochitl=*Govenia superba* (Ossenbach 2009)
cozolmecatl=p. 213 Hernandez Rerum/p. 765 Hernandez Historia=*Smilax spinosa* (*S. mexicana*)
cozticpatli=p. 236 Hernandez Rerum/p. 983 Hernandez Historia=*Thalictrum hernandezii* or *T. mexicanum* (Díaz 1976)
coztic patli acatlanensi/cozticpatli acatlanense=p. 235 Hernandez Rerum/p. 987 Hernandez Historia=*Thalictrum hernandezii* (de Batres et al. 2012; Ortiz de Montellano. 1990)
coztictzapotl/cozticzapotl=*Pouteria campechiana* (*Lucuma salicifolia*) (Díaz 1976)
coztic xihuitl/coztixihuitl=p. 303 Hernandez Rerum/p. 642 Hernandez Historia=*Vaccinium* sp.?

cozticxocotl=Spondias mombin (Dressler 1953)
coztihuisti palancapatli=Serjania mexicana (Díaz 1976)
coztomate/coztomatl=p. 203 Hernandez Rerum/p. 703 Hernandez Historia=*Physalis coztomatl* (Daunay et al. 2008; Díaz 1976; Farfán and Elferink 2010) or *Physalis* sp. (Estrada Lugo 1989) or *Solanum lycopersicum* (Díaz 1976)
cuachala=Amphipterygium adstringens (*Juliania adstringens*) (Díaz 1976)
cuachalalate=Amphipterygium adstringens (*Juliania adstringens*) (Díaz 1976)
cuachichil/cuachil/cuachile=Loeselia mexicana (*L. coccinea*) (Díaz 1976)
cuachilote=Parmentiera aculeata (P. edulis) (Díaz 1976)
cuahtzitzicaztli=Urera caracasana (Díaz 1976)
cuahuchichi=Penstemon campanulatus (Díaz 1976)
cuaicastle=Ceanothus caeruleus (Díaz 1976)
cualmecatl/cuamecatl=Serjania mexicana (*Paullinia mexicana*) (Díaz 1976)
cuamecate=Antigonon flavescens (Díaz 1976)
cuamuchil=Pithecellobium dulce (Díaz 1976)
cuamochitl=Pithecellobium dulce (Farfán and Elferink 2010)
cuanaxana=Calea ternifolia var. *hypoleuca* (*C. hypoleuca*) (Díaz 1976)
cuanaxonaxi=Croton ciliatoglandulifer (Díaz 1976)
cuapatli=Commelina tuberosa or *Tradescantia crassifolia* (Díaz 1976)
cuapopolchi=Piqueria trinervia (Díaz 1976)
cuaquil=Celtis caudata (Díaz 1976)
cuartoloti=Andira inermis (Díaz 1976)
cuastecomatl=Crescentia alata (Díaz 1976)
cuatecomate=Crescentia alata (Díaz 1976)
cuauchichi=Garrya ovata (Díaz 1976)
cuauchichic=Garrya laurifolia or *G. ovata* or *Loeselia mexicana* (*L. coccinea*) or *Penstemon campanulatus* (Díaz 1976)
cuauhayohtli=Parmentiera aculeata (*Crescentia aculeata*) (Díaz 1976)
cuauhayohuachili=Jatropha curcas (Díaz 1976)
cuauhtepatli/chupiri=FC=*Euphorbia calyculata* (Ortiz de Montellano 1975, 1990)
cuahtli=Eisenhardtia sp. (de Ávila Blomberg 2012)
cuauhtzapotl=Annona aff. *cherimola* (Zepeda and White 2008) or *A. squamosa* (Díaz 1976)
cuauhxihuitl=FC=*Senna occidentalis* (Estrada Lugo 1989)
cuauhzontli=Chenopodium berlandieri (*C. berlandieri* subsp. *nuttalliae*) (Dressler 1953)
cuaulote=Guazuma ulmifolia (*G. tomentosa*) (Díaz 1976)
cuaunacaxtle=Enterolobium cycloparpum (Díaz 1976)
cuauteco=Heterotheca inuloides (Díaz 1976)
cuautecomate=Crescentia cujete or *C. alata* (Díaz 1976)
cuautecomatl=Crescentia alata (Díaz 1976)
cuautepatli=Euphorbia calyculata (Díaz 1976)
cuauteteco=Heterotheca inuloides (Díaz 1976)
cuchalolate=Amphipterygium adstringens (*Juliania adstringens*) (Díaz 1976)
cucoztli=Thalictrum hernandezii (Díaz 1976)

cuculin=*Leptolyngbya tenuis* (*Phormidium tenue*) or *Chroococcus turgidus* (Godínez et al. 2001)
cucultzin=*Euphorbia seguieriana* (*E. campestris*) (Díaz 1976)
cuechiliz tomatl/cuechyliztomatl=p. 372 Hernandez Rerum/p. 712 Hernandez Historia=?
cuechxochitl=*Plumeria rubra* (*P. rubra*) (Díaz 1976)
cuecuetzpahtli/cuencuetzpatli (mischievous/unabashed remedy)=pl. 82 Badianus (C-B fol. 45v)=*Clinopodium macrostemum* (*C. laevigatum, Satureja macrostema*) (Díaz 1976; Guerra 1952; Linares and Bye 2013; Miranda and Valdés 1964; Reko 1947; Valdés Gutiérrez et al. 1992)
cuetlaxochitl/cuetlaxxochitl=p. 375 Hernandez Rerum/p. 959 Hernandez Historia=FC685d=*Euphorbia pulcherrima* (Díaz 1976; Estrada Lugo 1989; Ortiz de Montellano 1990)
cueyauh quilitl=p. 375 Hernandez Rerum=*Portulaca oleracea*?
cuezalmetl=FC=*Agave* sp. (Estrada Lugo 1989)
cuicuilticayotli=*Cucurbita ficifolia* (Díaz 1976)
cuicuitlapile=FC489a=*Stachytarpheta* sp.?
cuicuitlapilli=p. 248 Hernandez Rerum/p. 722 Hernandez Historia=?
cuitlacamotli=*Valeriana sorbifolia* (*V. toluccana*) (Díaz 1976)
cuitlacoche=*Ustilago maydis* (Díaz 1976)
cuitlaçotl tepecuaquilcensi/cuitlazotl=p. 229 Hernandez Rerum/p. 716 Hernandez Historia=?
cuitlacopalli=p. 48 Hernandez Rerum/p. 544 Hernandez Historia=*Bursera fagaroides* (*Elaphrium odoratum*) or *B. simaruba* (*B. gummifera*) (Díaz 1976)
cuitlapatli=*Valeriana edulis* subsp. *procer*a (Farfán and Elferink 2010) or *Valeriana mexicana* (Díaz 1976; Ocaranza 2011) or *Euphorbia maculata* (Díaz 1976)
cuitlaxochitl=*Euphorbia pulcherrima* (Díaz 1976)
cuitlazotl=*Dysphania ambrosioides* (*Chenopodium ambrosioides*) (Díaz 1976)
cujete=Crescentia cujete (Díaz 1976)
*cuiztapaçollin/cuixyapazollin*n=p. 236 Hernandez Rerum/p. 1002 Hernandez Historia=*Phaseolus* sp.?
cungariqua/cunguricua=p. 249 Hernandez Rerum/p. 1008 Hernandez Historia=Apocynaceae?
cungariqua/cunguricua=p. 249 Hernandez Rerum/p. 1009 Hernandez Historia=Apocynaceae?
cupitri/cupitin=p. 375 Hernandez Rerum/p. 1004 Hernandez Historia=?
curaqua/pingüica=p. 121 Hernandez Rerum/p. 1006 Hernandez Historia=*Arctostaphylos pungens*?
cutaqua=*Solandra maxima* (*S. nitida*) (Díaz 1976)
cutaquatzitziqui=*Solandra maxima* (*S. nitida*) or *Swartzia nitida* (Díaz 1976)
cutiriqui=*Calochortus purpureus* (*C. bonplandianus*) (Díaz 1976; Farfán and Elferink 2010)
cuyotomate=*Vitex mollis* or *V. pyramidata* (Díaz 1976)

ecapacle/ecapatli/hecapatli=p. 112 Hernandez Rerum/p. 412 Hernandez Historia=FC514=*Senna alata* (*Cassia alata*) or *C. occidentalis* (Díaz 1976; Estrada Lugo 1989)

ecuicuilli=FC=*Phaseolus coccineus* (Estrada Lugo 1989)

ecuxo=FC=*Nicotiana* sp. (Estrada Lugo 1989)

eeloquiltic/eloquilitic/eloquilitl=FC529=*Iresine calea* (Díaz 1976; Estrada Lugo 1989) or *Amaranthus* sp. (Estrada Lugo 1989) or *Bidens pilosa* (Estrada Lugo 1989) or *B. aurea* (*B. tetragona*) (Estrada Lugo 1989) or *Peperomia* sp. (Estrada Lugo 1989) or *Piper* sp. (Estrada Lugo 1989)

ekkixil=*Dolichandra unguis-cati* (*Bignonia unguis-cati*) (Díaz 1976)

eloçacatl/elozacatl (tender-ear-of-maize grass)=pl. 54, 110 Badianus (C-B fol. 31v, 59v)=*Rumex* sp. (Valdés Gutiérrez et al. 1992) or *Cynodon* sp.? (Guerra 1952; Reko 1947) or *Equisetum myriochaetum* (Bye and Linares 2013) or *Cynodon dactylon* (Miranda and Valdés 1964)

elote=*Zea mays* (Rivera Morales 1941)

elocapolin (tender-ear-of-maize cherry)=pl. 86 Badianus (C-B fol. 47v)=*Prunus serotina* subsp. *capuli* (Díaz 1976; Linares and Bye 2013; Reko 1947; Valdés Gutiérrez et al. *1*992) or *Ardisia* sp. (Guerra 1952; Miranda and Valdés 1964)

elochiquahuitl=*Magnolia schiedeana* (Zepeda and White 2008)

eloquiltic=FC497=*Amaranthus* sp.?

eloquilitl (edible herb of the tender corncob)=FC=*Bidens pilosa* (Díaz 1976; Estrada Lugo 1987 in Picó and Nuez 2000a)

elosuchil=*Magnolia macrohylla* var. *dealbata* (*M. dealbata*) (Díaz 1976)

eloxochitl/ēlō-xōchitl/eloxochiquauitl (tender-ear-of-maize flower)=pl. 60, 69, 70 Badianus (C-B fol. 34v, 39r, 39v)=FC=*Magnolia schiedeana* (Alcántara Rojas 2008; Guerra 1952; Miranda and Valdés 1964; Reko 1947; Zepeda and White 2008) or *Magnolia macrophylla* var. *dealbata* (*M. dealbata*) (Díaz 1976; Estrada Lugo 1989; Linares and Bye 2013) or *Magnolia* sp. (de Ávila Blomberg 2012)

elozacatl=*Cynodon dactylon* (*Capriola dactylon*) (Díaz 1976)

enguanche=*Bocconia arborea* (Díaz 1976)

epaxihuitl=*Croton vulpinus*? (Ocaranza 2011)

epazotl/epaçotl/epazote=FC=*Dysphania ambrosioides* (*Chenolpodium ambrosoides*) (Comas et al. 1995; Díaz 1976; Estrada Lugo 1989; Pico and Nuez 2000b; Ocaranza 2011; Ortiz de Montellano 1975, 1990) or *Chenopodium vulvaria* (*C. foetidum*) (Díaz 1976)

epitactli=FC=*Phaseolus coccineus* (Estrada Lugo 1989)

equimitl=FC=*Erythrina mexicana* (Díaz 1976)

eratihueni=*Cuphea jorullensis* (Díaz 1976)

estachahuit=*Conocarpus erectus* (Díaz 1976)

etenquilitl=FC=*Phaseolus vulgaris* (Estrada Lugo 1989)

etzpatli/ezptli/totonquixihuitl/totoncaxihuitl=*Croton gosypiifolius* (*C. sanguifluus*) or *C. cortesianus* (Comas et al. 1995)

etzapotl/etzaptl/quauhtzapotl=FC=*Annona* aff. *cherimola* (Zepeda and White 2008) or *Annona* sp. (Estrada Lugo 1989)

etzquahuitl=*Croton gossypiifolius* (*C. sanguifluus*) (Díaz 1976) *ezpahtli/ezpatli* (blood remedy) pl. 33 Badianus (C-B fol. 20r)=FC=*Croton draco* (Estrada Lugo 1989; Reko 1947) or *Croton gossypiifolius* (*C. sanguifluus*) (Díaz 1976) or *Bouvardia ternifolia* (Bye and Linares 2013) or *Jatropha dioica* (*J. spathulata*) or *Pterocarpus* sp. (Estrada Lugo 1989) or *Bouvardia* sp. (Guerra 1952)

ezquahuitl=*Croton gossypiifolius* (*C. sanguifluus*) (Ortiz de Montellano 1990) or *C. draco* (Díaz 1976) or *C. suberosus* (*C. cladotrichus*) (Díaz 1976)

hacayotli=*Cucurbita moschata* (Díaz 1976)

haceoitli=*Indigofera suffruticosa* (*I. anil*) (Díaz 1976)

hachoque=*Piper palmeri* (Díaz 1976)

hahuiyacxihuitl (fragrant herb)=pl. 107 Badianus (C-B fol. 58r)=*Lippia graveolens* (*L. berlandieri*) (Farfán and Elferink 2010; Guerra 1952; Reko 1947; Valdés Gutiérrez et al. 1992) or *L. mexicana* (Linares and Bye 2013) or *Alnus* sp. (Miranda and Valdés 1964)

hanchinol=*Heimia salicifolia* or *Nesaea syphilitica* (*Heimia syphilitica*) (Díaz 1976)

hatanal=*Celosia virgata* (Díaz 1976)

hauayche=*Ageratum littorale* (Díaz 1976)

hecapahtli/hecapatli/ehcapahtli (wind remedy)=pl. 33, 65, 70, 87, 90 Badianus (C-B fol. 20r, 37r, 39v, 48r, 49v)=*Senna occidentalis* (Guerra 1952; Miranda and Valdés 1964; Reko 1947)

hiacamote=*Manihot esculenta* (*M. utilissima*) (Díaz 1976)

hoaca xochitl=p. 377 Hernandez Rerum=Araceae?

hoauhquilitl=p. 269 Hernandez Rerum=*Chenopodium* sp. (Altamirano 1896)

hoauhtli=FC=*Amaranthus hypochondriacus* (Estrada Lugo 1989)

hoaxin=*Senna sophera* (*Cassia esculenta*) or *Leucaena esculenta* (Díaz 1976)

huaxui=*Leucaena esculenta* (Altamirano 1896)

hoehoetzontecomatl=*Aristolochia odoratissima* (Díaz 1976)

hoeiamatl=*Ficus congesta* (*F. fasciculata*) (Díaz 1976)

hoeipochotli=*Pseudobombax ellipticum* (*Bombax ellipticum*) (Altamirano 1896)

hoitzcolotli=p. 222 Hernandez Rerum=*Eryngium deppeanum* (*E. painteri*) (Díaz 1976)

hoitziloxitl=p. 51 Hernandez Rerum/p. 559 Hernandez Historia=*Myroxylon balsamum* var. *pereirae* (*Toluifera pereirae*) or *Myroxylon sonsonatense*? (Díaz 1976; Ocaranza 2011)

hoitzilziltenxochitl=p. 379 Hernandez Rerum=*Ranunculus* sp.?

hoitzit xilxochitl=p. 103 Hernandez Rerum=*Salvia* sp.?

hoitzitziltentli=*Lobelia laxiflora* (Díaz 1976)

hoitzitzilxochitl=*Loeselia mexicana* (*L. coccinea*) (Díaz 1976)

hoitzmamaxalli=*Acacia cornigera* (Díaz 1976)

hoitzocuitlapilli=*Disocactus flagelliformis* (*Aporocactus flagelliformis*) (Díaz 1976)

hoitzquahuitl=*Caesalpinia vesicaria* or *Haematoxylum campechianum* (Díaz 1976) or *Caesalpinia* sp. (Altamirano 1896)

hoixachinquahuitl=*Havardia albicans* (*Pithecellobium albicans*) (Altamirano 1896; de Batres et al. 2012; Díaz 1976)

holcuahuitl/holquahuitl=Castilla elastica (Altamirano 1896; Farfán and Elferink 2010; Ocaranza 2011)
holli (ulli)=pl. 78 Badianus (C-B Fol. 43v)=resin of *Castilla elastica* (Guerra 1952)
huauhtli=Amaranthus sp. (Farfán and Elferink 2010)
huacalxochitl/huacalxōchitl (huacal flower)=pl. 14, 30, 46, 70, 104 Badianus (C-B fol. 10v, 18v, 27v, 39v, 56v)=*Xanthosoma sagittifolium* (*X. roseum*), *Anthurium cordatum* (Guerra 1952; Reko 1947) or *Philodendron radiatum* var. *pseudoradiatum* (de Ávila Blomberg 2012; Díaz 1976; Miranda and Valdés 1964) or *Philodendron mexicanum* (Alcántara Rojas 2008; Bye and Linares 2013) or *Philodendron* sp. (Valdés Gutiérrez et al. 1992)
huacalazochitl=Philodendron aff. *mexicanum* (Zepeda and White 2008)
huacamote/huacamotl/huacamotli=Maniot esculenta (*M. aipi, M. utilissima*) (Díaz 1976)
huacanala=Morella cerifera (*Myrica xalapensis*) (Díaz 1976)
huachacata=Galphimia glauca (Díaz 1976)
huachichile=Loeselia scariosa (Díaz 1976)
huachipil=Alvaradoa amorphoides (Díaz 1976)
huamuch/huamuchi/huamuchil/huitzquahuitl/quamochitl/coacamachalli=Pithecellobium dulce (Comas et al. 1995; Díaz 1976) or *Pithecellobium unguis-cati* (*Inga unguis-cati*) or *Haematoxylum campechianum* (Comas et al. 1995)
huanacaxtle=Enterolobium cyclocarpum (Díaz 1976)
huanchal=Trophis racemosa (Díaz 1976)
huanini=Bourreria huanita (Díaz 1976)
huatari=Lophophora williamsii (Díaz 1976)
huauachinoli/huauchinal/huauchinol=Heimia salicifolia (Díaz 1976)
huauhtli=Amaranthus caudatus (*A. leucocarpus*) (Dressler 1953)
huauhtzontli/uauhtzontli (hair of the huauhquilitl)=*Chenopodium berlandieri* (*C. berlandieri* subsp. *nuttalliae*) (Dressler 1953; Picó and Nuez 2000b; Safford 1918)
huaxacan/huaxaxan=Guaiacum sanctum (Díaz 1976)
huehuelicaton=Matelea reticulata (*Gonolobus reticulatus*) (Díaz 1976)
hueitlachpaoaztli=Oenothera laciniata (Díaz 1976)
huetzcanixochitl (flower that habitually laughs)=pl. 67 Badianus (C-B fol. 38r)=*Zephyranthes* sp. (Guerra 1952; Miranda and Valdés 1964; Reko 1947) or *Z. fosteri* (Linares and Bye 2013) or *Z. brevipes* (Valdés Gutiérrez et al. 1992)
hueynacatztli/huēinacaztli (large ear)=pl. 70, 104 Badianus (C-B fol. 39v, 56v)=*Cymbopetalum penduliflorum* (Comas et al. 1995; de Ávila Blomberg 2012; Guerra 1952; Reko 1947) or *Enterolobium cyclocarpum* (Comas et al. 1995; Díaz 1976; Miranda and Valdés 1964)
huicahoatl=Quercus ellptica (*Q. lanceolata*) (Díaz 1976)
huichichile=Loeselia mexicana (*L. coccinea*) (Díaz 1976)
huichichilemetl=Xylosma flexuosa (Díaz 1976)
huichoquachaqua=Iostephane heterophylla (Díaz 1976)

huihuitz mallotic (full of needles)=pl. 52, 59 Badianus (C-B fol. 30v, 34r)=*Mentzelia hispida* (Díaz 1976; Guerra 1952; Reko 1947; Linares and Bye 2013; Valdés Gutiérrez et al. 1992)
huihuitz yocochizxihuitl/huīhuitzyoh cochizxihuitl (thorny sleep herb)=pl. 20 Badianus (C-B fol. 13v)=*Mimosa albida* (de Ávila Blomberg 2012; Díaz 1976; Guerra 1952; Reko 1947; Viesca and Aranda 1996) or *M. ervendbergii* (Bye and Linares 2013) or *Mimosa* sp. (Valdés Gutiérrez et al. 1992) or *Mimosa* sp. (Miranda and Valdés 1964)
huilote=Vitex mollis or *V. pyrimidata* (Díaz 1976)
huinacaxtle=Enterolobium cyclocarpum (Díaz 1976)
huiquilitl=Indigofera suffruticosa (*I. anil*) (Díaz 1976)
huitlacoche=Ustilago maydis (Díaz 1976)
huitzcolotl=Calliandra houstoniana var. *anomala* (Díaz 1976)
huitzcolohtli/uuitzcolohtli/huitzcoloxochitl/huitzcōlohtli (thorny curved)=pl. 33 Badianus (C-B fol. 20r)=*Berberis* sp. (de Ávila Blomberg 2012) or *Condalia* sp. or *Solanum* sp. (Guerra 1952; Reko 1947)
huitzihtzilxochitl (humming bird flower)=pl. 66, 92, 113 Badianus (C-B fol. 37v, 50v, 61r)=p. 78 Hernandez Rerum=*Loeselia mexicana* (*L. coccinea*) (Guerra 1952; Reko 1947) or *Salvia gesneriiflora* (Díaz 1976) or *Salvia* sp. (Linares and Bye 2013; Miranda and Valdés 1964; Valdés Gutiérrez et al. 1992)
huitziloxitl=Myroxylon balsamum or *M. balsamum* var. *pereirae* (Díaz 1976)
huitzitzillacotl=Zephyranthes sp. (de Batres et al. 2012)
huitzilzitzin/huitzitziltzin/huitzitzilxochitl=Loeselia mexicana (*L. coccinea*) (Díaz 1976; Viesca and Aranda 1996)
huitzmamaxalli=p. 86 Hernandez Rerum=*Acacia* sp.?
huitzpatli=Baccharis conferta (Díaz 1976)
huitzqua(h)uitl/huitz-cuahuitl (spiny tree)=pl. 68, 70, 78 Badianus (C-B fol. 38v, 39v, 43v)=*Haematoxylum brasiletto* (de Ávila Blomberg 2012; Díaz 1976; Farfán and Elferink 2010; Valdés Gutiérrez et al. 1992; Viesca and Aranda 1996) or *Condalia hookeri* (*C. obovata*) (Guerra 1952; Reko 1947) or *H. campechianum* (Linares and Bye 2013)
huitzquilitl (thorny edible herb)=p. 10, 73, 79, 90 Badianus (C-B fol. 8v, 41r, 44r, 49v)=*Cirsium ehrenbergii* (Bye and Linares 2013; de Ávila Blomberg 2012; Linares and Bye 2013; Reko 1947; Valdés Gutiérrez et al. 1992; Zapeda and White 2008) or *Cirsium* sp. (Guerra 1952; Miranda and Valdés 1964)
huitztomatzin=p. 108 Hernandez Rerum/p. 707 Hernandez Historia=*Solanum hernandesii* (Díaz 1976)
huixochi/huixachi=FC=*Acacia farnesiana* (Estrada Lugo 1989)
icçotl=FC387=Cycadaceae?
icçotl=FC691=?
ichcatl=p. 308 Hernandez=*Gossypium hirsutum* (*G. mexicanum*)
ichcatlepatli/ichcatlepahtli/ychcatlepatli=p. 442 Hernandez Rerum/p. 181 Hernandez Historia=*Cordia bullata* var. *globosa* (*C. globosa*) or *C. sonorae* (Díaz 1976) or *Croton* sp.

iczotli/izote=FC=*Yucca gigantea* (*Y. elephantipes*) (Dressler 1953; Estrada Lugo 1989) or *Y. aloifolia* (Estrada Lugo 1989; Farfán and Elferink 2010) or *Y. faxoniana* (*Y. australis*) (Díaz 1976) or *Y. filamentosa* (Estrada Lugo 1989) or *Yucca* sp. (Altamirano 1896; Estrada Lugo 1989)

ili/ilitl=*Betula* sp. (Siméon 2010)

illamatzapotl=*Annona muricata* (Díaz 1976)

ipazote=*Chenopodium vulvaria* (*C. graveolens*) or *Dysphania ambrosioides* (*Chenopodium ambrosioides*) (Díaz 1976)

itzauhyatl=*Artemisia mexicana* (Farfán and Elferink 2010; Ocaranza 2011)

itzcuinpatli/itzcuinpahtli/ytzcuinpahtli (dog remedy)=pl. 46, 62, 65, 66 Badianus (C-B fol. 27v, 35r, 37r, 37v)=FC632=*Plumbago pulchella* (Díaz 1976; Ortiz de Montellano 1975, 1990) or *Schoenocaulon oficinale* (*Veratrum officinale*) (Díaz 1976; Reko 1947) or *Roldana ehrenbergiana* (*Senecio canicidus*) (Altamirano 1896; Díaz 1976; Estrada Lugo 1989; Flores 1886–1888, 1:240; Guerra 1952; Ramírez and Alcocer 1902; Siméon 2010) or *Anticlea frigida* (*Stenanthium frigidum, Veratrum frigidum*)? (Reko 1947) or *Plumbago pulchella* (Miranda and Valdés 1964) or *Solanum* sp.? (de Ávila Blomberg 2012) or *Phytolacca icosandra* (Bye and Linares 2013)

itzcuinyoyotli=*Roldana ehrenbergiana* (*Senecio canicidus*) (Díaz 1976)

itzmaqua=*Laelila speciosa* (Ossenbach 2009)

itzmiquilitl (arrow-shaped edible herb)=FC464a=*Portulaca* sp. (Estrada Lugo 1987 in Pico and Nuez 2000a)

itztac quilitl (salty edible herb)=FC521=*Pistacia mexicana* (Estrada Lugo 1989) or Fabaceae

itztamatl=*Pseudobombax ellipticum* (*Bombax ellipticum*) (Díaz 1976)

itztauhyatl=FC=*Artemsia mexicana* (Ortiz de Montellano 1975)

ixpostli/ixpul=*Pinaropappus roseus* (Díaz 1976)

iyamolin=*Phytolacca icosandra* (Díaz 1976) or *Phytolacca octandra* (Altamirano 1896)

izcatl=*Gossypium hirsutum* (Dressler 1953)

izcuinpahtli=*Roldana ehrenbergiana* (*Senecio canicidus*) (Díaz 1976)

izote=*Yucca faxoniana* (*Y. australis*) (Díaz 1976)

izpacihuitl=*Croton dioicus* (Díaz 1976)

izquixochitl/izquixochiquauitl (fragrant flower tree)=p. 292 Sahagún=FC685a=*Bourreria huanita* (*B. formosa*) (Alcántara Rojas 2008; Altamirano 1896; Díaz 1976; Estrada Lugo 1989) or *B. litoralis* or *Bourreria* sp. (Estrada Lugo 1989)

iztacatzoyatl=*Senecio salignus* (Díaz 1976)

iztacayotli=*Cucurbita pepo* (Díaz 1976)

iztacaxixpatli=FC=*Persicaria punctata* (*Polygonum acre*) or *Polygonum aviculare* (Estrada Lugo 1989)

iztacchichicquauitl=FC=*Garrya laurifolia* (Estrada Lugo 1989)

iztacetl=FC=*Phaseolus coccineus* (Estrada Lugo 1989)

iztac nochnopalli=FC533=*Packera bellidifolia* (*Senecio vulneraria*)

itzacoanenepili=*Cissampelos pareira* (Díaz 1976)

iztac ocoxochitl=*Didymaea mexicana* (Díaz 1976)

iztac palancapatli=FC=*Packera bellidifolia* (*Senecio vulneraria*) (Díaz 1976; Estrada Lugo 1989)

iztac patli/iztacquauitl=FC563=*Pediomelum pentaphyllum* (*Psoralea pentaphylla*) (Díaz 1976; Estrada Lugo 1989; Farfán and Elferink 2010) *or Matelea reticulata* (*Gonolobus reticulatus*) (Díaz 1976) or *Dalea lutea* (*Parosela plumosa*) (Díaz 1976)

iztacpatli chichipiltic=*Pseudognaphalium canescens* (*Gnaphalium canescens*) (Farfán and Elferink 2010)

iztacpatlipipitzahoac=*Polygala verticillata* (Díaz 1976)

iztacquauhxiotl=*Pseudosmodingium perniciosum* (Díaz 1976)

iztacquauitl=FC=*Mariosousa willardiana* (*Acacia willardiana*) (Estrada Lugo 1989) or *Ceiba* sp. or *Maytenus phyllanthoides* (Estrada Lugo 1989)

iztactepetzacuxochitl=*Govenia liliacea* (Ossenbach 2009)

iztactlalamatl=*Desmodium triflorum* (*D. stipulaceum*) (Díaz 1976)

iztactzilacayotli=*Cucurbita pepo* (Díaz 1976)

iztaczapotl/iztactzapotl=*Casimiroa edulis* (del Pozo 1966; Díaz 1976) or *C. pubescens* (Díaz 1976)

iztaczazalic=FC=*Ipomoea orizabensis* (*I. longipedunculata*) (Estrada Lugo 1989)

iztamatl=*Pseudobombax ellipticum* (*Bombax ellipticum*) (Díaz 1976)

iztaquilitl/iztaquiltic=FC=*Portulaca oleracea* or *Suaeda suffrutescens or S. torreyana* (Estrada Lugo 1989)

iztauhyatl=FC567=*Artemisia mexicana* (Comas et al. 1995; Díaz 1976; Estrada Lugo 1989; Ortiz de Montellano 1990) or *A. longifolia* (Estrada Lugo 1989)

iztecauhticmixitl=FC=*Argemone mexicana* (Estrada Lugo 1989) or *Bocconia frutescens* (Estrada Lugo 1989) or *Commelina erecta* or *C. pallida* or *C. tuberosa* (Estrada Lugo 1989)

itztolin/itztolli=FC639b=*Cyperus* sp. (Estrada Lugo 1989)

macahuite=*Cordia bossieri* or *Ficus insipida* (*F. glabrata*) or *F. aurea* (*F. tecolutensis*) (Díaz 1976)

maçayelli/mazāyēlli (deer liver)=pl. 22 Badianus (C-B fol. 14v)=*Smilax* sp. (Bye and Linares 2013; de Ávila Blomberg 2012; Guerra 1952; Reko 1947; Valdés Gutiérrez et al. 1992)

maçaxocotl=FC417=*Spondias mombin*?

macazpipilloli=p. 388 Hernandez Rerum=*Quamoclit bracteata* (*Exogonium bracteatum*)?

macolcol=FC=*Agave* sp. (Estrada Lugo 1989)

macoztic metl=FC524=*Agave atrovirens*?

macpalxochi quahuitl=p. 383,459 Hernandez Rerum=*Chiranthodendron pentadactylon* (*C. platanoides*) (Reko 1947; Safford 1912)

macpalxóchitl/macpilxóchitl/mapilxóchitl/mācpalxōchitl=pl. 57 Badianus(C-B fol. 33r)=FC=*Cheiranthodendron pentacactylon* (de Ávila Blomberg 2012; Díaz 1976; Estrada Lugo 1989; Guerra 1952; Miranda and Valdés 1964; Ortis de Montellano 1990; Zepeda and White 2008)

macuil/macuili/maculiz=*Tabebuia heterophylla* (*T. pentaphylla*) (Díaz 1976)

macuilxochitl=p. 155 Hernandez Rerum/p. 649 Hernandez Historia=FC678a=*Tagetes erecta* or *T. patula* (Estrada Lugo 1989)
maguey=FC=*Agave americana* or *A. atrovirens* or *A. salmiana* (*A. cochlearis*) (Estrada Lugo 1989)
mahuitli=*Indigofera suffruticosa* (*I. anil*) (Díaz 1976)
malinalli/malinalxochitl/malīnalli (twisted [object])=pl. 18, 33, 108 Badianus (C-B fol. 12v, 20r, 58v)=*Muhlenbergia macroura* (Bye and Linares 2013; de Ávila Blomberg 2012; Díaz 1976; Farfán and Elferink 2010; Guerra 1952; Reko 1947; Miranda and Valdés 1964; Valdés Gutiérrez et al. 1992; Zepeda and White 2008)
mamaxtla/mā-māxtla (loincloths)=p. 59, 60 Badianus (C-B fol. 34r, 34v)=FC466d=p. 265 Sahagún=*Rumex* sp. (de Ávila Blomberg 2012; Miranda and Valdés 1964; Valdés Gutiérrez et al. 1992) or *Rumex mexicanus* (Díaz 1976) or *Iresine calea* (Guerra 1952; Linares and Bye 201; Reko 1947) or *Iresine* sp. (Estrada Lugo 1989)
mamaxtlaquilitl=FC=*Iresine* sp. (Estrada Lugo 1989)
mamaxtlanelhuatl (many roots)=pl. 75 Badianus (C-B fol. 42r)=*Iresine calea* (Guerra 1952)
manenepili=*Pleopeltis macrocarpa* (*Polypodium lanceolatum*) (Díaz 1976)
mapasuchil=*Chiranthodendron pentadactylon* (Díaz 1976)
mapilxochitl=FC=*Chiranthodendron pentadactylon* (Díaz 1976)
mastuerzo=FC=*Tropaeolum majus* (Estrada Lugo 1989; Pico and Nuez 2000b)
matilishuate=*Tabebuia heterophylla* (*T. pentaphylla*) (Díaz 1976)
matlacuahuitl=*Guaiacum sanctum* (Díaz 1976) or *G. officinale* (Comas et al. 1995)
matlalin/matlalli=FC=*Commelinia* sp.? (Altamirano 1896) or *Commelina erecta* or *C. tuberosa* or *Tradescantia* sp. (Estrada Lugo 1989)
matlaliztic/matlalitztic=*Commelina pallida* (Comas et al. 1995; Ortiz de Montellano 1990) or *C. coelestris* (Díaz 1976) or *C. diffusa* (Díaz 1976) or *C. erecta* (Díaz 1976) or *C. tuberosa* (Comas et al. 1995)
matlatlquahuitl=*Guaiacum sanctum* (Ortiz de Montellano 1990)
matlalxihuitl=FC=*Tradescantia crassifolia* (Estrada Lugo 1989; Rivera Morales 1941)
matlalxochitl/mātlalxōchitl (blue/green flower)=pl. 14 Badianus (C-B fol. 10v)=*Commelina coelestis* (Bye and Linares 2013; de Ávila Blomberg 2012; Díaz 1976; Miranda and Valdés 1964; Valdés Gutiérrez et al. 1992) or *C. pallida* (Reko 1947) or *Commelina* sp. (Guerra 1952)
matlaxihuitl=FC=*Commelina* sp. (Estrada Lugo 1989)
matzatli/matzali=p. 311 Hernandez Rerum=*Ananas comosus* (*A. sativus*) (Díaz 1976; Zepeda and White 2008)
mazaxochotli/mazacaxocotl/mazaxocotl=p. 384 Hernandez Rerum=*Spondias purpurea* (de Batres et al. 2012) or *S. mombin* (Estrada Lugo 1989)
matzapotl=*Annona cherimola* (Díaz 1976)
maxacocote=*Cyrtocarpoa procera* (Díaz 1976)
maxten=FC=*Allium* sp. (Estrada Lugo 1989)
mazacoanenepilli=*Passiflora suberosa* (Díaz 1976)
mazacoxocoyl=*Spondias* sp.? (Altamirano 1896)

mazatanes=*Valeriana ceratophylla* (Díaz 1976)
mazatlxocotl=*Spondias purpurea* (Díaz 1976)
mecacahoatl=*Theobroma cacao* (Díaz 1976)
mecapatli=*Smilax aristolochiifolia* (*Smilax medica*) (Díaz 1976; Ocaranza 2011; Ortiz de Montellano 1990)
mecapatli mecatlanensi/*mecapatli mecatlanense*=p. 304 Hernandez Rerum/p. 757 Hernandez Historia=*Ipomoea purpurea* (Díaz 1976)
mecapatli o zarzaparrilla=p. 288 Hernandez Rerum/p. 759 Hernandez Historia=*Smilax moranensis* (Díaz 1976)
mecaxochitl/*mecaxōchitl* (cord flower)=pl. 70, 104 Badianus (C-B fol. 39v, 56v)=p. 144 Hernandez Rerum/p. 750 Hernandez Historia=p. 287 Sahagún=FC=*Piper amalago* (Comas et al. 1995; de Ávila Blomberg 2012; Estrada Lugo 1989; Ortiz de Montellano 1990) or *Peperomia* sp. (Guerra 1952; Miranda and Valdés 1964; Reko 1947) or *Vanilla planifolia* (Díaz 1976; Estrada Lugo 1989; Linares and Bye 2013; Valdés Gutiérrez et al. 1992) or *Piper umbellatum* (*P. cuernavacanum*) or *Piper* sp. (Estrada Lugo 1989)
memeya/*memeyal*/*memealxihuitl*=FC556, 664=*Euphorbia* sp. (Estrada Lugo 1989)
memeyalquahuitl=*Protium heptaphyllum* (Díaz 1976)
memeyaxiuhtontli/*memeyaxihuitl*/*mē-mēya-xihuitl* (herb [with sap] flowing abundantly)=pl. 111 Badianus (C-B fol. 60r)=*Euphorbia maculata* (de Ávila Blomberg 2012; Díaz 1976; Guerra 1952; Miranda and Valdés 1964; Reko 1947; Valdés Gutiérrez et al. 1992) or *E. hirta* (Linares and Bye 2013) or *Euphorbia* sp. (Guerra 1952)
metl=p. 270 Hernandez Rerum/p. 1037 Hernandez Historia=*Agave atrovirens* (Comas et al. 1995; Díaz 1976; Dressler 1953)
metl coztli/*mecoztli*=p. 271 Hernandez Rerum/p. 1040 Hernandez Historia=*Agave* sp.?
metztliyçacauh (the moon's grass)=pl. 67 Badianus (C-B fol. 38r)=*Sprekelila formosissima*? (Guerra 1952; Miranda and Valdés 1964; Reko 1947) or *Polianthes tuberosa* (Linares and Bye 2013) or Liliaceae?
metzolle/*metzollin*=*Mammillaria bicolor* (Díaz 1976)
mexcalmetl=p. 272 Hernandez Rerum/p. 1043 Hernandez Historia=*Agave horrida* (Díaz 1976)
mexixquilitl/*mexix-quilitl* (mexixin edible herb)=pl. 34 Badianus (C-B fol. 20v)=*Lepidium sordidum* (*L. granulare*) (de Ávila Blomberg 2012; Díaz 1976; Miranda and Valdés 1964; Valdés Gutiérrez et al. 1992) or *L. virginicum* (Bye and Linares 2013) or *L. sativum* (Guerra 1952)
mexixquilitl (water cress potherb or golden saxifrage potherb)=p. 161 Hernandez=*Tropaeolum* sp. or *Lepidium sativum* (Reko 1947) or *L. virginicum* (Pico and Nuez 2000b)
mexocotl=p. 271 Hernandez Rerum/p. 1044 Hernandez Historia=FC=*Bromelia pinguin* (Ortiz de Montellano 1975, 1990) or *B. humilis* (Díaz 1976)
mexoxoctli/*mexoxochtli*=FC=*Agave sisalana* (*A. ridida* var. *sisalana*) (Estrada Lugo 1989)
mezcal=*Lophophora williamsii* (Díaz 1976)

mezquitl=*Prosopis juliflora* (Díaz 1976)
micaquahuitl=*Cordia brossieri* (Díaz 1976) or *Ipomoea murucoides* (Díaz 1976; Standley 1920–1926:1205)
micatlach pahoaztli=p. 386 Hernandez Rerum=?
mica xoxo coyollin/micaxoxocoioli (sorrel of the deaths)=p. 386 Hernandez Rerum=FC=*Oxalis hernandezii* (Estrada Lugo 1987 in Pico and Nuez 2000a)
miccauexotl=FC=*Salix* sp. (Estrada Lugo 1989)
miccaxoxocoyoli=FC=*Oxalis hernandezii* (Estrada Lugo 1989)
michiuauhtli (fish grain)=pl. 113 Badianus (C-B fol. 61r)=FC=*Amaranthus hybridus* (Miranda and Valdés 1964) or *A. cuentus* (*A. paniculatus*) (Estrada Lugo 1989) or *Chenopodium berlandieri* (*C. berlandieri* subsp. *nuttalliae*) (Díaz 1976; Guerra 1952)
michpatli=*Buddleia* sp.? (Altamirano 1896)
micle=*Jacobionia spicigera* (Díaz 1976)
miltomates=FC=*Physalis aequata* or *P. peruviana*, or *P. philadelphica* (Daunay et al. 2008; Estrada Lugo 1989) or *Solanum lycopersicum* (*Lycopersicon esculentum*) (Estrada Lugo 1989) or *P. angulata* (Estrada Lugo 1989)
miltomatl=p. 295 Hernandez Rerum/p. 701 Hernandez Historia=*Physalis ixocarpa* (Dressler 1953; Daunay et al. 2008)
mispatle=*Buddleja sessiliflora* (Díaz 1976)
misquicopalli=*Prosopis juliflora* (Díaz 1976)
mixitl=FC452b=*Datura stramonium* (Daunay et al. 2008; Díaz 1976; Estrada Lugo 1989)
mixixe=*Lepidum armoracia* (*L. intermedium*) (Díaz 1976)
mixtlaxihuitl=*Lopezia racemosa* (Díaz 1976)
mizquitl/mizquite/miquahuitl (moon to come forth from)=pl. 14 Badianus (C-B fol. 10v)=p. 455 Hernandez Rerum=FC423=*Prosopis juliflora* (Altamirano 1896; Díaz 1976; Estrada Lugo 1989; Guerra 1952; Miranda and Valdés 1964; Ortiz de Montellano 1990) or *Prosopis chilensis* (Reko 1947) or *Pithecellobium circinale* (*Inga circinalis, Mimosa circinalis*) (Díaz 1976; Estrada Lugo 1989) or *Prosopis laevigata* (*P. dulcis*) or *Prosopis* sp. (Estrada Lugo 1989)
mizquixochicopalli=*Bursera graveolens* (Díaz 1976)
moçoquilitl/mozoquilitl (sticky edible herb)=FC469b=*Ageratina deltoidea* (*Eupatorium deltoideum*) (Estrada Lugo 1987 in Pico and Nuez 2000a; Estrada Lugo 1989)
mocuepanixochitl (flower that habitually turns)=pl. 67 Badianus (C-B fol. 38r)=*Penstemon* sp. (Guerra 1952; Reko 1947) or Convolvulaceae (Linares and Bye 2013)
mohuitli/mohintle=p. 387 Hernandez Rerum/p. 236 Hernandez Historia=*Justicia spicigera* (Díaz 1976)
moictle=*Justicia spicigera* (Díaz 1976)
molle=p. 54 Hernandez Rerum/p. 281 Hernandez Historia=*Schinus molle*?
molonqui=*Cissus verticillata* (*C. sicyoides*) or *C. tiliacea* (*Vitis tiliacea*) (Díaz 1976)
moyotli/moytli/mozote=*Justicia spicigera* (Díaz 1976)

muchite/muchitl=*Pithecellobium dulce* (Díaz 1976)
muicle/muitle=*Justicia spicigera* (Díaz 1976)
nacaetolin/nacacetoli=FC=*Cyperus* sp. (Estrada Lugo 1989)
nacapuli=*Ficus pertusa* (*F. padifolia*) (Díaz 1976)
nacazcolotl=FC=*Caesalpinia coriaria* (Díaz 1976; Estrada Lugo 1989) or *Enterolobium cyclocarpum* (Altamirano 1896)
nacazul/nacazcul=*Datura wrightii* (*D. innoxia*) (Safford 1920, p. 549) or *D. stramonium* (Díaz 1976)
nahuitepuz/nahuitiput=*Verbesina crocata* (*Bidens crocata*) (Díaz 1976; Ocaranza 2011)
namole=*Phytolacca americana* (*P. decandra*), *P. octandra*, or *P. rugosa* (Díaz 1976)
nanacace=FC=*Byrsonima crassifolia* (*B. cotinifolia*) or *Senecio ovalifolius* (Estrada Lugo 1989)
nanacate/nanacatl=*Panaeolus sphinctr*inus (Díaz 1976) or *Inocybe* sp. or *Panaeolus* sp. or *Psathyrella* sp. or *Psilocybe* sp. or *Stropharia* sp. (Estrada Lugo 1989)
nanahuapatli=*Packera bellidifolia* (*Senecio vulneraria*) or *Solidago sempervirens* var. *mexicana* (Díaz 1976)
nance/nanche/nanchi=*Byrsonima crassifolia* (Díaz 1976)
nantzinquahuitl=*Byrsonima crassifolia* (Díaz 1976)
nantzinxocotl=*Byrsonima crassifolia* (Díaz 1976; Dressler 1953)
nanzenquahuitl=*Byrsonima crassifolia* (Díaz 1976)
nauiytepuzt=p. 189 Hernandez Rerum=*Smallanthus* sp.?
*nautzinxocot*l=*Malpighia* sp.? (Altamirano 1896)
necouhxochitl/neuc-xōchitl (honey flower)=pl. 97 Badianus (C-B fol. 53r)=*Plumeria rubra* (*P. acutifolia*) (de Ávila Blomberg 2012; Guerra 1952; Linares and Bye 2013; Valdés Gutiérrez et al. 1992)
nequametl=p. 273 Hernandez Rerum/p. 1051 Hernandez Historia=FC=*Agave mexicana* (Estrada Lugo 1989) or *Dasylirion* sp.?
nequametl=FC753=*Agave vera-cruz* (*A. mexicana*) (Díaz 1976) or *A. potatorum* (Estrada Lugo 1989)
nequen/henequen=FC=*Agave fourcroydes* or *A. sisalana* (Estrada Lugo 1989)
nere=*Hymenaea courabil* (Díaz 1976)
nesesehuilone=*Geranium mexicanum* (Díaz 1976)
nestamalxochitl=FC=*Ranunculus petiolaris* (Estrada Lugo 1989)
netzahyualxochitl=*Phyla scaberrima* (*Lippia dulcis*) (Díaz 1976)
nexehuac/nex-ēhua-c (it has turned ashen)=pl. 49 Badianus (C-B fol. 29r)=*Datura ceratocaula* (Bye and Linares 2013; de Ávila Blomberg 2012; Díaz 1976; Guerra 1952; Reko 1947; Valdés Gutiérrez et al. 1992) or *Datura* sp. (Furst 1995; Miranda and Valdés 1964)
nexmetl=FC=*Agave* sp. (Estrada Lugo 1989)
nexmitzi=*Artemisia mexicana* (Díaz 1976)
nextalpe/tlanoquizpatli=p. 178 Hernandez Rerum=*Guilleminea densa* or *Alternanthera sessilis* (*A. repens*) (Bye and Linares 1990)
nextamalxochitl=FC678d=*Ranunculus petiolaris* (Díaz 1976; Estrada Lugo 1989)
niamoli yamoli=*Phytolacca rugosa* (Díaz 1976)

niamolli=*Phytolacca octandra* (Díaz 1976)
niiche=*Coccoloba uvifera* (Díaz 1976)
nikte=*Plumeria rubra* (*P. acutifolia*) (Díaz 1976)
nixtamalxochitl=*Ranuncularis petiolaris* or *Tecoma stans* (Díaz 1976)
nocheznopalli/nopalnocheztli=pl. 38 Badianus (C-B fol. 22v)=p. 78 Hernandez Rerum=*Nopalea cochenillifera* (Reko 1947)
nocheznopalli/nopalnocheztli/nocheztli=p. 78 Hernandez Rerum/p. 943 Hernandez Historia=*Nopalea cochenillifera* (Dressler 1953)
nocheznopalli/nopalnocheztli/napalnocheztli/nopalnocheztli/nocheztli=p. 78, 459 Hernandez Rerum/p. 943, 945 Hernandez Historia=*Nopalea cochenillifera* (Dressler 1953)
nocheztli (blood of the nopal)=pl. 38 Badianus (C-B fol. 22v)=*Nopalea cochenillifera* (Guerra 1952)
nochtli (primary name)=pl. 106 Badianus (C-B fol. 57v)=*Opunti*a sp. (Reko 1947) or *O. ficus-indica* or *O. megacantha* or *O. streptacantha* (Dressler 1953) or *O. ficus-indica* or *Cylindropuntia imbricata* (*O. imbricata*) (Díaz 1976) or *Opuntia* sp. (Guerra 1952)
nocuana/nocuana-nui=*Panaeolus sphinctrinus* (Díaz 1976)
nonochto (insignificant little catci that come out of anthills)=p. 392 Hernandez Rerum=*Portulaca* sp.?
nonochton azcapa'yxua/nōnōchtōn āzcapan ixhua (insignificant little catci that come up around anthills)=pl. 47 Badianus (C-B fol. 28r)=*Portulaca pilosa* (de Ávila Blomberg 2012; Díaz 1976) or *Pereskiopsis* sp. or *Nopalxochia* sp. (Guerra 1952; Reko 1947)
nonoquitlitl=*Luffa operculata* or *Sechiopsis triqueter* (Díaz 1976)
nopal/nopalli/nohpalli (primary name)=pl. 90 Badianus (C-B fol. 49v)=FC=*Nopalea cochenillifera* or *Opuntia ficus-indica* or *O. megacantha* or *O. streptacantha* (Dressler 1953) or *O. tuna* or *O. ficus-indica* (Díaz 1976) or *Opuntia/Nopalea* sp. (Miranda and Valdés 1964) or *O. ficus-indica* or *O. maxima* (*O. amyclaea*) or *O. megarhiza* or *O. streptacantha* (Estrada Lugo 1989) or *O. microdasys* (Guerra 1952)
nopalnocheztli=*Opuntia ficus-indica* (Díaz 1976)
nopalxochicuezaltic/nepalxochcuezalticquizi/nopalxochitl/nochxochitl=p. 392, 457/p. 930, 931 Hernandez Historia=FC713b=*Disocactus phyllanthoides* (*Nopalxochia phyllanthoides*) or *D. ackermannii* (*Epiphyllum ackermanii*) (Díaz 1976; Estrada Lugo 1989)
noriten=*Agastache mexicana* (*Cedronella mexicana*) (Díaz 1976)
oaxin=*Leucaena esculenta* (Díaz 1976)
ocelometl=FC=*Agave* sp. (Estrada Lugo 1989)
oceloxochitl=FC727a=*Tigridia pavonia* (Díaz 1976; Dressler 1953; Estrada Lugo 19890; Safford 1912; Zepeda and White 2008) or *Tigridia* sp. (Estrada Lugo 1989)
ochmuul=*Triumfetta semitriloba* (Díaz 1976)
ococoxochitl=*Dahlia coccinea* (Díaz 1976)
ocolintequitcua=*Spigelia anthelmia* (Díaz 1976)

ocopetlatl=FC651=*Alsophila firma* (*Cyathea mexicana*) (Estrada Lugo 1989)
ocopetalatlyatlanuac=*Campyloneuron phylitidis* or *Niphidium crassifolium* (*Pleopeltis crassifolia, Polypodium crassifolium*) (Díaz 1976)
ocopiatztli=FC576=p. 222 Hernandez Rerum/p. 350 Hernandez Historia=*Eryngium foetidum* (not *E. aquaticum, E. beecheyanum,* or *E. painteri*)?
ocopiaztli=FC=*Polianthes geminiflora* (*Bravoa geminiflora*) or *Eryngium painteri* (Díaz 1976) or *Eryngium beecheyanum* (Díaz 1976; Estrada Lugo 1989)
ocopichin/ocopipin=*Cosmos diversifolius* (Díaz 1976)
ocoquilitl/ocoquiltic (resinous edible herb)=FC=p. 393 Hernandez Rerum/p. 358 Hernandez Historia=*Lactuca* spp. or *Sonchus oleraceus* (Estrada Lugo 1987 in Pico and Nuez 2000a) or *Launaea intybacea* (*Lactuca intybacea*) (Díaz 1976)
ocosuchil=*Didymaea mexicana* (Díaz 1976)
ocotl/ocote/ocotl/oxitl (pine)=pl. 70 Badianus (C-B fol. 39v) FC377=*Pinus teocote* (Altamirano 1896; Díaz 1976; Estrada Lugo 1989; Guerra 1952; Rivera Morales 1941; Standley 1920–1926:56) or *Pinus* sp. (Miranda and Valdés 1964)
ocotzoquauitl=FC=*Liquidambar styraciflua* (Díaz 1976; Ortiz de Montellano 1975; Standley 1920–1926:317) or *Pinus teocote* (Estrada Lugo 1989)
ocotzotl=FC=*Liquidambar styraciflua* (Estrada Lugo 1989; Ortiz de Montellano 1990) or *Pinus teocote* (Estrada Lugo 1989)
ocoxochitl (pine flower)=p. 14, 60, 75, 76, 99 Badianus (C-B fol. 10v, 34v, 42r, 42v, 54r)=FC=p. 288 Sahagún=FC=*Didymaea mexicana* (Díaz 1976; Estrada Lugo 1989; Guerra 1952; Martinez 1969:231; Miranda and Valdés 1964; Reko 1947)
ocoxote/ocozote=*Liquidambar styraciflua* (Díaz 1976)
octli=p. 101, 106, 111 (C-B fol. 55r, 57v, 60r)=wine from pulque (Guerra 1952)
ohuaxocoyolin/ohua-xocoyōlin (maize-stalk acidic plant)=pl. 17, 52, 59, 82 Badianus (C-B fol. 12r, 30v, 34r, 45v)=*Begonia gracilis* (Bye and Linares 2013; de Ávila Blomberg 2012; Reko 1947; Valdés Gutiérrez et al. 1992) or *Begonia* sp. (Furst 1995; Guerra 1952; Miranda and Valdés 1964)
olcacatzan/coçolmecatl=p. 212 Hernandez Rerum/p. 767 Hernandez Historia=*Smilax spinosa* (*S. mexicana*) or *S. pseudochina* (Comas et al. 1995; Díaz 1976) or *S. cordifolia* (Comas et al. 1995)
olcuahuitl=*Castilla elastica* (Díaz 1976)
olli/ulli/hulli/holli/ōlli (black gum)=pl. 53 Badianus (C-B. fol. 31r)=pl. 78 Badianus (C-B Fol. 43v)=p. 61, 304 Sahagún=*Castilla elastica* (de Ávila Blomberg 2012; Miranda and Valdés 1964; Ocaranza 2011; Reko 1947)
ololiuhqui/ololiuqui=FC=*Datura ceratocaula* (Daunay et al. 2008; Estrada Lugo 1989) or *Datura meteloides* (Estrada Lugo 1989; Safford 1920, p. 550) or *Turbina corymbosa* (*Rivea corymbosa*) (Comas et al. 1995; Díaz 1976; Estrada Lugo 1989; Furst 1995; Ortiz de Montellano 1990; Safford 1912; Schultes 1941, 1964; Wasson 1973, 1980) or *Helenium mexicanum* (Díaz 1976; Estrada Lugo 1989)
ololtic=FC=*Euphorbia seguieriana* (*E. campestris*) (Díaz 1976; Estrada Lugo 1989) or *Croton salutaris* (Estrada Lugo 1989) or *Euphorbia macropus* (*E. biformis*) (Estrada Lugo 1989)
ololticapitzalpatli=*Priva grandiflora* (*P. tuberosa*) (Díaz 1976)

olquauitl=FC=*Castilla elastica* (Díaz 1976)
omizochitl/omixochitl=p. 277 Hernandez Rerum=FC666, 715=*Polianthus tuberosa* (*P. gracilis*) (Díaz 1976; Estrada Lugo 1989; Ortiz de Montellano 1990)
oquechpatli=*Peperomia umbilicata* (Díaz 1976)
oquichtli cocaxochitl=p. 155 Hernandez Rerum/p. 647 Hernandez Historia=*Tagetes* sp.
otlatl=FC=*Bambusa* sp. (Estrada Lugo 1989)
oxotzin=*Brosimum alicastrum* (Díaz 1976)
oxte=*Malpighia souzae* (Díaz 1976)
oyametl/oiametl/oyamel=pl. 70 Badianus (C-B fol. 39v)=FC374=*Abies religiosa* (de Ávila Blomberg 2012; Díaz 1976; Estrada Lugo 1989; Guerra 1952; Miranda and Valdés 1964; Reko 1947; Standley 1920–1926:59)
ozote=*Ipomoea murucoides* (Díaz 1976; Standley 1920–1926:1205)
oztoam=*Byrsonima crassifolia* (Díaz 1976)
pac=*Solanum lycopersicum* (*Lycopersicon esculentum*) (Díaz 1976)
paccan=*Solanum americanum* (*S. nigrum*) (Díaz 1976)
pachiote=*Apeiba tibourboju* (Díaz 1976)
panxochitl=*Lobelia laxiflora* (Díaz 1976)
papaloquelite=*Porophyllum coloratum* (*P. seemannii*) (Díaz 1976 ; Estrada Lugo 1989) or *P. tagetoides* (Díaz 1976) or *P. viridiflorum* (Estrada Lugo 1989)
papaloquilitl/pāpālōquilitl (butterfly edible herb)=pl. 34, 46, 91, 99 Badianus (C-B fol. 20v, 27v, 50r, 54r)=*Porophyllum* [*viridiflorum*] (de Ávila Blomberg 2012) or *P. ruderale* subsp. *macrocephalum* (Bye and Linares 2013) or *Porophyllum* sp. (Valdés Gutiérrez et al. 1992) or *P. coloratum* (Guerra 1952)
papaloquilitl=FC480a=*Porophyllum coloratum* (*P. seemannii*) or *Porophyllum* spp. (Pico and Nuez 2000b) or *P. coloratum* (Reko 1947) or *P. coloratum* (*P. seemannii*) or *P. tagetoides* (Díaz 1976)
patachtli/pataxte=*Theobroma cacao, T. angustifolium*, or *T. bicolor* (Dressler 1953)
pati=p. 275 Hernandez Rerum/p. 1049 Hernandez Historia=*Yucca* sp.?
patimetl/patayametl=FC=*Agave* sp. (Estrada Lugo 1989)
patlahuactzitzicaztli/patlāhua-c tzītzicāztli (wide/broad nettle)=pl. 85 Badianus (C-B fol. 47r)=*Tournefortia hirsutissima*? (Reko 1947) or *T. suffruticosa* (Guerra 1952; Miranda and Valdés 1964) or *Wigandia* [*urens*?] (de Ávila Blomberg 2012) or *Wigandia urens* (*W. caracasana*) (Díaz 1976)
pauxochitl=*Lobelia laxiflora* (Díaz 1976)
peçotli=pl. 79 Badianus (C-B fol. 44r)=*Pachycereus* sp. (Guerra 1952)
pelon mexixquilitl=p. 161 Hernandez Rerum=*Tropaeolum majus* (Pico and Nuez 2000b)
pepetoaxin=*Leucaena leucocephala* (*L. glauca*) (Díaz 1976)
petlatollin/petatolli=FC642=*Typha latifolia* (Zepeda and White 2008) or *Cyperus* sp. (Estrada Lugo 1989)
petzapatli=?
petzicatl=FC=*Chamissoa altissima* (Estrada Lugo 1989)
peyotl/peyote=FC517=*Lophophora williamsii* (del Pozo 1966; Estrada Lugo 1989)
phehuame=*Aristolochia* sp.? (Ocaranza 2011)

piçietl/picietl/pycietl/piciete/piecete/pycicti/pītz-yetl (little fragrant tobacco or tobacco perfume)=pl. 54, 82 Badianus (C-B fol. 31v, 45v)=p. 173 Hernandez Rerum/p. 246 Hernandez Historia=cultivated *Nicotiana rustica* (Comas et al. 1995; de Ávila Blomberg 2012; Díaz 1976; Dressler 1953; Daunay et al. 2008; Estrada Lugo 1989; Furst 1995; Guerra 1952; Reko 1947) or *N. tabacum* (*N. mexicana*) (Comas et al. 1995; Díaz 1976; Estrada Lugo 1989; Ocaranza 2011)

piltzin tecouhxochitl chiyaua (greasy flower of the child-lord)=p. 62 Badianus (C-B fol. 35v)=*Tagetes* sp.?

piltzintecouhxochitl chiyaua/pil-tzin-tēuc-xōchitl (flower of the child-lord)=p. 62, 65, 67 Badianus (C-B fol. 35v, 37r, 38r)=*Lantana camara* (de Ávila Blomberg 2012; Díaz 1976; Linares and Bye 2013; Miranda and Valdés 1964; Valdés Gutiérrez et al. 1992) or *Zinnia pauciflora* (Díaz 1976) or *Bonellia macrocarpa* subsp. *macrocarpa* (*Jacquinia aurantiaca*) (Guerra 1952)

pinahuihuiztli=*Mimosa sensitiva* (Díaz 1976)

pioa=FC=*Ananas comosus* (*A. sativus*) (Estrada Lugo 1989)

pipilolxochitl=*Lobelia laxiflora* (Díaz 1976)

pipiloxihuitl=*Cestrum nocturnum* (Díaz 1976)

pipiloxochitl=*Lobelia laxiflora* (Díaz 1976)

pipitzahuac/pipitzahoa/pipitzauac=FC520=*Acourtia humboldtii* (*Perezia adnata*) (Comas et al. 1995; Díaz 1976; Estrada Lugo 1989; Ortiz de Montellano 1975, 1990) or *A. fruticosa* (*Perezia fruticosa*)? (Comas et al. 1995; Ocaranza 2011) or *Eupatorium sessilifolium* (Díaz 1976; Estrada Lugo 1989) or *Polygala verticillata* (Díaz 1976) or *A. cuernavacana* (*Perezia cuernavacana*) (Estrada Lugo 1989)

pita=FC=*Agave fourcroydes* or *A. sisalana* (Estrada Lugo 1989)

pitzauac=*Indigofera suffruticosa* (*I. anil*) (Díaz 1976)

pocayo=p.398 Hernandez Rerum=*Smilax* sp.?

pochotl/pochote=FC380=*Ceiba pentandra* (*Eriodendron occidentale*) (Díaz 1976; Estrada Lugo 1989) or *Pseudobombax ellipticum* (*Bombax ellipticum*) (Díaz 1976)

pom=*Protium copal* (Dressler 1953)

popoloquolite=*Porophyllum ruderale* subsp. *macrocephalum* (*P. macrocephalum*) (Díaz 1976)

popoyauh=FC=*Raphanus raphanistrum* (Estrada Lugo 1989)

popoyauhquilitl=p. 397 Hernandez Rerum/p. 240 Hernandez Historia=Brassicaceae?

potonxihuite=*Cestrum dumetorum* (Díaz 1976)

poxahoac=*Euphorbia seguieriana* (*E. campestris*) (Díaz 1976)

poçahualizxiuhtontli/pozahualizxiuhto[n]tli (insignificant little herb for swellings)=pl. 110 Badianus (C-B fol. 59v)=*Hieracium* sp. (Linares and Bye 2013) or *Echeandia* sp.(Valdés Gutiérrez et al. 1992)

pozaualizpatli=FC=*Croton cortesianus* (Estrada Lugo 1989)

pozual=*Euphorbia cortesianus* (Díaz 1976)

puchotl=FC=*Bombax ceiba* (Estrada Lugo 1989)

quachichil=*Loeselia mexicana* (*L. coccinea*) (Díaz 1976)

quachtlacalhuaztli/quachilacalhoaztli=FC564=*Mentzelia hispida* (Díaz 1976) or *Phaseolus* sp.?

quahayohuachtli=*Cassia grandis* (Comas et al. 1995) or *Jatropha curcas* (Comas et al. 1995)
quahizquixochitl=*Cordia elaeagnoides* (Díaz 1976)
quametl=p. 273 Hernandez Rerum/p. 1052 Hernandez Historia=*Agave* sp.?
quamiauatl/quamiahoatl=p. 258 Hernandez Rerum/p. 1034 Hernandez Historia=*Huperzia orizabae*? or *Huperzia dichotoma* (*Lyocopodium dichotomum*) (de Batres et al. 2012)
quamochitl=FC=*Pithecellobium dulce* (Díaz 1976; Ortiz de Montellano 1975, 1990)
quanacaquilitl (edible herb of the fowls)=FC=*Sonchus oleraceus* (Estrada Lugo 1987 in Picó and Nuez 2000a)
quantlapatzitzintl=*Senecio toluccanus* (Díaz 1976)
quapatlachtli=FC=*Theobroma bicolor* (Estrada Lugo 1989)
quapinolli=*Hymenaea courbaril* (Díaz 1976)
quapipillolxochitl=*Lonicera pilosa* (Díaz 1976)
quapopoltzin/quaupopoltzin=FC571=*Piqueria trinervia* (Díaz 1976; Estrada Lugo 1989)
quappachcintli=FC=*Zea mays* (Estrada Lugo 1989)
quappoquietl (woody smoking-cane)=pl. 62 Badianus (C-B fol. 35v)=*Schkuhria* sp. (Guerra 1952)
quaquauhtzontic/quaquaunton=p. 250 Hernandez Rerum/p. 131,132 Hernandez Historia=*Matelea reticulata* (*Gonolobus reticulatus*) (Díaz 1976)
quauchtlacalhuatztli=*Mentzelia hispida* (Díaz 1976)
quauhalahuac/cuauh-ālāhua-c (slippery [plant] of the forest)=pl. 106, 109 Badianus (C-B fol. 57v, 59r)=*Heliocarpus* sp. (de Ávila Blomberg 2012) or *Bartlettina sordida* (*Eupatorium sordidum*) (Guerra 1952; Reko 1947) or *Heliocarpus terebinthinaceus* (*Grewia terebinthinacea, H. reticulatus*) (Díaz 1976; Linares and Bye 2013; Miranda and Valdés 1964; Valdés Gutiérrez et al. 1992)
quauhayacachtli=*Lycianthes somniculenta* (*Solanum somniculentum*) (Díaz 1976)
quauhayohuachtli=p. 173 Hernandez Rerum/p. 173 Hernandez Historia=*Jatropha curcas* (Díaz 1976)
quauhayohuachtli=p. 87 Hernandez Rerum/p. 176 Hernandez Historia=*Cassia fistula* (Comas et al. 1995; Díaz 1976)
quauhayohuachtli=*Manihot esculenta* (*M. utilissima*) (Díaz 1976)
quauhayotli=*Parmentiera aculeata* (*Crescentia aculeata, P. edulis*) (Díaz 1976)
quauhcacahoatl=*Theobroma cacao* (Díaz 1976)
quauhcamotl/quauhcamotli=FC428, 440=*Manihot esculenta* (*M. dulcis, M. utilissima*) (Díaz 1976; Dressler 1953; Estrada Lugo 1989)
quauhchian=p. 399 Hernandez Rerum/p. 213 Hernandez Historia=*Priva mexicana* (*P. hispida*) (Díaz 1976)
quauhchichicpatli=p. 400 Hernandez Rerum/p. 580 Hernandez Historia=*Desmodium* sp.?
quauhchichioalli=*Rhus terebinthifolia* (Díaz 1976)
quauhchilli=*Bocconia arborea* or *Capsicum frutescens* (Díaz 1976)
quauhcoyloi=p. 401 Hernandez Rerum/p. 514 Hernandez Historia=*Bactris mexicana*?

quauheloquilitl=FC=*Ageratina deltoidea* (*Eupatorium deltoideum*) (Estrada Lugo 1989)

quauhhuitzihtzilxochitl (hummingbird-flower of the forest)=pl. 69 Badianus (C-B fol. 39r)=*Loeselia mexicana* (*L. coccinea*) (Díaz 1976; Linares and Bye 2013; Reko 1947; Valdés Gutiérrez et al. 1992) or *Cestrum* sp. (Miranda and Valdés 1964) or *Loeselia* sp. (Guerra 1952)

quauhichcatl=FC=*Bombax palmeri* (Estrada Lugo 1989) or *Ceiba pentandra* (Estrada Lugo 1989)

quauhizquiztli/quauhhizquiztli=p. 404 Hernandez Rerum/p. 124 Hernandez Historia=*Baccharis conferta* (Díaz 1976)

quauhmecapatli=p. 290 Hernandez Rerum/p. 761 Hernandez Historia=*Smilax aristolochiaefolia*?

quauhmecatl=p. 289 Hernandez Rerum/p. 763 Hernandez Historia=*Serjania mexicana* (Díaz 1976; Ocaranza 2011)

quauhpatlathli/quauhpatachtli/cacahoaquahuitl/cacaoapatli=pl. 89 Badianus (C-B fol. 49r)=p. 80 Hernandez Rerum/p. 912, 991 Hernandez Historia=*Theobroma bicolor* (Bye and Linares 1990) or *Theobroma* sp. (Furst 1995)

quauhtecomatl=*Crescentia cujete* or *Amphitecna macrophylla* (*Crescentia macrophylla*) (Díaz 1976)

quauhtepatli=FC=*Rhus glabra* (Estrada Lugo 1989)

quauhtepopopatli=*Yucca aloifolia* (Farfán and Elferink 2010) or *Y. faxoniana* (*Y. australis*) (Díaz 1976)

quauhtepoztli/quauhtepuztli=FC=*Buddleia* sp. (Estrada Lugo 1989) or *Guibourtia hymenaefolia* (*Copaifera hymenaeifolia*) or *Quercus* sp. (Estrada Lugo 1989)

[quauhtla] huitzquilitl/cuauhtlah huitzquilitl (thorny edible herb)=pl. 55, 73, 75 (C-B fol. 32r, 41r, 42r)=*Cirsium* sp. (de Ávila Blomberg 2012; Guerra 1952; Miranda and Valdés 1964; Reko 1947) or *Cirsium ehrenbergii* (Bye and Linares 2013; Valdés Gutiérrez et al. 1992)

quauhtlatlatzin=*Hura crepitans* (Díaz 1976; Ocaranza 2011) or *H. polyandra* (Díaz 1976; Estrada Lugo 1989)

quauhtlaxoxocoyolin/cuauhtlah xōxocoyōlin (acid plant of the forest)=pl. 9, 32 Badianus (C-B fol. 8r 19v)=*Begonia gracilis* (Bye and Linares 2013; de Ávila Blomberg 2012; Valdés Gutiérrez et al. 1992) or *Begonia* sp. (Guerra 1952)

quauhtla xoxocoyalin/cuauhtlah xōxocoyōlin (wild acid plant)=pl. 32 Badianus (C-B fol. 19v)=FC=*Oxalis tetraphylla* (Bye and Linares 2013; Miranda and Valdés 1964) or *O. corniculata* (de Ávila Blomberg 2012; Estrada Lugo 1987 in Picó and Nuez 2000a) or *Begonia* sp. (Reko 1947) or *O. herdandezii* (Valdés Gutiérrez et al. 1992)

quauhtlaxoxocoyalin (wild acid plant)=p. 474 Hernandez=*Oxalis latifolia*?

quauhtlaxoxocoyalin (wild acid plant)=*Oxalis acetosella* (*O. americana*) (Díaz 1976)

quauhtlepatli=p. 67 Hernandez Rerum/p. 183 Hernandez Historia=*Euphorbia calyculata* (Díaz 1976; Farfán and Elferink 2010)

quauhtlepatli=p. 122 Hernandez Rerum/p. 185 Hernandez Historia=?

quauhtlepatli=p. 122 Hernandez Rerum/p. 186 Hernandez Historia=*Lippia alba* (*L. geminata*) (Díaz 1976)

quauhtzahuatl=Ipomoea arborescens (*Convolvulus arborescens*)? (Ocaranza 2011)
quauhtzapotl=FC=*Annona cherimola* (Díaz 1976; Dressler 1953) or *Psidium guajava* (Díaz 1976) or *Ternstroemia tepezapote* (Díaz 1976)
quauhtzontic=Gonolobus prostratus (Díaz 1976)
quauhtzitzicaztli/cuauhtzitzicaztli/cuauhtzītzicāztli (forest nettle)=p. 85 Badianus (C-B fol. 47r)=*Urera baccifera* (de Ávila Blomberg 2012; Guerra 1952; Reko 1947) or *Urera caracasana* (Miranda and Valdés 1964) or *Urtica* sp.?
quauhuayohuachtli=Jatropha curcas (Díaz 1976)
quauhxicalli=Crescentia cujete (Díaz 1976)
quauhxilotl=Parmentiera aculeata (*Crescentia aculeata, P. edulis*) (Díaz 1976)
quauhxiotl=Bursera fagaroides (*Elaphrium fagaroides, E. odoratum*) or *Pseudosmodingium andrieuxii* or *P. perniciosum* (Díaz 1976)
quauhxiyotl/quauhxiotl (wild itch)=p. 81 Badianus (C-B Vol. 45r)=p. 90, 406 Hernandez Rerum/p. 549 Hernandez Historia=*Pseudosmodingium perniciosum* or *Parmentiera aculeata* (*P. edulis*)? (Reko 1947) or *Bursera lancifolia* or *B. longipes* (Miranda and Valdés 1964) or *Bursera* sp. (Guerra 1952)
quauhxiuhtic=FC=*Bursera multijuga* or *B. fagaroides* (Díaz 1976; Reko 1947) or *Bursera multijuga* (*Elaphrium multijugum*) or *B. fagaroides* (*E. odoratum*) (Estrada Lugo 1989)
quauhxochitl=FC=*Plumeria rubra* (Díaz 1976; Estrada Lugo 1989)
quauhxoxocoyli=FC478ab=*Oxalis frutescens* subsp. *angustifolia* (*O. angustifolia*) or *O. corniculata* (Estrada Lugo 1989)
quauhyaya[h]ual (vine-rings of the forest)=pl. 81, 95 Badianus (C-B fol. 45r, 52r)=*Smilax* sp. (Guerra 1952; Reko 1947) or *Lycianthes* sp. (Linares and Bye 2013) or Solanaceae?
quauhyetl=p. 403 Hernandez=wild *Nicotiana rustica* (Daunay et al. 2008) or *N. tabacum* (Díaz 1976)
quahuitl/quauhyyauhtli/yyauhtli (forest/to incense)=pl. 65, 70, 88, 91, 92, 95, 110 Badianus (C-B fol. 37r, 39v, 48v, 50r, 50v, 52r, 59v)=FC508a, 628=p. 160 Hernandez Rerum/p. 969 Hernandez Historia=p. 288 Sahagún=*Tagetes lucida* (Linares and Bye 2013; Valdés Gutiérrez et al. 1992) or *Chrysactinia mexicana* (Guerra 1952; Reko 1947)
quauhyzquixochitl (wild popcorn flower)=pl. 77 Badianus (C-B fol. 43r)=*Bourreria huanita* (Guerra 1952; Reko 1947) or *Cordia elaeagnoides* (Miranda and Valdés 1964)
quauitzquilitl (headed & prickly thistle)=FC=*Cirsium* sp. (Estrada Lugo 1987 in Picó and Nuez 2000a)
quaulinecapatli=Smilax aristolochiifolia (Díaz 1976)
quaumaychuachilli=Jatropha curcas (Díaz 1976)
quaumecapatl=Smilax aristolochiifolia (Díaz 1976)
quautecomatl=Crescentia alata (Díaz 1976)
quauxitol=Parmentiera aculeata (*P. edulis*) (Dressler 1953)
quauxiyotl=Bursera lancifolia or *B. longipes* (*Terebinthus longipes*) (Díaz 1976)
queppachcintli=Zea mays (Rivera Morales 1941)

quequexqui/quequixquic=FC=*Colocasia antiuquorum* (Estrada Lugo 1989) or *Xanthosoma violaceum* or *X. robustum* or *Xanthosma* sp. (Estrada Lugo 1989)
quetzalaylin (precious ylin)=pl. 53, 54, 116 (C-B fol. 31r, 31v, 62v)=?
quetzalhatzonyatl/quetzal-atzonyatl (plume-like atzonyatl)=pl. 94 Badianius (C-B fol. 51v)=*Barkleyanthus salicifolius* (de Ávila Blomberg 2012; Linares and Bye 2013) or *Senecio salignus* (Díaz 1976; Guerra 1952; Miranda and Valdés 1964; Reko 1947; Valdés Gutiérrez et al. 1992)
quetzala[h]uexotl/quetzalhuexotl/quetzalauexotl/quetzalāhuexōtl (precious willow water)=pl. 106 Badianus (C-B fol. 57v)=FC386, 578=*Salix lasiolepis* (de Ávila Blomberg 2012; Díaz 1976; Estrada Lugo 1989; Ortiz de Montellano 1990; Reko 1947) or *Salix paradoxa* (Linares and Bye 2013) or *Salix* sp. (Guerra 1952; Miranda and Valdés 1964; Valdés Gutiérrez et al. 1992)
quetzalichtli=p. 275 Hernandez Rerum=*Agave* sp.?
quetzalmizquitl (precious mesquite)=FC592=pl. 62, 89 Badianus (C-B fol. 35v, 49r)=*Prosopis chilensis* (de Ávila Blomberg 2012; Farfán and Elferink 2010; Guerra 1952; Reko 1947) or *Parkinsonia aculeata* (Díaz 1976; Miranda and Valdés 1964) or *Prosopis laevigata* (Linares and Bye 2013; Valdés Gutiérrez et al. 1992) or *Prosopis pubescens* or *Prosopis* sp. (Estrada Lugo 1989)
quetzalotlatl=FC=*Bambusa* sp. (Estrada Lugo 1989)
quetzalxoxouh capahtli/quetzalxoxouhqui (precious blue/green remedy)=pl. 43, 65 Badianus (C-B fol. 26r, 37r)=*Monnina* sp. cf. *M. salicifolia* or *Turbina corymbosa* (Díaz 1976; Guerra 1952; Reko 1947; Valdés Gutiérrez et al. 1992)
quetzalylin/quetzalytzlin/quetzalīlīn (precious alder)=pl. 69, 70, 83, 89, 115 (C-B fol. 39r, 39v, 46r, 49r, 62r)=*Alnus* sp. (de Ávila Blomberg 2012; Guerra 1952; Reko 1947;Valdés Gutiérrez et al. 1992) or *A. acuminata* subsp. *arguta* (Miranda and Valdés 1964) or *A. jorullensis* (Linares and Bye 2013)
queyauhquilitl=*Heliotropium curassavicum* (Díaz 1976)
quilamolli=*Dioscorea convolvulacea* or *Microsechium palmatum* (*M. helleri*) (Díaz 1976)
quilauacatl=FC=*Persea* sp. (Estrada Lugo 1989)
quiltonilli=FC464b=*Amaranthus hypochondriacus* (Estrada Lugo 1989)
quimichpatli=FC456b=*Buddleja sessiliflora* (Estrada Lugo 1989) or *Schoenocaulon officinale* (*Sabadilla officinarum*) (Díaz 1976 ; Estrada Lugo 1989) or *Roldana ehrenbergiana* (*Senecio canicidus*) (Díaz 1976) or *Veratrum* sp. (Altamirano 1896) or *Anticlea frigida* (*Stenanthium frigidum*) (Estrada Lugo 1989)
quimixpatle/quimixpatli=*Buddleja sessiliflora* (*B. verticillata*) (Díaz 1976)
sacatlaxcalli=*Cuscuta umbellata* (Díaz 1976)
sasafras=p. 61 Hernandez Rerum/p. 293 Hernandez Historia=*Bursera graveolens* or *Elaphrium pubescens* or *Sassafras albidum* (*S. officinalis*) (Díaz 1976)
sinicuiche/sinicuichi/sinicuilche=*Heimia salicifolia* (Díaz 1976; Wasson 1973, 1980)
suchilcahue=*Plumeria rubra* (*P. acutifolia*) (Díaz 1976)
tacanalquiltic=FC=*Artemisia ludoviciana* subsp. *redolens* (*A. redolens*) (Estrada Lugo 1989)
tacopaste/tacopatle/tacopaxtle=*Aristolochia anguicida* (*A. mexicana*) or *Aristolochia odoratissima* (Díaz 1976)

talayote=Gonolobus erianthus (Díaz 1976)
tamalayote/tamalayotl/tamalayotli=Cucurbita maxima (Díaz 1976)
taman=Gossypium hirsutum (*G. schottii*) (Díaz 1976)
tamatli=p. 118 Hernandez Rerum/p. 641 Hernandez Historia=?
tancazche=Zanthoxylum fagara (Díaz 1976)
tapate=Datura stramonium (Díaz 1976)
tatlauhcapatli=Geranium carolinianum (Farfán and Elferink 2010)
tchomechichi=Solanum americanum (*S. nigrum*)? (Ocaranza 2011)
teamoxtli (stone moss)=pl. 7, 29, 30, 58, 63, 80, 90, 91, 114 Badianus (C-B. fol. 7r, 18r, 18v, 33v, 36r, 44v, 49v, 50r, 61v)=bryophyte (Guerra 1952)
teachpaoaztli=Oenothera laciniata (Díaz 1976)
tecpatli/tecpalotl=FC=*Manfreda scabra* (*Agave saponaria*) or *Iostephane heterophylla* (Estrada Lugo 1989)
tecochitixihuitl (stone sleep herb)=pl. 76 Badianus (C-B fol. 42v)=*Boccconia* sp.? (Guerra 1952; Reko 1947)
tecolonochnopalli=FC=*Opuntia streptacantha* (Estrada Lugo 1989)
tecomahaca=p. 55 Hernandez Rerum/p. 554 Hernandez Historia=*Bursera tecomaca* (*Elaphrium tecomaca*) or *Protium heptaphyllum* (Díaz 1976)
tecomalyyac=Protium heptaphyllum (Díaz 1976)
tecomasuchil=Cochlospermum vitifolium (Díaz 1976)
tecomate/tecomatl=Crescentia alata or *C. cujete* (Díaz 1976)
tecomaxochitl=FC511=p. 408 Hernandez Rerum/p. 443 Hernandez Historia=*Datura* sp. (Daunay et al. 2008) or *Solandra maxima* (*Datura maxima*, *S. nitida*) (Díaz 1976; Farfán and Elferink 2010; Elferink et al. 1994; Zepeda and White 2008) or *Solandra guerrerensis* or *Solandra brevicalyx* (Farfán and Elferink 2010) or *Solandra maxima* (*S. hartwegii*) (Safford 1920, p. 560) or *Brugmansia arborea* (*Datura arborea*) or *Datura stramonium* (Díaz 1976) or *Amphitecna macrophylla* (Estrada Lugo 1989) or *Cochlospermum vitifolium* (*Maximilianea vitifolia*) or *Solandra grandiflora* or *Datura* sp. or *Swartzia guttata* (Estrada Lugo 1989)
tecomitl=FC=*Mammillaria polythele* (*M. tetracantha*) or *M. dolichacantha* (Estrada Lugo 1989)
teçonpahtli (remedy that grows on red porous volcanic rock)=pl. 10 Badianus (C-B vol. 8v)=*Asclepias linaria* (Bye and Linares 2013; Guerra 1952; Valdés Gutiérrez et al. 1992; Miranda and Valdés 1964)
tecopal/tecoaplli=Bursera copallifera (*B. jorullensis*) or *Rhus saxatilis* (Díaz 1976)
teçontzapotl=FC=*Manilkara zapota* (*Calocarpum mammosum*, *Lucuma mammosa*) (Estrada Lugo 1989)
tecopalli quahuitl/tecopalquahuitl=p. 47 Hernandez Rerum/p. 535 Hernandez Historia=*Protium copal* (Díaz 1976) or *Bursera* sp.?
tecopalquanhuitl pitzahoac=Bursera bipinnata (Díaz 1976)
tecopal quahuitl tepotztlani/tocopalquahuitl=p. 49 Hernandez Rerum/p. 537 Hernandez Historia=*Bursera* sp.?
tecote=Crescentia alata (Díaz 1976)

tecpatli (medicine used for broken bones)=pl. 41 Badianus (C-B fol. 25r)=p. 191 Hernandez Rerum=*Smallanthus* sp.? or *Helianthus* sp.? (Altamirano 1896) or Asteraceae (Miranda and Valdés 1964) or *Datura* sp. (Guerra 1952)
tecuampatli=*Marsdenia zimapanica* (Díaz 1976)
tecuitlatl (excreta of stone)=*Spirulina maxima* (*S. geitlerii*) (Ortiz de Montellano 1990) or *Spirulina* sp. (Farrar 1966)
tehuistle/tehuitzli/tehuxtle/tehuztle=*Sapindus saponaria* (Díaz 1976)
temahuiztiliquauitl (people-honoring tree)=pl. 68, 69 (C-B fol. 38v, 39r)=Melastomataceae (Guerra 1952; Reko 1947)
temamatlatzin (shine with many leaves/stone shining with leaves)=pl. 7, 36 Badianus (C-B fol. 7r, 21v)=bryophyte (Guerra 1952)
temecatl=*Cissus tiliacea* (*Vitis tiliacea*) (Díaz 1976; Ocaranza 2011) or *Ipomoea emetica* or *I. hederifolia* (Díaz 1976)
temecatl expurgante/temecatl purgante=p. 305 Hernandez Rerum/p. 775 Hernandez Historia=*Ipomoea hederifolia*?
temecatl yauhtepecensi=p. 200 Hernandez Rerum/p. 777 Hernandez Historia=?
temeicaxihuitl/temecaxihuitl=p. 410 Hernandez Rerum/p. 778 Hernandez Historia=?
tememetla/temēmetla/tememetlatl (gizzard)=pl. 38 Badianus (C-B fol. 22v)=FC507a=*Echeveria fulgens* (de Ávila Blomberg 2012; Estrada Lugo 1989; Guerra 1952; Reko 1947) or *E. gibbiflora* (Bye and Linares 2013; Díaz 1976; Valdés Gutiérrez et al. 1992) or *Echeveria* sp. (Estrada Lugo 1989)
temetl=FC=*Agave* sp. (Estrada Lugo 1989)
tempate=*Jatropha curcas* (Díaz 1976)
tempixque=*Pouteria campechiana* (*Vitellaria sphaerocarpa*) (Díaz 1976)
tempixquiztli=*Randia aculeata* (*R. mitis*) (Díaz 1976)
tempixtle=*Sideroxylon palmeri* (*Bumelia laetevirens*) or *Pouteria campechiana* (*Vitellaria sphaerocarpa*) or *Sideroxylon persimile* subsp. *subsessiliflorum* (*Bumelia subsessiliflora*) (Díaz 1976)
tempizquixtli=*Sideroxylon palmeri* (*Bumelia laetevirens*) or *Sideroxylon persimile* subsp. *subsessiliflorum* (Bumelia subsessiliflora) (Díaz 1976)
tenamaz-nanapoloa (carrying triplets)/*tenamazton/tlalamatl*=p. 129 Hernandez Rerum=*Scutellaria mexicana* (*Salazaria mexicana*) (Tucker and Janick 2016)
tencuanete/tenauanete=*Euphorbia calyculata* (Díaz 1976)
tenochitli=*Opuntia ficus-indica* (Díaz 1976)
tenopalli=FC=*Discocactus* sp. (*Epiphyllum* sp.) (Estrada Lugo 1989)
tenpachoaztli=*Selaginella sellowii* (*S. rupestris*) (Díaz 1976)
tentzonxochitl=*Calliandra houstoniana* var. *anomala* (Díaz 1976; Farfán and Elferink 2010) or *Bauhinia racemosa* (Díaz 1976)
teocaltzincense=*Caphalanthus occidentalis* (Díaz 1976)
teocomitl=FC754a=*Mammillaria polythele* (*M. tetracantha*) (Díaz 1976) or *Echinocactus* sp. (Estrada Lugo 1989) or *Ferocactus* sp. (Estrada Lugo 1989)
teocotl=FC=*Pinus teocote* (Díaz 1976; Estrada Lugo 1989)
teocuahuitl=*Cedrela* sp. (Altamirano 1896)
teoezquauitl/teōezcuahuitl (sacred blood tree)=pl. 68 Badianus (C-B fol. 38v)=*Croton [draco]* sp. (de Ávila Blomberg 2012; Miranda and Valdés 1964) or

Rauvolfia sp. (Valdés Gutiérrez et al. 1992) or *Mammea americana* (Guerra 1952)

teohamatl/teoamatl (sacred paper plant)=pl. 77, 78, 82 Badianus (C-B fol. 43r, 43v, 45v)=*Agave* sp. (*Spirulina* sp.) (Farrar 1966; Reko 1947)

teometl=FC524=*Agave atrovirens* (Díaz 1976; Estrada Lugo 1989) or *A. potatorum* (Díaz 1976) or *A. vivipata*? (Ocaranza 2011)

teonacaztli/ueinacaztli (sacred ear)=FC420, 686=p. 237 Sahagún=*Cymbopetalum penduliflorum* (Díaz 1976; Estrada Lugo 1989) or *Chiranthodendron pentadactylon* (Estrada Lugo 1989)

teonanacatl=*Psilocybe aztecorum* or *P. mexicana* (Farfán and Elferink 2010), *P. zapotecorum, P. semperviva,* or *Panaeolus sphinctrinus* (Díaz 1976; Schultes 1940) or *Salvia divinorum* (del Pozo 1966) or *Hylocereus undatus* (Díaz 1976) or *Inocybe* sp. or *Panaeolus* sp. or *Psathyrella* sp. or *Psilocybe* sp. or *Stropharia* sp. (Estrada Lugo 1989)

teonochtli/teō-nōchtli (sacred cactus)=pl. 28 Badianus (C-B fol. 17v)=*Pachycereus weberi* (Reko 1947) or *Stenocereus* sp. (de Ávila Blomberg 2012; Valdés Gutiérrez et al. 1992; Zepeda and White 2008) or *Isolatocereus dumortieri* (Bye and Linares 2013) or *Hylocereus undatus* (Estrada Lugo 1989) or *Pachycereus* sp. (Guerra 1952)

teopochotl=FC743=*Bombax* sp. (Estrada Lugo 1989) or *Ceiba* sp. (Estrada Lugo,1989)

teouaxin=FC=*Lysiloma acapulcense* (Estrada Lugo 1989)

teoyztaquilitl (sacred salt edible herb)=pl. 31, 46, 51, 75, 79 Badianus (C-B fol. 19r, 27v, 30r, 42r, 44r)=Brassicaceae (Guerra 1952; Miranda and Valdés 1964; Reko 1947)

tepachtli=*Selaginella sellowii* (*S. rupestris*) (Díaz 1976)

tepapaquiltiqu[h]auitl (tree making people happy)=pl. 69, 70, 91 Badianus (C-B fol. 39r, 39v, 50r)=*Hibiscus* sp. (Guerra 1952; Reko 1947)

tepaquiltixiuhtontli (insignificant little herb making people happy)=pl. 67 Badianus (C-B fol. 38r)=*Hibiscus* sp. (Reko 1947; Valdés Gutiérrez et al. 1992) or *Ipomoea cristulata* (Linares and Bye 2013) or *Pavonia schiedeana* (*Malache rosea*) (Guerra 1952)

teparisirangua=*Commelina tuberosa* (Díaz 1976)

tepate=*Datura stramonium* (Díaz 1976)

tepepapaloquilitl/tepēpālōquilitl (butterfly edible herb of the mountains)=pl. 103 Badianus (C-B fol. 56r)=*Porophyllum* sp.? (de Ávila Blomberg 2012; Miranda and Valdés 1964; Valdés Gutiérrez et al. 1992) or *Porophyllum punctatum* (Linares and Bye 2013)

tepeacxoyatl=*Asclepias verticillata* (Díaz 1976)

tepeamalacotl=FC545=*Solidago* sp. (Estrada Lugo 1989)

tepeamatl/texcalamatl (cliff yellow paper)=p. 81–82 Hernandez Rerum=*Ficus nymphaeifolia* (Díaz 1976) or *Populus* sp.

tepecenpoal xochitl=p. 156 Hernandez Rerum/p. 652 Hernandez Historia=*Tagetes* sp.?

tepechayotli=*Selaginella pallescens* (*S. cuspidata*) (Díaz 1976)

tepechian/tepechien/tepē-chian (mountain chia)=pl. 32, 61, 62, 63, 65, 76, 81 Badianus (C-B fol. 19v, 35r, 35v, 36r, 37r, 42v, 45r)=*Salvia [hispanica]* (de Ávila Blomberg 2012; Guerra 1952) or *Verbena* sp. (Reko 1947) or *Salvia occidentalis* (Díaz 1976) or *Salvia* sp. (Miranda and Valdés 1964)
tepecoacuilco=Galphimia paniculata (*G. humboldtiana*) (Díaz 1976)
tepecuilote=Cornus excelsa (Díaz 1976)
tepeguaje=Lysiloma acapulcense (Díaz 1976)
tepehoaxin=Senna pallida (*Cassia acapulcensis*) or *Lysiloma acapulcense* (Díaz 1976)
tepehoila capitzxochitl=p. 211 Hernandez Rerum=*Ipomoea* sp.?
tepehuaje=Leucaena pulverulenta or *Lysiloma acapulcense* (Díaz 1976)
tepeizquitl=Arctostaphylos pungens (Díaz 1976)
tepejilote=Chamaedorea elatior or *C. tepejilote* (Díaz 1976)
tepejiloxochitl=Calliandra houstoniana var. *anomala* (*C. anomala*) (Díaz 1976)
tepemizquitl=Lysiloma acapulcense (Díaz 1976)
tepenexcomitl=p. 410 Hernandez Rerum/p. 939 Hernandez Historia=*Ferocactus recurvus* (*F. latispinus*)
tepeoaxin=Lysiloma acapulcense (Díaz 1976)
tepepapaloquilitl (mountain butterfly edible plant)=pl. 103 Badianus (C-B fol. 56r)=*Porophyllum* sp. (Guerra 1952; Reko 1947)
tepesquitl=Arctostaphylos pungens (Díaz 1976)
tepetlachichicx xiuhuitl/tepetlachichicxihuitl=p. 192 Hernandez Rerum/p. 572 Hernandez Historia=?
tepetomatl=FC383, 585=*Arctostaphylos tomentosa* (Díaz 1976; Estada Lugo 1989; Ocaranza 2011) or *Physalis* sp. (Estrada Lugo 1989)
tepetuage=Cassia acapulcensis (Díaz 1976)
tepexiloxochitl=Calliandra houstoniana var. *anomala* (*C. anomala, C. grandiflora*) (Díaz 1976)
tepexonacatl=FC=*Allium* sp. (Estrada Lugo 1989)
tepeyauhtli/ yauhtli/yyauhtli montes=p. 167 Hernandez Rerum/p. 971 Hernandez Historia=*Chromolaena pulchella* (*Eupatorium pulchellum*)?
tepeylin (mountain alder)=pl. 70 Badianus (C-B fol. 39v)=*Alnus* sp. (Guerra 1952; Miranda and Valdés 1964; Reko 1947)
tepezapote=Phytolacca dioica or *Ternstroemia tepezapote* (Díaz 1976)
tepezentli/tepececentli=p. 223 Hernandez Rerum/p. 867 Hernandez Historia=Apocynaceae?
tepezquite=Arctostaphylos pungens (Díaz 1976)
tepicquilitl (edible herb found among the stones)=FC471b=?
tepitoton teamoxtli (small stone mosses)=pl. 30 Badianus (C-B fol. 18v)=bryophyte (Guerra 1952)
tepixiloxochitl=Calliandra houstoniana var. *anomala* (Farfán and Elferink 2010)
tepepote/tepepotl/tepepotli=Baccharis conferta or *B. salicina* (*B. glutinosa*) or *B. multiflora* or *Ephedra aspera* or *E. trifurca* (Díaz 1976)
tepoyan=Ehretia anacua (*E. scabra*) (Díaz 1976)

tepozan=pl. 81 Badianus (C-B fol. 45r)=*Buddleja* sp. (de Ávila Blomberg 2012) or *B. americana* or *B. cordata* or *Haplophyton cimicidum* (Díaz 1976)

tepoztlanensi=p. 31 Hernandez Rerum/p. 28 Hernandez Historia=*Dahlia* sp. aff. *D. coccinea*?

tequammaytl/tequammaitl/tegnammaytl (man-eater's hand)=pl. 10 Badianus (C-B fol. 8v)=*Smilax spinosa* (Miranda and Valdés 1964; Reko 1947) or *S. moranensis* (Díaz 1976) or *Dioscorea* sp. (Bye and Linares 2013; Guerra 1952; Valdés Gutiérrez et al. 1992)

tequa niytzoncal=p. 411 Hernandez Rerum=*Vittaria* sp.?

tequequetzal=FC656=*Selaginella lepidophylla* (Díaz 1976; Estrada Lugo 1989) or *Adiantum* sp. or *S. pilifera* (Estrada Lugo 1989)

tequixquiçacatl/tequizquizacatl (soda grass)=pl. 19 Badianus (C-B fol. 13r)=*Distichlis spicat*a (Bye and Linares 2013; Díaz 1976; Guerra 1952; Valdés Gutiérrez et al. 1992) or *Sporolobus* sp. (Guerra 1952; Miranda and Valdés 1964; Reko 1947)

teshuate=*Conostegia xalapensis* (Díaz 1976)

tetezhuatic/tē-tezhua-tic (smooth or polished)=pl. 40 Badianus (C-B fol. 24v)=*Conostegia* sp. or *Miconia* sp. (de Ávila Blomberg 2012; Guerra 1952; Reko 1947) or *Conostegia xalapensis* (Díaz 1976; Miranda and Valdés 1964)

tetlatia=*Toxicodendron radicans* (Díaz 1976; Ocaranza 2011)

tetzapotl (stone sapote)=pl. 57, 58 Badianus (C-B fol. 33r, 33v)=*Manilkara zapota* or *Annona squamosa* (Reko 1947) or *Pouteria sapota* (Linares and Bye 2013; Valdés Gutiérrez et al. 1992) or *Annona squamosa* (Guerra 1952)

tetzitzicaztli/te-tzīzicāztli (stone nettle)=pl. 85 Badianus (C-B fol. 47r)=*Jatropha* sp.? (Guerra 1952; Reko 1947) or *Cnidoscolus multilobus*? (de Ávila Blomberg 2012) or *Cnidoscolus* sp. (Linares and Bye 2013; Miranda and Valdés 1964; Valdés Gutiérrez et al. 1992)

tetzitzilin (making a stone-like ringing noise)=pl. 94 Badianus (C-B fol. 51v)=Cucurbitaceae (Guerra 1952; Reko 1947) or *Microsechium palmatum* (*M. helleri*) (Linares and Bye 2013) or *Cucurbita* sp. (Miranda and Valdés 1964; Valdés Gutiérrez et al. 1992)

tetzminohpalli/tetzminopalli (sedum nopal)=pl. 32 Badianus (C-B fol. 19v)=*Pereskia portulacifolia* (Guerra 1952; Reko 1947) or Crassulaceae (Miranda and Valdés 1964)

tetzmitl/tetzmetl/tetzmitic/tetetzmitic=pl. 19, 75, 79, 88, 90 Badianus (C-B fol. 13r, 42r, 44r, 48v, 49v)=FC496, 557, 754e=p. 171 Hernandez=*Sedum dendroideum* (Bye and Linares 2013; de Ávila Blomberg 2012; Díaz 1976; Estrada Lugo 1989; Guerra 1952; Reko 1947; Miranda and Valdés 1964; Valdés Gutiérrez et al. 1992)

tetzmicochitl (flower of the tetzmitl)=pl. 10, 60 Badianus (C-B fol. 8v, 34v)=see *tetzmitl*

tetzmolin/tetzmoli=FC654, 756=*Quercus fusiformis* or *Quercus* sp. (Estrada Lugo 1989)

teunacaztli=FC=*Cymbopetalum penduliflorum* (Estrada Lugo 1989)

texaltzapotl=*Annona cherimola* or *A. squamosa* (Díaz 1976) *texcalamacoztli/texcalāma-coztli* (yellow paper-tree of the boulders)=pl. 68 Badianus (C-B fol. 38v)=*Populus* sp. or *Ficus* sp. (de Ávila Blomberg 2012) or *Ficus petiolaris* (Guerra 1952; Linares and Bye 2013; Miranda and Valdés 1964; Reko 1947; Valdés Gutiérrez et al. 1992)

texcalamatichichic (bitter cliff paper)=pl. 78 Badianus (C-B fol. 43v)=*Ficus petiolaris* (Díaz 1976; Farfán and Elferink 2010; Guerra 1952; Reko 1947)

texcalamatl=*Ficus petiolaris* (*F. jaliscana*) or *F. nymphaeifolia* (Díaz 1976)

texcalpachtli=*Selaginella sellowii* (*S. rupestris*) (Díaz 1976)

texhuatl=p. 413 Hernandez Rerum=Melastomataceae?

texixiuhtli=*Heliotropium angiospermum* (*H. parviflorum*) (Díaz 1976)

texiotl/texyotl/texiutl (stone rash)=pl. 38, 90 Badianus (C-B fol. 22v, 49v)=FC755a=*Sedum dendroideum* (Díaz 1976; Estrada Lugo 1989) or *S. oxypetalum* (Díaz 1976; Miranda and Valdés 1964) or *S. bourgaei*? (Guerra 1952; Reko 1947) or *Sedum praealtum* subsp. *parvifolium* (Bye and Linares 2013; Valdés Gutiérrez et al. 1992)

texocotl=FC=*Crataegus gracilior* (*C. pubescens*) (Dressler 1953) or *Crataegus mexicana* (Altamirano 1896)

texochitl yamanqui/yamanquitexochitl (soft stone-flower)=pl. 64 Badianus (C-B fol. 36v)=*Selaginella* sp. (Guerra 1952; Reko 1947) or *S. lepidophylla* (Díaz 1976; Valdés Gutiérrez et al. 1992)

texocotl/texocutl=FC=*Crataegus mexicana* (Díaz 1976; Estrada Lugo 1989) or *C. gracilior* (*C. pubescens*) (Estrada Lugo 1989)

texoxochitl=*Calliandra houstoniana* var. *anomala* (*C. anomala, C. grandiflora*) (Díaz 1976)

texoxocoili/texoxocoytli (hard sorrel)=FC587=*Oxalis* sp. (Estrada Lugo 1987 in Pico and Nuez 2000a; Estrada Lugo 1989) or *Begonia gracilis* (Farfán and Elferink 2010)

teyacua=*Chiranthodendron pentadactylon* (Díaz 1976)

teyyayhtli=*Asplenium trichomanes* (Díaz 1976)

tezcapatli=*Senecio praecox* (Díaz 1976)

tezhaotl=*Melastoma* sp. (Altamirano 1896)

tezompotonic/"tzonpotonic"=p. 191 Hernandez Rerum/p. 135 Hernandez Historia=*Blumea viscosa* (*Conyza lyrata*)

tezonpatli=*Asclepias linaria* (Díaz 1976; Ortiz de Montellano 1990; Reko 1947)

tezonzapotl/tezontzapote (lava sapote)=*Manilkara zapota* (Díaz 1976; Dressler 1953) or *Mammea americana* (Díaz 1976)

tezuatl/tezoatl=FC=*Conostegia xalapensis* or *Miconia laevigata* (Estrada Lugo 1989)

theometl=p. 274 Hernandez Rerum/p. 1046 Hernandez Historia=*Agave atrovirens*?

tianquiz=*Alternanthera sessilis* (*A. repens*) or *Plumbago pulchella* (Díaz 1976)

tianquizpepetla=*Alternanthera sessilis* (*A. repens*) (Díaz 1976; Farfán and Elferink 2010) or *A. pungens* (*Illecebrum achyrantha*)? (Ocaranza 2011)

titilamatl=*Pseudobombax ellipticum* (*Bombax ellipticum*) (Díaz 1976)

tizexochitl=*Plumeria rubra* (*P. acutifolia*) (Díaz 1976)

tlacacahoatl=*Arachis hypogaea* (Díaz 1976)
tlacacamohtli/*tlāca-camohtli* (man tuber)=pl. 48, 79 Badianus (C-B fol. 28v, 44r)=*Ipomoe*a sp. (de Ávila Blomberg 2012; Miranda and Valdés 1964; Valdés Gutiérrez et al. 1992) or *I. batatas* (Bye and Linares 2013) or *I. purga* or *Operculina pteripes* (*O. alatipes*) (Guerra 1952; Reko 1947)
tlacametl=p. 274 Hernandez Rerum/p. 1042 Hernandez Historia=FC=*Agave atrovirens* (Estrada Lugo 1989) or *A. potatorum* (Díaz 1976; Estrada Lugo 1989) or *A. salmiana* (Estrada Lugo 1989)
tlacapolin=*Gaultheria myrsinoides* (*Pernettya ciliata*) (Díaz 1976)
tlacapulin=*Rhamnus serrata* (Ortiz de Montellano 1990)
tlacazolauacatl=FC=*Persea* sp. (Estrada Lugo 1989)
tlacehuiloni=*Geranium hernandesii*? (Ocaranza 2011)
tlaceoitli/*tlaceuilli*=FC=*Indigofera suffruticosa* (*I. anil*) (Díaz 1976; Estrada Lugo 1989)
tlachichinoa=*Moussonia deppeana* (*Isoloma deppeanum, Kohleria deppeana*) or *Tournefortia capitata* or *T. hirsutissima* (Díaz 1976)
tlachichinoapatlahoac=p. 292 Hernandez Rerum/p. 366 Hernandez Historia=*Heliotropium angiospermum* (*H. parviflorum*) (Díaz 1976)
tlachichinoapatlahoac=p. 292 Hernandez Rerum/p. 367 Hernandez Historia=*Tournefortia mexicana* (Díaz 1976)
tlachichinoapatlahoac=p. 292 Hernandez Rerum/p. 368 Hernandez Historia=*Tournefortia volubilis* (Díaz 1976)
tlachichinoa xiuitl=FC=*Plumbago zeylanica (P. scandens*) (Estrada Lugo 1989) or *Tournefortia hirsutissima* (Estrada Lugo 1989)
tlachichinol=*Plumbago pulchella* (Díaz 1976)
tlachichinole=*Stevia linoides* (Díaz 1976)
tlachicon=*Curatella americana* (Díaz 1976)
tlachinole=*Croton draco* or *Urera caracasana* (Díaz 1976)
tlachinoltetzmitl=FC754f=*Moussonia deppeana* (*Kohleria deppeana*) (Díaz 1976; Estrada Lugo 1989) or *Croton draco* or *Sedum* sp. (Estrada Lugo 1989)
tlachipili=FC=*Crotalaria* sp. (Estrada Lugo 1989)
tlachpahuaztic=*Polygala verticillata* (Díaz 1976)
tlacoamatl/*tlacametl* (stalk plant)=pl. 53 Badianus (C-B fol. 31r)=p. 274, 884 Hernandez/p. 1042 Hernandez Historia=*Agave* sp.?
tlacoamatl (stalk paper)=pl. 53 Badianus (C-B fol. 31r)=*Morus* sp. (Guerra 1952; Reko 1947) or *Ehretia tinifolia* or *Morus celtidifolia* (Díaz 1976) or *Morus* sp. (Miranda and Valdés 1964)
tlacochich=*Pseudognaphalium conoideum* (*Gnaphalium conoideum*) (Díaz 1976)
tlacochichic=*Pseudognaphalium canescens* (*Gnaphalium canescens*) (Díaz 1976; Farfán and Elferink 2010) or *Blumea viscosa* (*Conyza canescens*) or *Pseudognaphalium canescens* (*Gnaphalium conoideum*) (Díaz 1976)
tlacoecapahtli/*tlacohecapahtli*/*tlacoecapatli* (shrubby wind-remedy)=pl. 88 Badianus (C-B fol. 48v)=*Litsea glaucescens* (Linares and Bye 2013) or *Senna occidentalis* (*Cassia occidentalis*) (Guerra 1952; Miranda and Valdés 1964; Reko 1947)

tlacoizquixochitl=*Cordia alba* (*C. dentata*) (Díaz 1976)

tlacopatli/tlacopatle=FC=*Aristolochia* sp. (Farfán and Elferink 2010) or *A. anguicida* (*A. mexicana*) (Díaz 1976; Estrada Lugo 1989; Ocaranza 2011) or *A. subclausa* (Díaz 1976) or *Iostephane heterophylla* or *Viguiera excelsa* (Estrada Lugo 1989)

tlacopopotl/tepopotl (stem broom)=pl. 108 Badianus (C-B fol. 58v)=FC757a=*Flaveria angustifolia* (Farfán and Elferink 2010; Guerra 1952; Reko 1947) or *Arundinella hispida* (Estrada Lugo 1989) or *Marina diffusa* (*Dalea diffusa*) (Díaz 1976; Miranda and Valdés 1964) or *Schkuhria pinnata* (Linnares and Bye 2013) or *Baccharis multiflora* (Estrada Lugo 1989)

tlacoxihuitl/tlacoxiutl=p. 416 Hernandez Rerum=FC550=*Bocconia arborea* (Estrada Lugo 1989) or *Bouvardia ternifolia* (Díaz 1976; Estrada Lugo 1989)

tlacoxilo[h]xochitl/tlacō-xīlō-xōchitl (shrubby maize-ear-flower)=pl. 33, 35, 78 Badianus (C-B. fol. 20r, 21r, 43v)=*Calliandra houstoniana* var. *anomala* (*C. anomala, C. grandiflora*) (Bye and Linares 2013; de Ávila Blomberg 2012; Díaz 1976; Farfán and Elferink 2010; Guerra 1952; Miranda and Valdés 1964; Valdés Gutiérrez et al. 1992)

tlacoxiloxochitl=p. 104 Hernandez Rerum/p. 455, 456, 457 Hernandez Historia=*Calliandra houstoniana* var. *anomala* (*C. anomala, C. grandiflora*) (Díaz 1976; Farfán and Elferink 2010; Ortiz de Montellano 1990; Reko 1947)

tlacoxochitl=FC701b=*Bouvardia ternifolia* (Díaz 1976; Farfán and Elferink 2010; Viesca and Aranda 1996) or *Calliandra houstoniana* var. *anomala* (*C. anomala, C. grandiflora*) (Díaz 1976)

tlacoyzquixochitl/tlacoizquixochitl (shrubby popcorn-flower)=pl. 14, 30, 60, 70 Badianus (C-B fol. 10v, 18v, 34v, 39v)=Boraginaceae or *Bourreria huanita* or *Cordia alba* (Reko 1947; Valdés Gutiérrez et al. 1992) or *Cordia elaeagnoides* (Linares and Bye 2013) or *Cordia alba* (Guerra 1952)

tlacozazalic=FC=*Bidens* sp. (Estrada Lugo 1989) or *Fuchsia thymifolia* (Estrada Lugo 1989)

tlaelpatli/apitzalpatli=p. 285 Hernandez Rerum/p. 6 Hernandez Historia=*Spermacoce tenella* (Díaz 1976) or *Priva* sp.?

tlahchinolpan yxuaxihuitl (herb that comes up in burned land)=pl. 41 Badianus (C-B fol. 25r)=Poaceae (Guerra 1952) or Cyperaceae?

tlahçolpahtli/tlazolpatli/azcapan yxhuatlahcolpahtli (ill smelling medicine)=pl. 20, 76 Badianus (C-B fol. 13v, 42v)=*Datura stramonium* (Guerra 1952; Reko 1947) or *Datura* sp. (Furst 1995)

tlahçoteoçacayl/tlahçolteoçacatl/tlahcolteocacatl /tlazolteozacatl (grass of Tlahcolteotl)=pl. 75, 80 Badianus (C-B fol. 42r, 44v)=*Sobralia macrantha* (Guerra 1952; Miranda and Valdés 1964; Reko 1947) or *Tigridia violacea*?

tlahcuilolquauitl/tla-hcuilō-l-cuahuitl (written tree/wood)=pl. 69 Badianus (C-B fol. 39r)=*Dalbergia* sp. (de Ávila Blomberg 2012; Miranda and Valdés 1964; Valdés Gutiérrez et al. 1992), Fabaceae (Guerra 1952; Reko 1947), or *D. brownei* (*D. amerimnum*) (Díaz 1976) or *Haematoxylon brasiletto* (Linares and Bye 2013)

tlahnextia xiuhtonli (little herb that shines)=pl. 29, 31 Badianus (C-B fol. 18r, 19r)=*Eryngium carlinae* (Bye and Linares 2013) or *Eryngium* sp. (Guerra 1952) *tlahoelilocaquahuitl*=p. 455 Hernandez Rerum/p. 555 Hernandez Historia=?

tla[h]tla[h]cotic/tlatlacotic (having many stems/sticks)=pl. 38, 45, 92, 94, 98 Badianus (C-B fol. 22v, 27r, 50v, 51v, 53v)=*Bidens odorata* (Bye and Linares 2013) or *Bidens* sp. (Miranda and Valdés 1964) or *Baccharis brachylaenoides* var. *ligustrina* (*Baccharis saligna*) (Guerra 1952) or *Brosimum alicastrum* or *Zinnia* sp. or *Scabiosa fruticulosa*

tlakatsitsintli=*Cordyceps capitata* (Díaz 1976)

tlalacxoyatl=p. 419 Hernandez Rerum/p. 42 Hernandez Historia=*Asclepias linaria* (Díaz 1976; Ortiz de Montellano 1990) or *A. tuberosa* (Díaz 1976)

tlalahoacatl=p. 420 Hernandez Rerum/p. 90 Hernandez Historia=*Tagetes coronopifolia* (*T. multifida*) or *Stevia linoides* (Díaz 1976)

tlalahoehoetl=*Selaginella sellowii* (*S. rupeastris*) (Díaz 1976)

tlalalxixquilitl=*Osmorhiza mexicana* (Díaz 1976)

tlalamatl=FC572=p. 70 Hernandez Rerum/p. 261 Hernandez Historia=*Desmodium orbiculare* or *D. angustifolium* (Estrada Lugo 1989)

tlalamatl=p. 306 Hernandez Rerum/p. 264 Hernandez Historia=*Desmodium orbiculare* (Díaz 1976)

tlalamatl=p. 420 Hernandez Rerum=*Capsicum* sp.?

tlalantlacuitlapilli=*Ipomoea jalapa* (*Batatas jalapa*)? (Ocaranza 2011)

tlalaquequetzal=p. 194 Hernandez Rerum=*Achillea millefolium*?

tlalauhcapatli=*Geranium mexicanum* (Díaz 1976)

tlal[h]aueuetl/tlalhahuehuetl/tlalhuaueuetl (earth cypress)=pl. 8, 84, 86, 89, 113 Badianus (C-B fol. 7v, 46v, 47v, 49r, 61r)=*Agastache mexicana* (*Cedronella mexicana*) (Bye and Linares 2013; Díaz 1976; Guerra 1952; Linares and Bye 2013; Miranda and Valdés 1964; Reko 1947; Valdés Gutiérrez et al. 1992) or *Acalypha* sp. (Miranda and Valdés 1964; Valdés Gutiérrez et al. 1992) or *Verbena* sp. (Guerra 1952)

tlalayote/tlalayotic/tlalayotli/tlatlalayotli/tlalayotli/tlallayotli=FC488a, 544=*Asclepias linaria* (Estrada Lugo 1989) or *Gonolobus erianthus* or *G. pedunculatus* (Díaz 1976; Estrada Lugo 1989) or *G. nummularius* (Díaz 1976) or *G. parviflorus* (Estrada Lugo 1989)

tlalcacahoatl/tlalcacahuatl=*Arachis hypogaea* or *Theobroma cacao* (Díaz 1976)

tlalcacaloxochitl=FC671=*Plumeria rubra* (*P. acutifolia*) (Díaz 1976; Estrada Lugo 1989)

tlalcacahuatl chimalhuacani chalcensi/tlalcacahoatl chimalhuacan=p. 281 Hernandez Rerum/p. 919 Hernandez Historia=?

tlalcacapol/tlālcācapol (earth cherries)=pl. 89 Badianus (C-B fol. 49r)=*Rhamnus serrata*? or *Rhamnus* sp. (de Ávila Blomberg 2012) or *Malpighia glabra* (Guerra 1952; Miranda and Valdés 1964; Reko 1947) or *R. serrata* (Díaz 1976) or *R. humboldtiana* (*Karwinskia humboldtiana*) (Guerra 1952; Linares and Bye 2013; Valdés Gutiérrez et al. 1992)

tlalcacauatl/tlalcacahuatl/tlalcacahoatl=p. 253 Hernandez Rerum/p. 917 Hernandez Historia=FC495=*Arachis hypogaea* (Dressler 1953; Estrada Lugo 1989; Farfán and Elferink 2010)

tlalcapoli=FC658=*Rhamnus humboldtiana* (*Karwinskia humboldtiana*) (Estrada Lugo 1989) or *Rhamnus serrata* (Estrada Lugo 1989)
tlalcapolin/tlalcapulin=p. 230 Hernandez Rerum/p. 906 Hernandez Historia=FC=*Rhamnus humboldtiana* (*Karwinskia humboldtiana*) (Standley 1920–1926:717) or *R. serrata* or *Chimaphila umbellata* (Díaz 1976) or *Gaultheria myrsinoides* (*Pernettya ciliata*) (Estrada Lugo 1989)
tlalchamallin/tlalchamolin=p. 420 Hernandez Rerum/p. 928 Hernandez Historia=?
tlalchichinboaxihuitl=*Tournefortia mexicana*? (Ocaranza 2011)
tlalchichicpatli=p. 421 Hernandez Rerum/p. 579 Hernandez Historia=*Desmodium* sp.?
tlalchichinoa patlahoal=p. 292 Hernandez Rerum=*Solidago* sp.
tlachichinole=*Moussonia deppeana* (*Isoloma deppeanum, Kohleria deppeana*) (Díaz 1976)
tlachinole=Nesaea syphilitica (Heimia syphilitica) (Díaz 1976)
tlalchipili=FC551=*Crotalaria* sp. or *Frasera albicaulis* var. *nitida* (*Swertia nitida*) (Díaz 1976; Estrada Lugo 1989)
tlacocol=*Polygala verticillata* (Díaz 1976)
tlalhecapa[h]tli (earth wind medicine)=pl. 65, 79, 92 Badianus (C-B fol. 37r, 44r, 50v)=*Cassia* sp.? (Guerra 1952)
tlacuilolquauitl=FC=*Philodendron sagittifolium* (*P. sanguineum*) (Estrada Lugo 1989)
tlalhuaxin (earth huaxin)=pl. 57 Badianus (C-B fol. 33r)=*Mimosa* sp. (Guerra 1952; Reko 1947; Valdés Gutiérrez et al. 1992) or *Mimosa aculeaticarpa* (*Mimosa aculeaticarpa* var. *biuncifera*) (Linares and Bye 2013)
tlalhoaxin=*Senna occidentalis* (*Cassia occidentalis*) or *Chamaecrista nictitans* var. *mensalis* (*Cassia leptadenia*) (Díaz 1976)
tlalixtli=*Ceanothus caeruleus* (Díaz 1976)
tlalizquitl=*Cordia elaeagnoides* (Díaz 1976)
tlallaala/tlaltzacutli=FC=p. 419 Hernandez Rerum/p. 381 Hernandez Historia=*Sphaeralcea angustifolia* (Estrada Lugo 1989)
tlallaalam=*Sphaeralcea angustifolia* (*Malva angustifolia*) (Díaz 1976)
tlallaala alia=p. 419 Hernandez Rerum/p. 382 Hernandez Historia=*Malvastrum spicatumi?*
tlallantlacuacuitlapilli/tlallantlacacuitlapilli=p. 238 Hernandez Rerum/p. 726 Hernandez Historia=?
tlallantlacuacuitlapilli/tlallantlacacuitlapilli=p. 238 Hernandez Rerum/p. 727 Hernandez Historia=?
tlallantlacuacuitlapilli/tlallantlacacuitlapilli=p. 239 Hernandez Rerum/p. 728 Hernandez Historia=?
tlallantlacuacuitlapilli/tlallantlacacuitlapilli=p. 239 Hernandez Rerum/p. 729 Hernandez Historia=*Marsdenia mexicana* (Díaz 1976) or *Gonolobus* sp.?
tlallayotli=p. 201 Hernandez Rerum/p. 166 Hernandez Historia=*Gonolobus nummularius* or *G. pedunculatus* (Díaz 1976)
tlalmatzalin hocxotzincensi=p. 294 Hernandez Rerum=*Brazoria arenaria*?
tlalocopetate/tlalocopetatl=*Coriaria ruscifolia* subsp. *microphylla* (*C. atropurpurea*) (Díaz 1976)

tlalocoxochitl=*Drymaria ovata* (Díaz 1976)

tlalmizquitl (earth mesquite)=pl. 40, 113 Badianus (C-B fol. 24v, 61r)=FC547=*Dalea foliolosa* (Bye and Linares 2013) or *Prosopis* sp. (Estrada Lugo 1989) or *Prosopis juliflora*? or *Hoffmannseggia glauca* (*H. densiflora*) (Guerra 1952; Reko 1947) or *Dalea* sp. (Miranda and Valdés 1964; Valdés Gutiérrez et al. 1992)

tlalquequetzal/tlāl-quē-quetzal (earth plume-headdress)=pl. 39, 62, 78 Badianus (C-B fol. 24r, 35v, 43v)=FC656=*Achillea millefolium* (Bye and Linares 2013; de Ávila Blomberg 2012; Díaz 1976; Estrada Lugo 1989; Farfán and Elferink 2010; Guerra 1952; Ortiz de Montellano 1990; Reko 1947; Valdés Gutiérrez et al. 1992) or *Adiantum trapeziforme*? (Ocaranza 2011) or *Pleopeltis macrocarpa* (*Polypodium lanceolatum*) (Díaz 1976; Estrada Lugo 1989) or *Adiantum* sp. (Estrada Lugo 1989)

tlaltzacuitli=*Sphaeralcea angustifolia* (*Malva angustifolia*) (Díaz 1976)

tlaltzilacayotli=p. 206 Hernandez Rerum/p. 164 Hernandez Historia=?

tlaltzitzicaztli=*Urtica dioica* (Díaz 1976)

tlalyetl/tlalietl/tlally yetl=FC=p. 423 Hernandez Rerum/p. 247 Hernandez Historia=*Erigeron glacialis* (*E. scaposus*) (Estrada Lugo 1989)

tlalxiloxochitl=*Calliandra houstoniana* var. *anomala* (Farfán and Elferink 2010)

tlalyetl/tlalietl/tlalyetl =FC560, 658=*Erigeron longipes* (*E. scaposus*) (Díaz 1976) or *Cunila* sp. (Estrada Lugo 1989)

tlamacatzcatzotl/tlamacazatzotl/tlamacazcacootl=*Calliandra houstoniana* var. *anomala* (*C. anomala, C. grandiflora*) (Díaz 1976)

tlamacazqui ipapa=FC=*Huperzia dichotoma* (*Lycopodium dichotomum*) (Díaz 1976; Estrada Lugo 1989)

tlamapaquelite=*Piper sanctum* (Díaz 1976)

tlamexcaltzin=*Argemone mexicana* or *A. ochroleuca* (Díaz 1976)

tlanchalagua=*Zeltnera stricta* (*Erythraea stricta*) (Díaz 1976)

tlanchichinole=*Moussonia deppeana* (*Isoloma deppeanum, Kohleria deppeana*) (Díaz 1976)

tlanchichinolli=*Plumbago zeylanica* (*P. scandens*) (Díaz 1976)

tlanenpo[h]poloua xiuhtontli/xiuhtontli tlanenpopoloua/tlanempopolo hua xiuhtontli (little herb that squanders its fortune)=pl. 58, 88 Badianus (C-B fol. 33v, 48v)=*Asclepias oenotheroides* (*A. longicornu*) (Reko 1947) or *Argemone mexicana* (Guerra 1952; Reko 1947) or *Asclepias* sp. (Linares and Bye 2013; Valdés Gutiérrez et al. 1992) or *Argemone* sp. (Miranda and Valdés 1964)

tlanepaquelite/tlanepaquelitl=*Piper auritum* or *P. sanctum* (Díaz 1976)

tlanextia qua[h]uitl/tlanexti quauitl (bright tree)=pl. 69, 75, 86 Badianus (C-B fol. 39r, 42r, 47v)=Iridaceae? or Amaryllidaceae?

tlanextiaxi(h)uitl=pl. 75, 80, 96, 116 Badianus (C-B fol. 42r, 44v, 53r, 62v)=*Eryngium* sp. (Guerra 1952)

tlahnextia xiuhtontli/tlahnextia/tlanextia/tlahnextli/tlanextia xiuhtontli/tlanextiyxiuh (little herb that shines)=pl. 29, 31, 36, 54, 55, 82, 106 Badianus (C-B fol. 18r 19r, 21v, 31v, 32r, 45v, 57v)=*Polygala* sp. (Bye and Linares 2013; Valdés Gutiérrez et al. 1992) or *Eryngium* sp. (Guerra 1952; Reko 1947; Valdés Gutiérrez et al. 1992) or *Eryngium carlinae* (Bye and Linares 2013)

tlanoquizpatli/nextalpe=p. 178 Hernandez Rerum=*Guilleminea densa* or *Alternanthera sessilis* (*A. repens*) (Bye and Linares 1990)
tlaolcapulin/tlaolcapuli=FC427=*Prunus* sp. (Estrada Lugo 1989)
tlaolli=p. 242, 243 Hernandez Rerum/p. 874, 876 Hernandez Historia=*Zea mays* (Díaz 1976; Dressler 1953)
tlapa/tlapatl=FC=*Datura stramonium* (Díaz 1976; Estrada Lugo 1989; Furst 1995) or *Datura* sp. (Guerra 1952) or *Ricinus communis* (Estrada Lugo 1989)
tlapahtli/tlapahtil/tapatl=pl. 22, 89 Badianus (C-B fol. 14v, 49r)=p. 278 Hernandez Rerum=FC450, 451=*Datura stramonium* (Daunay et al. 2008; Díaz 1976; Miranda and Valdés 1964; Ocaranza 2011; Ortiz de Montellano 1990; Reko 1947; Safford 1920) or *Datura* sp. (Guerra 1952)
talpal=FC=*Bourreria* sp. (Estrada Lugo 1989)
tlapalachiyotl/tlapalāchiotl (paint annatto)=pl. 78 Badianus (C-B fol. 43v)=*Bixa orellana* (de Ávila Blomberg 2012; Díaz 1976; Farfán and Elferink 2010; Guerra 1952; Miranda and Valdés 1964; Reko 1947)
tlapalcacauatl/tlapalcacahuatl/tlapalcacahoatl (red cacao)=pl. 68, 70 Badianus (C-B fol. 38v, 39v)=*Theobroma cacao* (de Ávila Blomberg 2012; Díaz 1976; Furst 1995; Linares and Bye 2013; Guerra 1952; Miranda and Valdés 1964; Reko 1947; Valdés Gutiérrez et al. 1992)
tlapalcoçatli coaxochitl=p. 156 Hernandez Rerum/p. 650 Hernandez Historia=*Tagetes patula*?
tlapalezpatli=p. 91 Hernandez Rerum/p. 628 Hernandez Historia=FC=*Croton gossypiifolius* (*C. sanguifluus*) or *Eysenhardtia polystachya* (Díaz 1976; Estrada Lugo 1989) or Fabaceae?
tlapalezquauitl=FC=*Jatropha dioica* (*J. spathulata*) (Estrada Lugo 1989)
tlapalizquixochitl=FC685c=*Bourreria* sp.?
tlapalnochnopalli=FC=*Opuntia* sp.?
tlapaloaxin=*Leucaena pulverulenta* (Díaz 1976)
tlapaltecacayatl=p. 155 Hernandez Rerum/p. 648 Hernandez Historia=*Tagetes patula*
tlaquilin (frilled flower, primary name)=pl. 22, 83, 89 Badianus (C-B fol. 14v, 46r, 49r)=p. 279 Hernandez Rerum=*Mirabilis jalapa* (de Ávila Blomberg 2012; Díaz 1976; Guerra 1952; Miranda and Valdés 1964; Ortiz de Montellano 1990; Reko 1947; Valdés Gutiérrez et al. 1992) or *Ipomoea stans* (Bye and Linares 2013)
tlatlalamatic=p. 426 Hernandez Rerum/p. 266 Hernandez Historia=*Cuphea angustifolia* (Díaz 1976)
tlatlanquaye/tlā-tlancuā-yeh/tlatlancuaya (possessor of several knees)=pl. 8, 22, 34, 36, 50, 52, 54, 66, 80, 81, 82, 83 (C-B fol. 7v, 14v 20v, 21v, 29v, 30v, 31v, 37v, 44v, 45r, 45v, 46r)=FC588=*Iresine calea* (Bye and Linares 2013; de Ávila Blomberg 2012; Díaz 1976; Estrada Lugo 1989; Guerra 1952; Ortiz de Montellano 1990) or *Iresine diffusa* (*I. celosioides*) (Díaz 1976) + pl. 73 Badianus (C-B fol. 41r)=*Peperomia galioides* (de Ávila Blomberg 2012; Díaz 1976; Linares and Bye 2013; Miranda and Valdés 1964; Valdés Gutiérrez et al. 1992) or *Iresine* sp. (Miranda and Valdés 1964; Valdés Gutiérrez et al. 1992)

tlatlaolton (insignificant little maize kernels)=pl. 46, 66 Badianus (C-B fol. 27v, 37v)=*Porophyllum coloratum* (Bye and Linares 2013; Reko 1947) or *Porophyllum* sp. (Guerra 1952; Reko 1947; Valdés Gutiérrez et al. 1992)

tlatlauhcapatli =FC491=*Achimenes erecta* (*A. coccinea*) or *Geranium carolinianum* (Díaz 1976)

tlatlauhquitezontzapotl=FC=*Manilkara zapota* (*Achras sapota, Lucuma mammosa*) (Estrada Lugo 1989)

tlatlauhqui=FC=*Ceiba pentandra* or *Geranium carolinianum* (Estrada Lugo 1989)

tlatlauhqui amoxtli (red amoxtill)=pl. 59, 90 Badianus (C-B fol. 34r, 49v)=?

tlatlauhquipatli=FC=*Justicia* aff. *aurea* (Zepeda and White 2008) or *Geranium carolinianum* or *G. alchemilloides* or *G. mexicanum* (Estrada Lugo 1989) or *Justicia* sp. (Estrada Lugo 1989)

tlatoc nochtli/tla-tōc-nōchtli (planted cactus)=pl. 90 Badianus (C-B fol. 49v)=*Opuntia [ficus-indica]* (de Ávila Blomberg 2012) or *O. microdasys* (Guerra 1952; Miranda and Valdés 1964), or *O. streptacantha* (Zepeda and White 2008) or *O. tomentosa* (Linares and Bye 2013) or *Opuntia* sp. (Valdés Gutiérrez et al. 1992)

tlatocnocnopalli=FC=*Opuntia* sp. (Estrada Lugo 1989)

tlaltzacutli=FC=*Sphaeralcea angustifolia*?

tlatzan=*Persea americana* (*P. gratissima*) (Díaz 1976)

tlatzcan/tlazcantic (cypress, primary name)=pl. 86 Badianus (C-B fol. 47v)=FC373, 374=p. 98 Hernandez Rerum=*Cupressus lusitanica* var. *benthamii* (*C. benthamii, C. lindleyi*) (plus *Tillandsia usneoides*) (Altamirano 1896; de Ávila Blomberg 2012; Díaz 1976; Estrada Lugo 1989; Guerra 1952; Linares and Bye 2013; Miranda and Valdés 1964; Reko 1947; Siméon 2010; Valdés Gutiérrez et al. 1992) or *Juniperus poblana* (*Cupressus thurifera*) (Estrada Lugo 1989)

tlavepaquelite=*Piper sanctum* (Díaz 1976)

tlaxcapan=*Ipomoea stans* (Díaz 1976)

tlaxtloxochitl=*Calliandra houstoniana* var. *anomala* (*C. anomala, C. grandiflora*) (Díaz 1976)

tlayapaloni/tla-yapalō-ni (it habitually makes things black)=pl. 8, 86, 113 Badianus (C-B fol. 7v, 47v, 61r)=*Cissus verticillata* (*C. sicyoides*) (Bye and Linares 2013; de Ávila Blomberg 2012; Díaz 1976; Guerra 1952; Miranda and Valdés 1964; Valdés Gutiérrez et al. 1992) or Bromeliaceae (Guerra 1952)

tlayapaloni xiuhtontli (it habitually makes things black)=pl. 80 Badianus (C-B fol. 44v)=*Plantago* sp. (de Ávila Blomberg 2012; Miranda and Valdés 1964; Valdés Gutiérrez et al. 1992) or Bromeliaceae (Reko 1947) or *Plantago australis* (*Plantago australis* subsp. *hirtella*) (Linares and Bye 2013) or *Cissus verticillata* (*C. sicyoides*) (Guerra 1952)

tlazalxihuitl=*Datura stramonium* (Farfán and Elferink 2010)

tlazolpahtli=*Datura stramonium* or *D. meteloides* (Viesca and Aranda 1996)

tlepatli=p. 429 Hernandez Rerum/p. 189 Hernandez Historia=FC=*Plumbago pulchella* (Díaz 1976; Farfán and Elferink 2010; Ortiz de Montellano 1975, 1990) or *P. scandens* (Díaz 1976) *tlepatli*=p. 102 Hernandez Rerum/p. 192 Hernandez Historia=*Tournefortia hirsutissima* or *T. mexicana* (Díaz 1976; Farfán and Elferink 2010)

tlaquilitl=*Senna bacillaris* (*Cassia bacillaris*) (Díaz 1976)

tletlemaitl=FC=*Plumbago pulchella* (Ortiz de Montellano 1975) or *Euphorbia* sp. or *Plumbago* sp. (Estrada Lugo 1989)

tlilamatl=*Ficus maxima* (*F. guadalajarana*) or *Ficus aurea* (*F. tecolutensis*) (Díaz 1976)

tlilpoton=*Eryngium carlinae* (Díaz 1976)

tlilpotonqui=*Eryngium beecheyanum* (Díaz 1976)

tlilticchien=FC=*Salvia hispanica* (Estrada Lugo 1989)

tliltzapotl/tlilzapotl=p. 430 Hernandez Rerum/p. 270 Hernandez Historia=*Diospyros nigra* (*D. digyna, D. ebenaster*) (Díaz 1976; Dressler 1953)

tlilxochitl/tlīl-xōchitl (black flower)=pl. 70, 104 Badianus (C-B fol. 39v, 56v)=FC668=p. 38 Hernandez Rerum=*Vanilla planifolia* (*V. fragrans*) (Comas et al. 1995; de Ávila Blomberg 2012; Díaz 1976; Dressler 1953; Estrada Lugo 1989; Farfán and Elferink 2010; Guerra 1952; Miranda and Valdés 1964; Ocaranza 2011; Ossenbach 2009; Reko 1947; Valdés Gutiérrez et al. 1992) or *V. mexicana* (*Epidendrum vanilla*) (Estrada Lugo 1989)

tlipoton=*Eryngium foetidum* (Farfán and Elferink 2010)

tocaxihuitl=*Momordica charantia* (Díaz 1976)

tochacaxihuitl=*Cunila polyantha* (*C. longiflora*) (Díaz 1976)

tochomitl=*Moussonia deppeana* (*Isoloma deppeanum, Kohleria deppeana*) (Díaz 1976)

tohmiyoxihuitl (fleece herb)=pl. 111 Badianus (C-B fol. 60r)=*Packera bellidifolia* (Linares and Bye 2013)

to[h]toloctzin (small/revered one that repeatedly bows its head)=pl. 99 Badianus (C-B fol. 54r)=*Rubus pumilus* (Linares and Bye 2013; Valdés Gutiérrez et al. 1992) or *Cayaponia racemosa* (Miranda and Valdés 1964)

toliaman=FC=*Cyperus* sp. (Estrada Lugo 1989)

tollon=*Typha latifolia* (Zepeda and White 2008)

tolmimilli=FC641=*Cyperus* sp. (Estrada Lugo 1989) or *Schoenoplectus* sp.

toloa/toloua/tolohua=pl. 22, 42 Badianus (C-B fol. 14v, 23v)=FC518=*Datura* sp. (Daunay et al. 2008; Estrada Lugo 1989) or *Datura stramonium* (Estrada Lugo 1989)

toloatzin/toloazin=FC=*Datura stramonium* (Daunay et al. 2008; Díaz 1976; Farfán and Elferink 2010)

toloatzin/toloache=FC=*Datura wrightii* (*D. innoxia*) (Farfán and Elferink 2010; Safford 1920, p. 549)

tolohua xihuitl/tolouazihuitl/tōloā-xihuitl (herb that lowers its head)=pl. 41, 49, 57, 64, 65, 86 Badianus (C-B fol. 25r, 29r, 33r, 36v, 37r, 47v)=*Datura/Brugmansia* sp. (de Ávila Blomberg 2012; Furst 1995; Reko 1947) or *D. stramonium* (Bye and Linares 2013; Díaz 1976; Miranda and Valdés 1964; Valdés Gutiérrez et al. 1992) or *D. wrightii* (*D. meteloides*) (Bye and Linares 2013; Díaz 1976; Guerra 1952; Miranda and Valdés 1964; Valdés Gutiérrez et al. 1992)

tolompatl/tolonpatl=*Ipomoea purga* (Díaz 1976)

tolpatlactli/tōl-patlactli (wide rush)=pl. 29, 79 Badianus (C-B fol. 18r, 44r)=FC640=*Typha* sp. (Bye and Linares 2013; Miranda and Valdés 1964; Valdés Gutiérrez et al. 1992) or *Typha angustifolia* (Reko 1947) or *T. latifolia* (de Ávila Blomberg 2012; Díaz 1976; Guerra 1952; Reko 1947; Zepeda and White 2008) or *Cyperus* sp. (Estrada Lugo 1989)

tolpatle/tolpatli=*Cyperus esculentus* (Díaz 1976)

tomahuactlacopatli=p. 42 Hernandez Rerum/p. 789 Hernandez Historia=*Aristolochia* sp.?

tomame=*Physalis peruviana* or *P. pubescens* (Díaz 1976)

tomate/tomates=FC=*Solanum lycopersicum* (*Lycopersicon esculentum*) or *Physalis ixocarpa* (Díaz 1976; Estrada Lugo 1989) or *P. angulata* or *P. coztomatl* or *P. pubescens* (Daunay et al. 2008; Estrada Lugo 1989) or *P. philadelphica* (Estrada Lugo 1989)

tomatl=*Physalis ixocarpa* (Dressler 1953; Daunay et al. 2008) or *Solanum lycopersicum* or *Saracha* sp. (Jenkins 1948) or *Solanum lycopersicum* (Díaz 1976)

tomazq'tl/tomazquitl (tomato popcorn)=pl. 69 Badianus (C-B fol. 39r)=FC=*Arbutus* sp. (de Ávila Blomberg 2012) or *Arctostaphylos* sp. (Guerra 1952; Reko 1947) or *Arctostaphylos pungens* (Díaz 1976; Linares and Bye 2013; Miranda and Valdés 1964; Valdés Gutiérrez et al. 1992)

tomiyo xihuitl=Asteraceae (Reko 1947)

tompilillin=*Melothria scabra* (Díaz 1976)

tonacaxochitl (human-sustenance flower)=pl. 99 Badianus (C-B fol. 54r)=*Amphilophium buccinatorium* (*Distictis buccinatoria*) (Guerra 1952; Reko 1947; Zepeda and White 2008) or *Amphilophium crucigerum* (*Pithecoctenium echinatum*) (Miranda and Valdés 1964)

tonalahuate=*Persea americana* (*P. gratissima*) (Díaz 1976)

tonalaxochitl=*Oenothera laciniata* (Díaz 1976)

tonalchichicaq (bitter edible herb of dry lands)=FC=*Oenothera pubescens* (Estrada Lugo 1987 in Pico and Nuez 2000a)

tonalchichicaquilitl=FC=*Oenothera laciniata* (Díaz 1976; Estrada Lugo 1989) or *Hieracium mexicanum* (Estrada Lugo 1989)

tonalxihuitl (bitter edible herb of the summer)=FC660b=*Stevia salicifolia* (Estrada Lugo 1989; Ortiz de Montellano 1990) or *Veronica americana* (Farfán and Elferink 2010)

tonalxihuitl/tonalxiuitl=FC574=p. 251 Hernandez Rerum/p. 61 Hernandez Historia=*Cosmos diversifolius* var. *dahlioides* (*Bidens dahlioides*) or *Oenothera laciniata* or *Senecio helodes* (*S. purpurascens*) or *Stevia salicifolia* or *Veronica americana* (Díaz 1976; Estrada Lugo 1989) or *Oenothera rosea* (*O. rubra*) or *O. laciniata* (*O. sinuata*) (Estrada Lugo 1989)

tonalxochitl ocoitucensi=p. 251 Hernandez Rerum/p. 59 Hernandez Historia=*Ludwigia* sp.?

tonalxochitl (sun warmth flower/flower of the dry warm season)=pl. 67 Badianus (C-B fol. 38r)=*Mirabilis* sp. (Guerra 1952) or *Zinnia elegans*?

tonalxochitl (sun warmth flower/flower of the dry warm season)=FC670d=*Amphilophium crucigerum* (*Pithecoctenium echinatum*) (Díaz

1976; Estrada Lugo 1989; Reko 1947) or *Bletia coccinea* (Estrada Lugo 1989; Ocaranza 2011; Ossenbach 2009) or *Prosthechea vitellina* (*Epidendrum vitellinum*) (Estrada Lugo 1989)

tonalxuitl=*Stevia salicifolia* (Díaz 1976)

tonatiuhyxiuh (sun herb)=pl. 41, 61, 75, 76, 79, 82, 87, 97, 116 (C-B fol. 25r, 35r, 42r, 42v, 44r, 45v, 48r, 53r, 62v)=*Abutilon* sp. (Guerra 1952)

tonatiuh yxiuh ahhauachcho (the sun's dewy herb)=pl. 42, 51 Badianus (C-B fol. 25v, 30r)=Malvaceae or Brassicaceae (Miranda and Valdés 1964) or *Abutilon* sp. or *Malvaviscus* sp. (Guerra 1952)

tonatiuh yxiuh pepetlaca (the sun's herb that shimmers)=pl. 51 Badianus (C-B fol. 30r)=*Oxalis tetraphylla* (Valdés Gutiérrez et al. 1992) or Malvaceae (Guerra 1952; Reko 1947) or *Marsilea mollis* (Bye and Linares 2013) or Brassicaceae (Guerra 1952)

toonchichi/tonchichi=*Solanum nigrum* (Díaz 1976; Farfán and Elferink 2010)

topoçan/tepozán=pl. 81 Badianus (C-B fol. 45r)=FC577=*Buddleja americana* (Díaz 1976; Guerra 1952; Miranda and Valdés 1964; Reko 1947) or *B. cordata* (Díaz 1976)

torna-loco=*Datura ceratocaula* (Safford 1920)

torote=*Bursera fagaroides* (*Elaphrium fagaroides, E. odoratum*) or *Bursera tecomaca* (*Elaphrium tecomaca*) or *Jatropha cardiophylla* (Díaz 1976)

totecyxiuh/to-tēuc i-xiuh/totec ixoxochiuh (our lord's herb)=pl. 110 Badianus (C-B fol. 59v)=FC=*Heliotropium* sp. (de Ávila Blomberg 2012; Guerra 1952; Miranda and Valdés 1964; Reko 1947; Valdés Gutiérrez et al. 1992) or *Nama undulata* (Linares and Bye 2013) or *Heliotropium curassavicum* (Estrada Lugo 1989) or *Tournefortia* sp. (Guerra 1952)

totecyxochitl=*Sesuvium portulacastrum* (Díaz 1976)

totocaplllin/totocapolin=p. 432 Hernandez Rerum/p. 908 Hernandez Historia=?

totocuit laxil/totocuitlatl=p. 433 Hernandez Rerum/p. 734 Hernandez Historia=*Psittacanthus calyculatus* (Díaz 1976)

totolcuitatzapotl=FC416=*Diospyros nigra* (*D. ebenaster, D. digyna*) (Díaz 1976; Estrada Lugo 1989)

totolicxitl=*Artemisia ludoviciana* subsp. *redolens* (*A. redolens*) (Díaz 1976)

totolinyxochitl=*Bocconia arborea* (Díaz 1976)

totomatic=*Margaranthus solanaceus* or *Nicandra physalodes* (Díaz 1976)

totoncapatli=p. 115 Hernandez Rerum/p. 1023 Hernandez Historia=*Galphimia glauca* (*Thryallis glauca*) (Díaz 1976)

totoncaxihuitl/totoncaxiuitl=FC526=*Senna occidentalis* (*Cassia occidentalis*) or *Senna alata* (*Cassia alata*) (Díaz 1976; Estrada Lugo 1989; Ortiz de Montellano 1975, 1990)

totoncaxoxocoyollin=*Begonia gracilis* (Díaz 1976; Farfán and Elferink 2010)

totonquixochitl/totōnqui xōchitl (hot/fever flower)=pl. 67 Badianus (2 plants) (C-B fol. 38r)=*Ruellia* sp. (de Ávila Blomberg 2012) or *Mimulus cardinalis* or *Ipomoea* sp. (Reko 1947; Valdés Gutiérrez et al. 1992) or *Ipomoea conzattii* (*Exogonium conzatii*) (Díaz 1976; Miranda and Valdés 1964) or *Penstemon roseus* (Linares

and Bye 2013) or *Cordia sebestena* (Valdés Gutiérrez et al. 1992) or *Distimake tuberosus* (*Ipomoea tuberosa*) (Guerra 1952)
totopsotz/totopzots=Malvastrum coromandelianum (Díaz 1976)
tototzapotl=Sideroxylon tepicense (*S. mexicanum*) (Díaz 1976)
totoyczitl=p. 256 Hernandez Rerum/p. 1000 Hernandez Historia=*Merremia tuberosa (Ipomoea tuberosa*) (Díaz 1976)
toxcuitlapilxochitl/toxcuitlapulxochitl=Lobelia laxiflora (Díaz 1976)
toxichec cimarron=Mikania cordifolia (Díaz 1976)
toxob=Caesalpinia vesicaria (Díaz 1976)
tozancuitlaxcolli=FC=*Canna indica* (Díaz 1976; Estrada Lugo 1989)
tozcuitlapil xochitl=p. 282 Hernandez Rerum/p. 735 Hernandez Historia=*Canna indica*?
tozcuitlapil xochitl=p. 282 Hernandez Rerum/p. 736 Hernandez Historia=*Maranta arundinacea*?
tozpatli=Dorstenia contrajerva (Díaz 1976)
tragorigano=p. 196 Hernandez Rerum/p. 230 Hernandez Historia=*Clinopodium macrostemum* (Díaz 1976)
tsayaltsay=Anoda cristata (Díaz 1976)
tsayuntsal=Mentzelia hispida (Díaz 1976)
tsayuntsay=Priva lappulacea (Díaz 1976)
tsocosnichat=Dioscorea floribunda (Díaz 1976)
tsots-k-ab=Mentzelia hispida (Díaz 1976)
tsots'kab=Mentzelia aspera (Díaz 1976)
tsotz=Brosimum alicastrum (Díaz 1976)
tsulubtock=Bauhinia subrotundifolia (*B. spathacea*) (Díaz 1976)
ts'itsil-xiu=Erigeron canadensis (Díaz 1976)
ts'ui-che=Pithecellobium unguis-cati (Díaz 1976)
ts'uts'uk-al'=Cissampelos pareira (Díaz 1976)
tuche=Clinopodium macrostemum (*Calamintha macrostema, Satureja macrostema*) (Díaz 1976)
tuk'=Acrocomia aculeata (*A. mexicana*) (Díaz 1976)
tulasuchil=Tecoma stans (Díaz 1976)
tumutsali=Tagetes liucida (Díaz 1976)
tutsuy-kiu=Mirabilis jalapa (Díaz 1976)
tuzpatli=Dorstenia contrajerva (Díaz 1976)
tzacatepahtli=Asphodelus ramosus (Díaz 1976)
tzacouhxochitl (glue flower)=pl. 97 Badianus (C-B fol. 53r)=*Bletia campanulata* (Guerra 1952; Miranda and Valdés 1964) or *B. coccinea* (Miranda and Valdés 1964) or *Catasetum maculatum* or *Prosthechea pastoris* (*Encyclia pastoris, Epidendrum pastoris*)?
tzacutli/tzacuhtli/tzautli=p. 283 Hernandez Rerum/p. 377 Hernandez Historia=FC665=*Prosthechea pastoris* (*Encyclia pastoris, Epidendrum pastoris*) (Díaz 1976; Estrada Lugo 1989; Ossenbach 2009) or *Bletia* sp. or *Govenia* sp. or *Trichocentrum cebolleta* (*Oncidium cebolleta*) (Berdan 2007) or *Bletia campanulata* (Estrada Lugo 1989)

tzacuxilotl=*Arpophyllum spicatum*?
tzacuxochitl/tzacuxochitlxiuitl=FC721b=*Bletia campanulata* (Díaz 1976; Estrada Lugo 1989; Ossenbach 2009; Ortiz de Montellano 1990) or *Bletia coccinea* (Díaz 1976; Estrada Lugo 1989) or *Prosthechea vitellina* (*Epidendrum vitellinum*)
tzaguangueni=p. 148 Hernandez Rerum/p. 567 Hernandez Historia=*Artemisia ludoviciana* subsp. *redolens* (*A. redolens*) (Díaz 1976)
tzalmuy=*Annona squamosa* (Díaz 1976)
tzanaicxitl=*Asplenium trichomanes* (Díaz 1976)
tzaponochnopalli=FC=*Opuntia ficus-indica* (Díaz 1976; Estrada Lugo 1989)
tzaponochtli=*Opuntia ficus-indica* (Díaz 1976)
tzapotl/tzapoquauitl/tzapotes/zapotes=FC411=*Manilkara zapota* (*Achras sapota*) (Díaz 1976; Estrada Lugo 1989; Zepeda and White 2008) or *Casimiroa edulis* (Zepeda and White 2008) or *Pouteria dominigensis* (*Lucuma dominigensis*) (Díaz 1976) or *Diospyros nigra* (*D. digyna*) or *Manilkara zapota* (*Achras sapota, Lucuma mammosa*) or *Sideroxylon* sp. (Estrada Lugo 1989)
tzatzanaco=*Celtis caudata* (Díaz 1976)
tzatzapalli=FC=*Zea mays* (Estrada Lugo 1989; Rivera Morales 1941)
tzatzayanalquiltic=FC558=*Rhodosciadium tuberosum* (*Deanea tuberosa*) (Díaz 1976; Estrada Lugo 1989; Farfán and Elferink 2010)
tzatzequa=*Urera baccifera* (Farfán and Elferink 2010)
tzatzupu=*Sapindus saponaria* (Díaz 1976)
tzauc-xōchitl (glue flower)=pl. 97 Badianus (C-B fol. 53r)=*Bletia* or *Laeli*a sp. (de Ávila Blomberg 2012)
tzauhzilotl/tzauxilotl=*Arpophyllum spicatum* (Díaz 1976; Ossenbach 2009)
tzacouhxochitl/tzauxochitl (glue flower)=pl. 97 Badianus (C-B. fol. 53r)=p. 433 Hernandez Rerum=Orchidaceae (Reko 1947) or *Bletia campanulata* or *Prosthechea pastoris* (*Epidendrum pastoris*) or *Catasetum maculatum*
tzayanalquititl/tzaianalquitl/tzauāna-l-quilitl (quelite cut in pieces/torn edible herb)=pl. 79 Badianus (C-B fol. 44r)=*Rhodosciadium tuberosum* (*Deanea tuberosa*) (de Ávila Blomberg 2012; Díaz 1976; Guerra 1952; Pico and Nuez 2000b; Miranda and Valdés 1964; Reko 1947)
tzayentzal=*Priva lappulacea* (Díaz 1976; Martinez 1969:502)
tzazalicpatli/zazalicpatli=p. 434 Hernandez Rerum/p. 835 Hernandez Historia=?
tzicatzontecomatl=p. 254 Hernandez Rerum/p. 445 Hernandez Historia=*Spermacoce tenella* (Díaz 1976)
tzicozapotl=*Manilkara zapota* (Dressler 1953)
tzicxochitl=*Laelia anceps* (Ossenbach 2009)
tzihuaccopalli/tzīhuac-copalli (tzihuac-like copal tree)=pl. 88 Badianus (C-B fol. 48v)=*Bursera* sp. (de Ávila Blomberg 2012; Miranda and Valdés 1964; Reko 1947; Valdés Gutiérrez et al. 1992) or *Bursera bipinnata* (Linares and Bye 2013) or *Discocactus flagelliformis* (*Nyctocereus serpentinus*) (Guerra 1952)
tzilacayometl=FC=*Agave* sp. (Estrada Lugo 1989)
tzilacayotli=FC=*Cucurbita ficifolia* (Díaz 1976; Estrada Lugo 1989; Farfán and Elferink 2010) or *Melothria pendula* (*M. guadalupensis*) or *M. scabra* (Díaz 1976)

tzinacanatlapatli=p. 435 Hernandez Rerum=*Passiflora coriacea*?
tzinacancuitlaquahuitl=p. 58 Hernandez Rerum/p. 560 Hernandez Historia=*Mimosa* sp.?
tzipipatli/hueipatli=FC=*Senecio sessilifolius* (*S. cardiophyllus*) (Estrada Lugo 1989)
tziritzecua=*Prosopis juliflora* (Díaz 1976)
tzitzcatzquilitl/tzitzicazquilitl (edible nettle)=FC467b=*Urtica urens* (Estrada Lugo 1987 in Pico and Nuez 2000a)
tzitzicazquilitl=*Cnidoscolus urens* (*Jatropha urens*) (Díaz 1976; Estrada Lugo 1989)
tzitzicaztli=FC510=*Cnidoscolus urens* (*Jatropha urens*) (Díaz 1976;Estrada Lugo 1989) or *Gronovia scandens* (Estrada Lugo 1989)
tzitzicion=*Gymnosperma glutinosum* (*Selloa glutinosa*) (Díaz 1976)
tzictli=FC=*Manilkara zapota* (*Achras sapota*)?
tzitzicton/tzītzictōn (insignificant little chewing gums)=pl. 46, 64 Badianus (C-B fol. 27v, 36v)=*Gymnosperma glutinosum* (*Selloa glutinosa*) (Bye and Linares 2013; Guerra 1952; Linares and Bye 2013; Reko 1947; Valdés Gutiérrez et al. 1992) or *Asclepias* [*notha*?] (de Ávila Blomberg 2012)
tzitzicton (insignificant little chewing gums)=pl. 64 Badianus (C-B fol. 36v)=*Asclepias* [*notha*?] (de Ávila Blomberg 2012)
tzitziqui=*Tagetes lucida* (Díaz 1976)
tzitziquilitl (dark green edible herb)=FC468b=*Bidens pilosa* (Estrada Lugo 1989) or *Tagetes* sp. (Estrada Lugo 1987 in Pico and Nuez 2000a) or *Erigeron pusillus* (Díaz 1976; Estrada Lugo 1989) or *T. tenuifolia* (*T. peduncularis*) or *Zinnia elegans* (Estrada Lugo 1989)
tzocoijoyo=*Magnolia mexicana* (*Talauma mexicana*) (Díaz 1976)
tzocuxochitl=*Bletia campanulata* (Díaz 1976)
tzompachtzin (precious hairy, woolly one)=pl. 76 Badianus (C-B fol. 42r)=*Solanum donianum* (*S. verbascifolium*) (Guerra 1952; Reko 1947) or *Sedum* sp. (Miranda and Valdés 1964)
tzompantle/ztompantli/tzompacuahuitl/tzompanquahuitl/tzompanquauitl=FC689a=*Erythrina americana* (Altamirano 1896; del Pozo 1966; Díaz 1976; Estrada Lugo 1989) or *E. coralloides* (Díaz 1976) or *E. mexicana* (Estrada Lugo 1989) or *Escobedia* sp. (Estrada Lugo 1989)
tzompilihuitzpatli=*Mimulus rupestris* (Díaz 1976)
tzompotonic=*Pseudognaphalium conoideum* (*Gnaphalium conoideum*) (Díaz 1976)
tzonayotli=*Luffa acutangula* or *L. cylindrica* or *L. operculata* (Díaz 1976)
tzonpilihuizpahtli (cold in the head medicine)=pl. 42 Badianus (C-B fol. 23v)=?
tzonpilihuiz xihuitl (thorny cold-in-the-head plant/catarrh herb)=pl. 24, 51 Badianus (C-B fol. 15v, 30r)=*Mimulus rupestris* (Díaz 1976; Miranda and Valdés 1964; Valdés Gutiérrez et al. 1992) or *Agastache mexicana* (*C. mexicana*) (Guerra 1952; Reko 1947) or *Galium mexicanum* (Bye and Linares 2013) or *Galium* sp. (Guerra 1952)
tzonpotonic=*Blumea viscosa* (*Conyza lyrata*)?
tzontecomanacatl=FC457b=*Amanita muscaria* (Díaz 1976; Estrada Lugo 1989)
tzontecpatli=*Jatropha cordata* (Díaz 1976)

tzonxochitl=Calliandra *houstoniana* var. *anomala* (*C. anomala, C. grandiflora*) (Díaz 1976)
tzonpelchichixochitl/tzopel-chi-chil-xochitl=*Malvaviscus arboreus* or *Kosteletzkya pentacarpos* (*Hibiscus pentacarpos*) (Díaz 1976)
tzopelicacococ (sweetness-spicy)=pl. 35 Badianus (C-B fol. 21r)=*Phyla scaberrima* (*Lippia dulcis*) (Guerra 1952; Miranda and Valdés 1964; Reko 1947)
tzontecomatl tzopilotl=*Swietenia mahagoni* (de Batres et al. 2012)
tzontecomaxochitl=FC=*Calliandra houstoniana* var. *anomala* (*C. anomala*) or *Stanhopea tigrina* (Estrada Lugo 1989)
tzopiliquauitl=*Swietenia mahagoni* (Díaz 1976)
tzopilotl/tzopilotlizontecomatl=*Swietenia humilis* (Díaz 1976)
tzopilotlzonte=Swietenia *mahagoni* (Díaz 1976)
tzopotonic=*Pseudognaphalium canescens* (*Gnaphalium canescens*) (Díaz 1976)
tzupelicxuihuitl=*Phyla scaberrima* (*Lippia dulcis*) (Díaz 1976)
tzotzocaxihuitl (wart herb)=pl. 96 Badianus (C-B fol. 52r)=*Euphorbia helioscopia* (Guerra 1952; Reko 1947)
tzotzotlani (glistening plant)=pl. 65 Badianus (C-B fol. 37r)=*Solanum donianum* (*S. verbascifolium*) (Guerra 1952; Reko 1947)
uacalxochitl=FC709=*Philodendron tripartitum* (*P. affine*) or *P. sagittifolium* (Estrada Lugo 1989)
uaxin/uaxi=FC422=*Leucaena leucocephala* (*L. glauca*) (Díaz 1976) or *L. esculenta* (Estrada Lugo 1989)
uauhquilitl/uauhtli=FC462=*Amaranthus cruentus* (*A. paniculatus*) (Estrada Lugo 1989) or *A. caudatus*?
ueinacaztli/ueynacaztli=FC=*Chiranthodendron pentadactylon* or *Cymbopetalum penduliflorum* or *Enterolobium cyclocarpum* (Estrada Lugo 1989)
ueipatli=FC=*Solanum donianum* (*S. verbascifolium*) (Estrada Lugo 1989)
[h]uelicpahtli (delicious remedy)=pl. 55 Badianus (C-B fol. 32r)=*Ipomoea purga* (Bye and Linares 2013; Miranda and Valdés 1964; Valdés Gutiérrez et al. 1992) or *Operculina* sp. (Guerra 1952; Reko 1947) or *Ipomoea* sp. (Guerra 1952)
ueuei itzontecon=FC=*Gomphrena serrata* (*G. decumbens*) (Estrada Lugo 1989; Farfán and Elferink 2010) or *Helenium mexicanum* (Díaz 1976)
uexotl=FC=*Salix* sp. (Estrada Lugo 1989)
uitzitzilmetl=FC=*Agave xylonacantha* (Estrada Lugo 1989)
uitzitzilxochitl/uitzilxochitl=FC702a=*Loeselia mexicana* (*L. coccinea*) or *Myroxylon balsamum* var. *pereirae* (Díaz 1976; Estrada Lugo 1989)
uitzocuitlapilli=FC758a=*Disocactus flagelliformis* (*Aporocactus flagelliformis*) (Díaz 1976; Estrada Lugo 1989)
uitzocuitlapilxiuitl=*Bocconia frutescens* (Díaz 1976)
uitzquauitl=FC390=*Caesalpinia echinata* (Díaz 1976; Estrada Lugo 1989) or *Haematoxylum brasiletto* or *H. campechianum* (Estrada Lugo 1989)
uitzquilitl (prickly edible herb)=FC471a=*Cirsium mexicanum* (Díaz 1976; Estrada Lugo 1987 in Pico and Nuez 2000a; Estrada Lugo 1989) or *Cnicus* sp. (Estrada Lugo 1989)

ulquahuitl/ulquauoitl/ulli=FC=*Castilla elastica* (Comas et al. 1995; Estrada Lugo 1989)
utlatl=FC=*Baccharis* sp. (Estrada Lugo 1989)
utspeki/utsuppek/u-tsuts-pec/utsubilul=*Tabernaemontana amygdalifolia* (Díaz 1976)
uxitl=*Pinus teocote*? (Ocaranza 2011)
uzte=*Malpighia souzae* (Díaz 1976)
vexotl=*Salix pentandra*? (Ocaranza 2011)
xaan/xan=*Sabal yapa* (*Inodes japa*) (Díaz 1976)
xahualyztic=p. 437 Hernandez Rerum=Fabaceae?
xahuique=*Anticlea frigida* (*Stenanthium frigidum, Veratrum frigidum*) (Díaz 1976)
xalacotli/xalacocotli=FC726=*Arracacia atropurpurea* (Díaz 1976; Estrada Lugo 1989)
xalatlacense=*Solanum lycopersicum* (*Lycopersicum esculentum*) (Díaz 1976)
xaltomatl/xāl-tomatl (sand tomato)=pl. 14, 30, 79, 106 Badianus (C-B fol. 10v, 18v, 44r, 57v)=FC485, 537=*Jaltomata procumbens* (*Saracha jaltomata*) (Bye and Linares 2013; de Ávila Blomberg 2012; Díaz 1976; Estrada Lugo 1989; Guerra 1952; Ocaranza 2011; Valdés Gutiérrez et al. 1992) or *Physalis mollis* (Estrada Lugo 1989)
xalxocotl/xacocotl/xalcocotl=FC419=*Psidium guajava* (*Psidium pomiferum*) (Altamirano 1896; Díaz 1976; Dressler 1953; Estrada Lugo 1989; Guerra 1952; Ortiz de Montellano 1975, 1990; Reko 1947; Zepeda and White 2008)
xanabmucuy=*Euphorbia parviflora* (*E. pilulifera*) (Díaz 1976)
xanacol/xanacotl=*Pseudobombax ellipticum* (*Bombax ellipticum*) (Díaz 1976)
xanatl=*Morus celtidifolia* (*M. microphylla*) (Díaz 1976)
xate=*Argemone mexicana* (Díaz 1976)
xauay=*Euphorbia parviflora* (*E. pilulifera*) (Díaz 1976)
xaxachichitl=*Calea zacatechichi* (Díaz 1976)
xaxaxacotic/xaxaxaoti=*Galphimia glauca* (Díaz 1976)
xaxim=*Leucaena leucocephala* (*L. glauca*) (Díaz 1976)
xaxocotl/xalxocotl/xāl-xocotl (sand sour fruit)=pl. 53, 79 Badianus (C-B fol. 31r, 44r)=*Psidium guajava* (Bye and Linares 2013a; de Ávila Blomberg 2012; Miranda and Valdés 1964; Valdés Gutiérrez et al. 1992)
xaxocoyol=FC=*Oxalis* sp. (Estrada Lugo 1987 in Pico and Nuez 2000a)
xaxox=*Solanum donianum* (*S. verbascifolium*) (Díaz 1976)
xayillol/xayil-lol=*Anredera baselloides* (*Boussingaultia baselloides*) (Díaz 1976)
xcantiris=*Acacia farnesiana* (Díaz 1976)
xcunche=*Pseudobombas ellipticum* (*Bombax ellipticum*) (Díaz 1976)
xeigia=*Prunus serotina* subsp. *capuli* (*P. capuli*) (Díaz 1976)
xexihuitl (plant growing in stony places)=pl. 65 Badianus (C-B fol. 37r)=*Arracacia* sp.? (Guerra 1952; Reko 1947)
xezetzi=*Artemisia ludoviciana* subsp. *redolens* (*A. redolens*) (Díaz 1976)
xhalalnal/xhotalnan=*Celosia virgata* (Díaz 1976)
xiaxiu=*Evolvulus alsinoides* (Díaz 1976)
xicalan=*Bursera copallifera* (*Bursera jorullensis, Elaphrium jorullense*) (Díaz 1976)

xicama=p. 309 Hernandez Rerum/p. 522 Hernandez Historia=FC481a=*Pachyrhizus erosus* (*P. angulatus, P. palmatilobus*) (Díaz 1976; Dressler 1953; Estrada Lugo 1989)
xicamatic=*Ipomoea jicama* (Díaz 1976)
xichocopal=*Bursera linanoe* (*B. aloexylon*) (Díaz 1976)
xicora=*Lophophora williamsii* (Díaz 1976)
xicotzapotl=FC415=*Manilkara zapota* (*Achras sapota, Lucuma mammosa*) (Díaz 1976; Estrada Lugo 1989; Standley 1920–1926:1119)
xihuicxan/xihuixcan=*Pseudobombax ellipticum* (*Bombax ellipticum*) (Díaz 1976)
xihuitl tonalco mochiua hahuachcho (plant which comes up in the dry season and gets covered with dew)=pl. 12 Badianus (C-B fol. 9v)=*Mirabilis* sp.? (Bye and Linares 2013; Guerra 1952; Reko 1947)
xik-che=*Zanthoxylum fagara* (Díaz 1976)
xikin=*Calea zacatechichi* (Díaz 1976)
xilochitl=FC=*Pachira insignis, Calliandra houstoniana* var. *anomala* (*C. grandiflora*) (Farfán and Elferink 2010) or *Pseudobombax ellipticum* (*Bombax ellipticum*)?
xilometl=FC=*Agave* sp. (Estrada Lugo 1989)
xilotl/xilote=*Zea mays* (Rivera Morales 1941)
xiloxochicuahuitl=*Calliandra houstoniana* var. *anomala* (*C. anomala, C. grandiflora*) (Díaz 1976)
xiloxochitl (milky corn silk flower)=pl. 89 Badianus (C-B fol. 49r)=FC693=FC=*Pseudobombax ellipticum* (*Bombax ellipticum*) (Alcántara Rojas 2008; Díaz 1976; Estrada Lugo 1989; Miranda and Valdés 1964; Zepeda and White 2008) or *Calliandra houstoniana* var. *anomala* (*C. anomala, C. grandiflora*) (Díaz 1976; Estrada Lugo 1989; Guerra 1952; Reko 1947) or *Caesalpinia pulcherrima* (Díaz 1976) or *Pachira aquatica* (Díaz 1976) or *P. insignis* (Estrada Lugo 1989)
ximene=*Persea americana* (*P. gratissima*) (Díaz 1976)
xiopatla=*Gaulthera acuminata* (*G. nitida*) (Díaz 1976)
xiopatli=*Senna occidentalis* (*Cassia occidentalis*) (Díaz 1976)
xiote=*Pseudosmodingium perniciosum* (Díaz 1976)
xitomatl/xiomate=*Solanum lycopersicum* (Díaz 1976; Dressler 1953)
xitrus=*Cardiospermum halicacabum* (Díaz 1976)
xiuehcapahtli/xiuhhecapatli/xiuhehcapahtli/xihuecapahtli/xiuhecapatli/xiuh-ehēca-pahtli (plant wind medicine)=pl. 7, 65, 66, 75, 95, 99, 113 Badianus (C-B fol. 7r, 37r, 37v, 42r, 52r, 54r, 61r)=*Senna occidentalis* (*Cassia occidentalis*) (de Ávila Blomberg 2012; Díaz 1976; Guerra 1952; Reko 1947; Miranda and Valdés 1964; Valdés Gutiérrez et al. 1992) or *Senna septemtrionalis* (Bye and Linares 2013; Linares and Bye 2013)
xiuh[h]amolli (herbaceous soap)=pl. 11 Badianus (C-B fol. 9r)=*Ipomoea murucoides* (de Batres et al. 2012; Guerra 1952) or *Ipomoea* sp. (Bye and Linares 2013; Reko 1947; Miranda and Valdés 1964; Valdés Gutiérrez et al. 1992)
xiuheloquilitl/xiuhhēlōquilitl (green, tender edible plant)=pl. 108 Badianus (C-B fol. 58v)=*Bidens aurea* (*B. tetragona*) (Farfán and Elferink 2010; Guerra 1952;

Reko 1947) or *Galinsoga* sp. (de Ávila Blomberg 2012) or *Bidens pilosa* (Linares and Bye 2013)

xiuhpahtli (herbaceous remedy/turquoise remedy)=pl. 67, 108 Badianus (C-B fol. 38r, 58v)=*Tigridia pavonia* (Linares and Bye 2013; Miranda and Valdés 1964; Reko 1947; Valdés Gutiérrez et al. 1992) or *Tigridia* sp. (Guerra 1952)

xiuhquilitl=FC=*Zapoteca formosa* (*Calliandra gracilis*) or *Indigofera suffruticosa* (*I. anil*) (Estrada Lugo 1989)

xiuhtlemaytl (herbaceous incense burner)=pl. 85 Badianus (C-B fol. 47r)=*Pectis* sp.? (Guerra 1952; Miranda and Valdés 1964; Reko 1947)

xiuhtontli (small plant)=pl. 33, 52, 57, 64, 80 Badianus (C-B fol. 20r, 30v, 33r, 36v, 44v)=?

xiuquilitl/xiuquilitlpitzahoac=*Indigofera suffruticosa* (Altamirano 1896; Díaz 1976; Dressler 1953)

xoalacatl=*Heliocereus speciosus* (Díaz 1976)

xochicacahoatl=*Theobroma cacao* (Díaz 1976)

xochicamotli=FC=*Ipomoea batatas* (Estrada Lugo 1989)

xochicintli=FC=*Zea mays* (Estrada Lugo 1989; Rivera Morales 1941)

xochicopalli=p. 49 Hernandez Rerum/p. 539 Hernandez Historia=?

xochinacaztli/xochinacastle (ear flower)=p. 30 Hernandez Rerum=*Cymbopetalum penduliflorum* (Bye and Linares 1990; Díaz 1976; Farfán and Elferink 2010; Safford 1910, 1912)

xochiocotzol/xōchi-oco-tzotl=pl. 28, 34, 65, 92, 104 Badianus (C-B fol. 17v, 20v, 37r, 50v, 56v)=*Liquidambar styraciflua* (*L. macrophylla*) (de Ávila Blomberg 2012; Díaz 1976; Estrada Lugo 1989; Guerra 1952; Miranda and Valdés 1964; Reko 1947)

xochiocotzoquahuitl/xochiocotzotl quauhxihuitl=p. 56 Hernandez Rerum/p. 359 Hernandez Historia=FC=*Liquidambar styraciflua* (Comas et al. 1995; Díaz 1976; Estrada Lugo,1989; Ortiz de Montellano 1975)

xochipali/xochipalli/xochipatli=FC=*Cosmos sulphureus* (Díaz 1976; Estrada Lugo 1989) or *Bonellia macrocarpa* subsp. *macrocarpa* (*Jacquinia aurantiaca*) (Estrada Lugo 1989)

xochipitzahoac=*Polygala scoparia* or *P. verticillata* (Díaz 1976)

xococquauitl/xoco-c cuahuitl (sour tree)=pl. 69 Badianus (C-B fol. 39r)=*Conostegia xalapensis* (Guerra 1952; Reko 1947; Linares and Bye 2013; Miranda and Valdés 1964; Valdés Gutiérrez et al. 1992) or *Leandra* sp. or *Miconia* sp. (de Ávila Blomberg 2012)

xococotl=FC=*Rumex patientia* (Estrada Lugo 1989)

xocoitztli=*Solanum angustifolium* (*S. cornutum*) (Díaz 1976)

xoconochnopalli/xoconochtli=FC=*Cylindropuntia imbricata* (*Opuntia imbricata*) (Díaz 1976; Estrada Lugo 1989)

xocoatl=*Theobroma cacao* (Comas et al. 1995; Zepeda and White 2008)

xocotl=*Spondias purpurea* (Dressler 1953)

xocoxochitl=*Pimenta dioica* (*P. officinalis*) (Díaz 1976; Ocaranza 2011)

xolometl=p. 275 Hernandez Rerum/p. 1048 Hernandez Historia=*Agave* sp.?

xoloxiltic=*Piqueria trinervia* (Díaz 1976)

xomalin/tomali/tomalli (primary name)=pl. 110 Badianus (C-B fol. 59v)=FC=*Juncus* sp. (Guerra 1952; Miranda and Valdés 1964; Reko 1947) or *Cyperus* sp. (Estrada Lugo 1989)
xometontli=Senna occidentalis (*Cassia occidentalis*) (Díaz 1976)
xonacatl=FC=*Allium* sp. (Estrada Lugo 1989)
xonecuili=Heimia salicifolia (Díaz 1976)
xonequilitl=Gymnosperma glutinosum (*Selloa glutinosa*) (Díaz 1976)
xonitlzal=Piqueria trinervia (Díaz 1976)
xopancelia=Selaginella lepidophylla or *S. sellowii* (*S. rupestris*) (Díaz 1976)
xoxocoiolcuecuepoc (flowering sorrel)=FC=*Oxalis* sp. (Estrada Lugo 1987 in Pico and Nuez 2000a)
xoxocoiolhuiuila (creeping sorrel)=*Oxalis* sp.?
xoxocoiol hoihoilan (very bitter)=p. 440 Hernandez Rerum=FC=*Oxalis* sp. (Estrada Lugo 1987 in Pico and Nuez 2000a)
xoxocoyolhuiuila=FC=*Oxalis stricta* (Estrada Lugo 1989)
xoxocoyoli=FC=*Oxalis* sp. or *O. violacea* (Estrada Lugo 1989)
xoxocoyolcuecuepoc=FC=*Biophytum dendroides* (*Oxalis dendroides*) or *Oxalis* sp. (Estrada Lugo 1989)
xoxocoyoltic/xoxocoioltic =FC540=*Oxalis albicans* (Estrada Lugo 1987 in Pico and Nuez 2000a; Estrada Lugo 1989) or *Oxalis frutescens* subsp. *angustifolia* (*O. angustifolia*) (Díaz 1976; Estrada Lugo 1989) or *O. verticillata* (Estrada Lugo 1989)
xoxocquauitl=Conostegia xalapensis (Díaz 1976)
xoxonitzal/xoxonitztac=Piqueria trinervia (Díaz 1976)
xoxouhcapa[h]tli (blue/green remedy/ blue medicine)=FC=*Datura* sp. (Daunay et al. 2008; Estrada Lugo 1989) or *Caesalpinia cirsta* (Estrada Lugo 1989)
xoxouhcapahtli (blue/green remedy/blue medicine)=pl. 43, 63 Badianus (C-B fol. 26r, 36r)=FC=*Caesalpinia crista* (Díaz 1976; Estrada Lugo 1989) or *Lepidium virginicum* (Linares and Bye 2013) or *Lepidium* sp. (Valdés Gutiérrez et al. 1992) or *Hyptis* sp. (Miranda and Valdés 1964) or *Turbina corymbosa* (*Rivea corymbosa*) (Guerra 1952)
xoxohqui matlalxochitl=pl. 14 Badianus (C-B fol. 10v)=*Commelina* sp. (Guerra 1952)
xoxouhquipahtli (blue/green remedy/blue medicine)=pl. 22 Badianus (C-B fol. 14v)=*Caesalpinia pulcherrima* (Bye and Linares 2013) or *Caesalpinia crista* (Valdés Gutiérrez et al. 1992) or *Hyptis* sp.?
xoy=Melampodium divaricatum (Díaz 1976)
xumetl=Sambucus canadensis (*S. mexicana*) (Díaz 1976; Ortiz de Montellano 1990)
xunalixase=Croton ciliatoglandulifer (Díaz 1976)
xunequitl/xunequitzio=Gaudichaudia cynanchoides (Díaz 1976)
yacaquilitl=Cuphea jorullensis (Díaz 1976)
yamancapatli=Euphorbia serpens (*E. radicans*) (Farfán and Elferink 2010)
yamanquipatlis/yamanquipahtli/yamanquipatli (gentle or weak medicine)=pl. 78 Badianus (C-B fol. 43v)=p. 138 Hernandez Rerum=FC=*Manihot angustiloba* (Estrada Lugo 1989; Guerra 1952; Reko 1947) or *Asclepias verticillata* (Díaz

1976; Estrada Lugo 1989; Miranda and Valdés 1964) or *Manihot* sp. (Estrada Lugo 1989) or M. *rubricaulis*

yamanquitexochitl/texochitl yamanqui (soft stone-flower)=pl. 64 Badianus (C-B fol. 36v)=*Selaginella* sp. (Reko 1947) or *S. lepidophylla* (Miranda and Valdés 1964)

yamate=*Asclepias subulata* (Díaz 1976)

yapaxiutl=FC=*Amaranthus* sp. or *Croton diocus* (Estrada Lugo 1989)

yauhtli=*Tagetes lucida* (Díaz 1976; Farfán and Elferink 2010; Reko 1947)

yaxhalache=*Euphorbia tithymaloides* subsp. *parasitica* (*Pedilanthus itzaeus*) (Díaz 1976)

ychcatlepatli/ichcatlepatli=p. 442 Hernandez=*Solanum scandens*?

ycxitecuan=*Schaffneria nigripes* (*Phyllitis nigripes*) (Díaz 1976)

yecapixtla=*Psittacanthus calyculatus* (Díaz 1976)

yecuxoton=FC=*Nicotiana tabacum* (Estrada Lugo 1989)

yehcapaztzin=*Senna occidentalis* (*Cassia laevigata*) (Díaz 1976)

yehuanitzoz=*Serjania mexicana* (Díaz 1976)

yepacihuitl=*Croton dioicus* or *C. vulpinus* (Díaz 1976)

yetl=*Nicotiana* spp. (Daunay et al. 2008; Dressler 1953)

yiamoli/yiamolli/yamole=FC461b,759b=*Phytolacca octandra* (Díaz 1976; Estrada Lugo 1989) or *P. americana* (*P. decandra*) or *P. longespica* or *P. rugosa* (Estrada Lugo 1989)

yichcayo=FC=*Rhodosciadium tuberosum* (*Deanea tuberosa*) (Estrada Lugo 1989) or *Gossypium* sp. (Estrada Lugo 1989)

yietl=FC=*Nicotiana tabacum* (*N. mexicana*) (Estrada Lugo 1989)

ylin/īlīn (alder)=pl. 84, 87 Badianus (C-B fol. 46v, 48r)=*Alnus* sp. (de Ávila Blomberg 2012; Guerra 1952; Miranda and Valdés 1964; Reko 1947; Valdés Gutiérrez et al. 1992)

yllamatlantli=p. 443 Hernandez Rerum=*Solanum lycopersicum*?

yollopatli (heart-medicine)=*Magnolia mexicana*?

yolo chichiltic=p. 445 Hernandez Rerum/p. 68 Hernandez Historia=*Chromolaena collina* (*Eupatorium collinum*)?

yolloxochitl/iolloxochitl/yoloxochitl/yollochicuahuitl/yōlloh-xōchitl (heart flower)=pl. 59, 60, 98, 104 Badianus (C-B fol. 34r, 34v, 53v, 56v)=p. 40 Hernandez Rerum=FC=*Magnolia mexicana* (*Talauma mexicana*) (Alcántara Rojas 2008; de Ávila Blomberg 2012; del Pozo 1966; Díaz 1976; Estrada Lugo 1989; Farfán and Elferink 2010; Guerra 1952; Linares and Bye 2013; Miranda and Valdés 1964; Ocaranza 2011; Ortiz de Montellano. 1990; Valdés Gutiérrez et al. 1992; Zepeda and White 2008) or *M. virginiana* (*M. glauca*) (Estrada Lugo 1989)

yolochiachitl=*Psoralea glandulosa* (Díaz 1976)

yolochichi/yolochichil/yolochichiltic=*Chromolaena collina* (*Eupatorium collinum*) (Díaz 1976)

yolochichitl=*Chromolaena collina* (*Eupatorium collinum*) or *Ageratina petiolaris* (*Eupatorium petiolare*) (Díaz 1976)

yoloxiltic=*Piqueria trinervia* (Díaz 1976)

yoloxochitl=*Chromolaena collina* (*Eupatorium collinum*) or *Magnolia mexicana* (*Talauma mexicana*) or *M. macrophylla* (Díaz 1976)

yoxanitztac/yoxonitzal=*Piqueria trinervia* (Díaz 1976)

yoyotl/yoyote/yoyotli=*Cascabela thevetia* (*Thevetia neriifolia*) (del Pozo 1966) or *Cascabela thevetioides* (*Thevetia yccotli*) (Díaz 1976)

ytacchyatl/yztaccihuatl=p. 208 Hernandez Rerum/p. 896 Hernandez Historia=*Erigeron* sp.?

ytzaccococtlacotli/yztaccococtlacotl=p. 450 Hernandez Rerum/p. 625 Hernandez Historia=?

ytzcuimpatli (dog medicine)=p. 307 Hernandez Rerum=*Schoenocaulon officinale* (*Veratrum officinale*)?

ytzticpatli=p. 116 Hernandez Rerum=*Manihot* sp.?

ytzuquatzitziqui=*Solandra maxima* (*Datura maxima*) (Díaz 1976)

yumate=*Asclepias subulata* (Díaz 1976)

yuripitzqua/yuriripitaqua=*Desmodium orbiculare* (Díaz 1976)

yxiayaual=FC=*Scutellaria caerulea* (Estrada Lugo 1989)

yyacxihuitl=p. 448 Hernandez Rerum/p. 852 Hernandez Historia=*Cestrum* sp.?

yyamolin=*Phytolacca octandra* (Díaz 1976)

yyauhitl/yiauhtli/iyauhtli/yyahhitl (to incense)/*quauhyyauhtli* (forest incense)=pl. 91 Badianus (C-B fol. 50r)=FC508a, 628=p. 160 Hernandez Rerum/p. 969 Hernandez Historia=p. 288 Sahagún=*Tagetes lucida* (Díaz 1976; Estrada Lugo 1989; Guerra 1952; Ortiz de Montellano 1990)

yyauhtzin=*Crataegus laevigata* (*C. oxyacantha*) (Díaz 1976)

yzhoatomatl=*Solanum lycopersicum* (Díaz 1976)

yzquixochitl/īzqui-xōchitl (popcorn flower)=pl. 69, 70, 78, 80 Badianus (C-B fol. 39r, 39v, 43v, 44v)=*Bourreria huanita* (de Ávila Blomberg 2012; Farfán and Elferink 2010; Guerra 1952; Linares and Bye 2013; Valdés Gutiérrez et al. 1992) or *Bourreria* sp. (Miranda and Valdés 1964)

yztachuitzquahuitl/iztāc huitzcuahuitl (white spiny tree)=pl. 109 Badianus (C-B. fol. 58v, 59r)=*Haematoxylum campechianum* (de Ávila Blomberg 2012; Linares and Bye 2013; Valdés Gutiérrez et al. 1992) or *H. brasiletto* (Miranda and Valdés 1964) or *Condalia hookeri* (*C. obovata*) (Guerra 1952)

yztac ocoxochitl (white pine flower)=pl. 7, 54, 61, 74, 79, 92 Badianus (C-B fol. 7r, 31v, 35r, 42v, 44r, 50v)=*Didymaea mexicana* (Miranda and Valdés 1964; Reko 1947)

yztacpahtli (white medicine)=pl. 25 Badianus (C-B fol. 16r)=*Mimosa* sp. (Bye and Linares 2013; Guerra 1952)

yztacquauitl (white tree)=pl. 68 Badianus (C-B fol. 38v)=*Acacia [coulteri]* (de Ávila Blomberg 2012) or *Eysenhardtia polystachya* (Linares and Bye 2013) or Fabaceae (Guerra 1952; Miranda and Valdés 1964)

yztacquauhxiotl/iztāc cuahuitl=p. 57 Hernandez Rerum/p. 551 Hernandez Historia=*Bursera fagaroides* (*Elaphrium fagaroides*) (Díaz 1976)

yztacquauhxiotl=p. 57 Hernandez Rerum/p. 552 Hernandez Historia=*Bursera tecomaca* (*Elaphrium tecomaca*)?

yztactalamatl=p. 452 Hernandez Rerum/p. 258 Hernandez Historia=*Desmodium cinereum* (Díaz 1976)
yztauhyattl/yztauhtyatili/iztauhyatl (salty water)=pl. 43 Badianus (C-B fol. 26r)=p. 288 Sahagún=*Artemisia mexicana* (de Ávila Blomberg 2012; Díaz 1976; Guerra 1952; Miranda and Valdés 1964)
zazalic/zazalicpatli=FC=*Mentzelia hispida* (Farfán and Elferink 2010; Ocaranza 2011) or *Bidens* sp. (Estrada Lugo 1989) or *Perymenium pringlei* (Estrada Lugo 1989)
zacamatlalin/çacamatlalin/zaca-mātlalin (grassy matlalin)=pl. 75, 87 Badianus (C-B fol. 42r, 48r)=*Commelina* sp. (Guerra 1952) or *Tradescantia* sp. (de Ávila Blomberg 2012; Guerra 1952; Valdés Gutiérrez et al. 1992) or *Commelina coelestis* (Linares and Bye 2013; Miranda and Valdés 1964)
zacacamotontin=*Juncus balticus* subsp. *mexicanus* (*J. mexicanus*) (Díaz 1976)
zacachichi/zacachichic=*Calea zacatechichi* or *Conyza filaginoides* (Díaz 1976)
zacahuitztli=*Cenchrus echinatus* (Díaz 1976)
zacamalinalli=FC=*Baccharis* sp. or *Elymus* sp. (Estrada Lugo 1989)
zacamatlalin=*Commelina coelestris* (Díaz 1976)
zacanahuiteputz=*Verbesina crocata* (Díaz 1976)
zacacotl=*Cassia grandis* (*C. braziliana*) or *C. fistula* (Díaz 1976)
zacapipillolxochitl=*Lonicera pilosa* (Díaz 1976)
zacapolin=*Cenchrus tribuloides* (de Batres et al. 2012; Díaz 1976)
zacatechichi=*Calea zacatechichi* (Díaz 1976)
zacateteztli=FC=*Panicum* sp. (Estrada Lugo 1989)
zacatlamiahoalli=*Erigeron affinis* (Díaz 1976)
zacatlaxcale/zacatlaxcalli=FC=*Cuscuta tinctoria* (Díaz 1976; Estrada Lugo 1989) or *C. americana* or *C. odontolepis* (Estrada Lugo 1989)
zayoliscan/zayolitzcan/zayolizan/zayolizcan/zayolizcan=*Buddleja americana* or *B. cordata* (Díaz 1976)
zasale/zazalic=*Mentzelia hispida* (Díaz 1976)
zazaltzin=*Perymenium rude* (Díaz 1976)
zazamil=*Cordia alba* (*C. dentata*) (Díaz 1976)
zazanaczochitl=*Aristolochia odoratissima* (Díaz 1976)
zihoapactli/zihuatlapatl=*Montanoa tomentosa* (Díaz 1976)
zihuatlpatl=*Buddleja sessiflora* (*B. verticillata*) (Díaz 1976)
zoapatle/zoapatli=*Montanoa floribunda* or *M. frutescens* or *M. tomentosa* (Díaz 1976)
zohzoyat/zohzoyatic/zohzōyātic (full of palm leaves)/*çoçoyatic/zozoyatic*=pl. 102 Badianus (C-B fol. 55v)=p. 144 Hernandez Rerum/p. 855 Hernandez Historia=*Anticlea frigida* (*Stenanthium frigidum, Veratrum frigidum*) (Comas et al. 1995; Díaz 1976; Estrada Lugo 1989; Miranda and Valdés 1964; Ocaranza 2011; Zepeda and White 2008) or *Schoenocaulon caricifolium* (*S.coulteri*) (de Ávila Blomberg 2012; Ortiz de Montellano 1990) or *Echeandia mexicana* (Linares and Bye 2013) or *Agave* sp. (Valdés Gutiérrez et al. 1992) or *Schoenocaulon* sp. or *Zygadenus* sp.(Estrada Lugo 1989) or *Schoenocaulon officinale* (*Veratrum sabadilla*) (Estrada Lugo 1989) or *Anticlea* sp. (Guerra 1952)

zolcoanenepilli=Physalis nicandroides (Díaz 1976)
zompantle=Buddleja americana or *Erythrina coralloiides* (Díaz 1976)
zopilocuahuitl=Swietenia humilis (Díaz 1976)
zopilote/zopilotl=Swietenia humilis (*S. cirrhata*) or *S. macrophylla* (Díaz 1976)

Abbreviations (16th Century Sources and Commentaries)

Badianus=Emmart, E.W. 1940. *The Badianus Manuscript (Codex Barberini, Latin 241)*. Baltimore: Johns Hopkins Press.

C-B=Clayton, M., L. Guerrini, and A. de Avila. 2009. *Flora: The Aztec herbal.* London: The Royal Collection.

FC or FCillustration#=Sahagún, B. de. 1963. *Florentine Codex. General history of the things of New Spain. Book 11—Earthly things.* Transl. C. E. Dibble and A. J. O. Anderson. Salt Lake City: University of Utah.

Hernandez Historia=Hernandez, F. 1942. *Historia de las Plantas de Nueva España.* Ed. I. Mexico: Ochoterena. Imprenta Univ.

Hernandez Rerum=Hernandez, F., F. Celsi, F. Colonna, B. Deversini, J. Faber, J. Greuter, V. Mascardi, N.A. Recchi, and J. Terentius. 1651. *Rerum medicatum Novae Hispaniae Thesaurus, seu, Plantarum animalium mioneralium Mexicanorum historia.* Romae: Vitalis Mascardi.

Sahagún=Martinez, J. L. 1981. *El México antiguo, selección y reordenación de la Historia general de las cosas de Nueva España de fray Bernardino de Sahagún y de los informantes indígenas.* Caracas, Venezuela: Biblioteca Ayacucho.

Literature Cited (Post 16th Century Commentaries)

Alcántara Rojas, B. 2008. In Nepapan Xochitl: The power of flowers in the works of Sahagún. In *Colors between two worlds: The Florentine Codex of Bernardino de Sahagún*, ed. L.A. Waldman, 106–132. Florence: Villa I Tatti.

Altamirano, F. 1896. Historia natural aplicada de los antiguos mexicanos. *Anales del Instituto Médico Nacional México* 2: 261–272.

Berdan, F. F. 2007. The technology of ancient Mesoamerican mosaics: An experimental investigation of alternative super glues. FAMSI (Foundation for the Advancement of Mesoamerican Studies). http://www.famsi.org/reports/06015/06015Berdan01.pdf. Accessed 31 Jan 2014.

Bye, R.A., and E. Linares. 1990. Mexican market plants of the 16th century. I. Plants recorded in *Historia Natural de Nueva España*. *Journal of Ethnobiology* 10: 151–168.

Bye, R.A., and E. Linares. 2013. Códice de la Cruz-Badiano: Medicine préhispánica. Primera parte. *Arqueología Mexicana* 50: 7–91.

Comas, J., E. González, A. López Austin, G. Somolinos, and C. Viesca. 1995. *El mestizaje cultural y la medicina novohispanica del siglo XVI*. València: Universitat de València-C.S.I.C.

Daunay, M.-C., H. Laterrot, and J. Janick. 2008. Iconography and history of the Solanaceae: Antiquity to the 17th century. *Horticultural Reviews* 34: 1–111.

de Ávila Blomberg, A. 2012. Yerba del coyote, veneno del perro: La evidencia lexica para identificar plantas en el Códice de la Cruz Badiano. *Acta Botanica Mexicana* 100: 489–526.

de Batres, L.D.P., C.A.B. Alfaro, and J. Ghaemghami. 2012. Mesoamerica aesthetics: Horticultural plants in hair and skin care. *Chronica Horticulturae* 52 (2): 12–15.

del Pozo, E.C. 1966. Aztec pharmacology. *Review of Pharmacology* 6: 9–18.

Díaz, J.L. 1976. *Índice y sinonimia de las plantas medicinales de México*. México: Instituto Mexicano para el Estudio de las Plantas.

Dominguez, S.X.A. 1969. Algunos aspectos químicos y farmacológicos de sustancias aisladas de las plantas descritas en el Códice Badiano (Libellus de Medicinalibus Indorum Herbis).*Revista de la Sociedad Química de México* 13 (2): 85–89.

Dressler, R.L. 1953. The Pre-Columbian cultivated plants of Mexico. *Botanical Museum Leaflets* 16: 115–172.

Elferink, J.G.R., J. Flores, and C.D. Kaplan. 1994. The use of plants and other natural products for malevolent practices among Aztecs and their successors. *Estudio de Cultura Náhuatl (México: UNAM Instituto de Investigaciones Históricas)* 24: 27–47.

Estrada Lugo, E. 1989. *El Códice Florentino: Su información etnobotánica*. México: olegio de Postgraduados, Institución de Enseñanza e Investigación en Ciencias Agricolas, Montecillo.

Farfán, J.A.F., and J.G.R. Elferink. 2010. *Ethnobotany and Aztec sexuality*. Muenchen: Lincom Europa.

Farrar, W.V. 1966. Tecuitatl: A glimpse of Aztec food technology. *Nature* 211: 341–342.

Flores, F.A. 1886–1888. *Historia de la medicina en México*. México: Secretaría de Pomento.

Furst, P. 1995. This little books of herbs: Psychoactive plants as therapeutic agents in the Badianus Manuscript of 1552. In *Ethnobotany: Evolution of the discipline*, ed. R. Schultes and S. von Reis, 108–130. Portland: Dioscorides Press.

Godínez, J.L., M.M. Ortega, G. Garduño, M.G. Oliva, and G. Vilaclara. 2001. Traditional knowledge of Mexican continental algae. *Journal of Ethnohistory* 21: 57–88.

Guerra, F. 1952. Indice etimologico. In *Libellus de medicinalibus Indorum herbis. El manuscrito pictórico mexicano-latino de Martín de la Cruz y Juan Badiano de 1552. Estudio. Texto y Versión*, ed. M. de la Cruz, J. Badiano, and F. Guerra, 237–256. México: Editorial Vargas Rea y El Diario Español.

Jenkins, J.A. 1948. The origin of the cultivated tomato. *Economic Botany* 2: 379–392.

Lentz, D.L., M.D. Pohl, J.L. Alvarado, S. Tarighat, and R. Bye. 2008a. Sunflower (*Helianthus annuus* L.) as a pre-Columbian domesticate in Mexico. *Proceedings of the National Academy of Sciences of the United States of America* 105: 6232–6237.

Lentz, D.L., M.D. Pohl, and R. Bye. 2008b. Reply to Rieseberg and Burke, Heiser, Brown, and Smith: Molecular, linguistic, and archaeological evidence for domesticated sunflower in pre-Columbian Mesoamerica. *Proceedings of the National Academy of Sciences of the United States of America* 105: E49–E50.

Linares, E., and R. Bye. 2013. Códice de la Cruz-Badiano: Medicine préhispánica. Segunda parte. *Arqueología Mexicana* 51: 7–93.

Martinez, M. 1969. *Las plantas medicinales de Mexico*. 5th ed. Mexico: Andres Botas.

Miranda, F., and J. Valdés. 1964. Comentarios botánicos. In *Libellus de medicinalibus indorum herbis manuscrito Azteca de 1552 segun traduccion Latina de Juan Badiano version Espanola con estudios y comentarious por diversos autores*, ed. M. de la Cruz, 243–284. México: Instituto Mexicano del Seguro Social.

Ocaranza, F. 2011. *Historia de la medicina en México*. 2nd ed. México: Cien de Mexico, Cauhtémoc.

Ochoterena-Booth, H. 2000. Identity and current ethnobotanical knowledge of Francisco Hernández's "cicimatic". *Journal of Ethnobiology* 20: 61–71.

Ortiz de Montellano, B. 1975. Empirical Aztec medicine. *Science* 188: 215–220.

Ortiz de Montellano, B. 1990. *Aztec medicine, health, and nutrition*. New Brunswick: Rutgers University Press.

Ossenbach, C. 2009. Orchids and orchidology in Central America. 500 years of history. *Landesteriana* 9: 1–268.

Picó, B., and F. Nuez. 2000a. Minor crops of Mesoamerica in early sources (I). *Leafy vegetables. Genetic Resources and Crop Evolution* 47: 527–540.

Picó, B., and F. Nuez. 2000b. Minor crops of Mesoamerica in early sources (II). *Herbs used as condiments. Genetic Resources and Crop Evolution* 47: 541–552.

Ramírez, J., and G.V. Alcocer. 1902. *Sinonimia vulgar y científica de las plantas mexicanas*. Mexico: Oficina Tipográfica de la Secretaría de Fomento.

Reko, B.P. 1947. Nombres bótanicos del manuscrito Badiano. *Boletín de la Sociedad Botánica de México* 5: 23–43.

Rivera Morales, I. 1941. Ensayo de interacción botánica del libro X de la historia de Sahagún. *Anales del Instituto de Biología de la Universidad Nacional Autónoma de México, Serie botánica*12: 439–488.

Safford, W.E. 1910. The sacred ear-flower of the Aztecs: Xochinacaztli. *Annual Report of the Smithsonian Institution* 1910: 427–431.

Safford, W.E. 1912. Sacred flowers of the Aztecs. *Volta Review* 14: 87–93.

Safford, W.E. 1918. *Chenopodium nuttalliae,* a food plant of the Aztecs. *Journal of the Washington Academy of Science* 8: 521–527.

Safford, W.E. 1920. Daturas of the Old World and New: An account of their narcotic properties and their use in oracular and initiatory ceremonies. *Annual Report of the Smithsonian Institution*1920: 537–567, pl. 1–13.

Schultes, R.E. 1940. Teonanacatl: The narcotic mushroom of the Aztecs. *American Anthropologist* 42: 429–443.

Schultes, R.E. 1941. *A contribution to our knowledge of Rivea corymbosa, the narcotic ololinqui of the Aztecs*. Cambridge, MA: Botanical Museum of Harvard University.

Schultes, R.E. 1957. The genus *Quararibea* in Mexico and the use of its flowers as a spice for chocolate. *Botanical Museum Leaflets* 17: 247–264.

Schultes, R.E. 1964. The nomenclature of two Mexican narcotics. *Taxon* 13: 65–66.

Siméon, R. 2010. *Diccionario de la lengua Nahuatl o Mexicana*. México: Siblo Veintiuno.

Sorensen, P.D. 1970. The dahlia: An early history. *Arnoldia* 30: 121–138.

Standley, P.C. 1920–1926. *Trees and shrubs of Mexico*, Contributions of the U. S. National Herbarium. Vol. 23.

Tucker, A.O., and J. Janick. 2016. Identification of phytomorphs in the Voynich Codex. *Horticultural Reviews* 44: 1–64.

Valdés Gutiérrez, J., H. Flores Olivera, and H. Ochoterena-Booth. 1992. La botánica en el Codice de la Cruz. In *Estudios actuales sobre el libellus de medicinalibus Indorum herbis*, ed. J. Kumate, M.E. Pineda, C. Viesca, J. Sanfilippo, I. de la Peña Páez, J. Valdéz Gutiérrez, H. Flores Olvares, H. Ochoterena-Booth, and Z. Lozoyal, 129–180. México: Secretaria de Salud.

Viesca, C., and A. Aranda. 1996. Las alteraciones del sueño en el *Libellus de medicinalibus indorum herbis*. *Estudios de Cultura Náhuatl* 26: 145–161.

Wasson, R.G. 1973. The role of ‘flowers’ in Nahuatl culture: A suggested interpretation. *Botanical Museum Leaflets* 23: 305–324.

Wasson, R.G. 1980. *The wondrous mushroom: Mycolatry in Mesoamerica*. New York: McGraw-Hill.

Zepeda, G.C., and O.L. White. 2008. Herbolaria ye pintura mural: Plantas medicinales en los murals del Convento del Divino Salvador de Malinalco, Estado de México. *Polibotanica* 25: 173–199.

Appendix B: Plants Used by the Nahua, Arranged by Family

Plants Used By the Nahua, Arranged by Family

Plants cited as being used by the Nahua, arranged by family with updated nomenclature in black. Plant families and species unique to the *Voynich Codex* are in red. Taxa seen in both the *Voynich Codex* and 16th century codices are in blue. For References, see Appendix A.

Some species of bacteria and algae have no Nahuatl name recorded (Godínez et al. 2001).

BACTERIA

Chroococcaceae
Chroococcus turgidus (Kützing) Nägeli

Nostocaceae
Nostoc commune Vaucher ex Bornet & Flahault

Phormidiaceae
Leptolyngbya tenuis (Gomont) Anagnostidis & Komárek (*Phormidium tenue* Gomont)

Spirulinaceae
Spirulina maxima (Setchell & N.L.Gardner) Geitler (*S. geitleri* Auct.)

"ALGAE"

Chlorellaceae
Chlorella sp.

Prasiolaceae
Prasiola mexicana J. Agardh

"FUNGI"

Amanitaceae
Amanita muscaria (L.) Lam.

Inocybaceae
Inocybe sp.

Ophiocordycipitaceae
Cordyceps capitata (Holmsk.) Link

Psathyrellaceae
Psathyrella sp.

A. O. Tucker, J. Janick, *Flora of the Voynich Codex*,
https://doi.org/10.1007/978-3-030-19377-5

Strophariaceae
Panaeolus campanulatus (L.) Quél.
Panaeolus sphinctrinus (Fr.) Quél.
Psilocybe aztecorum R.Heim
Psilocybe cordispora R.Heim
Psilocybe hoogshagenii R.Heim
Psilocybe mexicana R.Heim
Psilocybe sempervivae R.Heim & Cailleux
Psilocybe zapotecorum R.Heim
Stropharia sp.

Ustilaginaceae
Ustilago maydis (DC.) Corda

LYCOPHYTES

Huperziaceae/Lycopodiaceae
Huperzia beiteliana Mickel
Huperzia dichotoma (Jacq.) Trevis. (*Lycopodium dichotomum* Jacq.)
Huperzia orizabae (Underw. & F.E.Lloyd) Holub
Huperzia reflexa (Lam.) Trevis

Selaginellaceae
Selaginella lepidophylla (Hook. & Grev.) Spring
Selaginella pallescens (C.Presl) Spring [*S. cuspidata* (Link) Link]
Selaginella pilifera A.Braun
Selaginella sellowii Hieron. [*S. rupestris* (L.) Spring]

PTERIDOPHYTES

Adiantaceae
Adiantum capillus-veneris L.
Adiantum trapeziforme L.
Pellaea ternifolia (Cav.) Link

Anemiaceae
Anemia mexicana Klotzsch

Aspleniaceae
Asplenium trichomanes L.
Schaffneria nigripes Fée [*Phyllitis nigripes* (Fée) Kuntze]

Cyatheaceae

Alsophila firma (Baker) D.S.Conant (*Cyathea mexicana* Schltdl. & Cham.)

Equisetaceae
Equisetum myriochaetum Schltdl. & Cham.

Hymenophyllaceae
Trichomanes ekmanii Wess. Boer.

Marsileaceae
Marsilea mollis B.L.Rob. & Fernald

Ophioglossaceae
Ophioglossum palmatum L.

Osmundaceae
Osmunda regalis L. var. *spectabilis* (Willd.) A.Gray

Polypodiaceae
Campyloneurum phyllitidis (L.) C.Presl
Niphidium crassifolium (L.) Lellinger [*Pleopeltis crassifolia* (L.) T. Moore, *Polypodium crassifolium* L.]
Pleopeltis macrocarpa (Bory ex Willd.) Kalf. (*Polypodium lanceolatum* L.)

Pteridaceae
Ceratopteris pteridoides (Hook.) Hieron.

Schizaeaceae
Schizaea elegans (Vahl) Sw.

Vittariaceae
Vittaria sp.

GYMNOSPERMS

Cupressaceae
Cupressus lusitanica Mill. var. *benthamii* (Endl.) Carrière [*C. benthamii* Engl., *C. lindleyi* Klotzsch ex Endl. subsp. *benthamii* (Endl.) Silba]
Juniperus poblana (Martínez) R.P.Adams (*Cupressus thurifera* Kunth)

Ephedraceae
Ephedra aspera Engelm. ex S.Watson
Ephedra trifurca Torr. ex S.Watson

Pinaceae

Abies religiosa (Kunth) Schltdl. & Cham.
Picea sp.
Pinus ayacahuite C.Ehrenb. ex Schltdl.
Pinus teocote Schied. ex Schltdl. & Cham.
Pseudotsuga sp.

Taxodiaceae
Taxodium mucronatum Ten. (*T. huegelii* C. Lawson)

Zamiaceae
Dioon edule Lindl.

ANGIOSPERMS

Acanthaceae
Dyschoriste sp.
Justicia aurea Schltdl.
Justicia spicigera Schltdl. (*Jacobinia mohintli* Benth. & Hook.f.)
Ruellia sp.
Ruellia tuberosa L.

Adoxaceae
Sambucus canadensis L. (*S. mexicana* C. Presl ex DC.)

Aizoaceae
Sesuvium portulacastrum (L.) L.

Alismataceae
Sagittaria macrophylla Zucc.

Alstroemeriaceae
Bomarea edulis (Tussac) Herb. [*B. hirtella* (Kunth.) Herb.]

Altingiaceae
Liquidambar styraciflua L. (*L. macrophylla* Oerst.)

Amaranthaceae
Alternanthera pungens Kunth (*Illecebrum achyrantha* L.)
Alternanthera sessilis (L.) R.Br. ex DC. (*A. repens* J.F.Gmel.)
Amaranthus caudatus L. (*A. leucocarpus* S. Watson)
Amaranthus cruentus L. (*A. paniculatus* L.)
Amaranthus hybridus L.
Amaranthus hypochondriacus L.
Amaranthus palmeri S. Watson

Amaranthus spinosus L.
Celosia argentea L.
Celosia virgata Jacq.
Chamissoa altissima (Jacq.) Kunth
Gomphrena serrata L. (*G. decumbens* Jacq.)
Guilleminea densa (Willd. ex Schult.) Moq.
Iresine calea (Ibantz) Standl.
Iresine diffusa Humb. & Bonpl. ex Willd. (*I. celosioides* L.)
Suaeda suffrutescens S. Watson
Suaeda torreyana S. Watson

Amaryllidaceae
Sprekelia formosissima (L.) Herb. (*Amaryllis formosissima* L.)
Zephyranthes brevipes Standl.
Zephyranthes carinata Herb.
Zephyranthes fosteri Traub

Anacardiaceae
Amphipterygium adstringens (Schltdl.) Standl. (*Juliania adstringens* (Schltdl.) Schltdl.)
Anacardium sp.
Cyrtocarpa edulis (Brandegee) Standl.
Cyrtocarpa procera Kunth
Pistacia mexicana Kunth
Pseudosmodingium andrieuxii (Baill.) Engl.
Pseudosmodingium perniciosum (Kunth) Engl.
Rhus glabra L.
Rhus saxatilis DC. [unresolved name]
Rhus terebinthifolia Schltdl. & Cham.
Schinus molle L.
Spondias mombin L. (*S. lutea* L.)
Spondias purpurea L.
Toxicodendron radicans (L.) Kuntze (*Rhus radicans* L.)

Annonaceae
Annona cherimola Mill.
Annona muricata L.
Annona reticulata L.
Annona squamosa L.
Cymbopetalum penduliflorum (Dunal) Baill.

Apiaceae
Arracacia atropurpurea (Lehm.) Benth. & Hook.f. ex Hemsl.
Arracacia trífida Coult. & Rose in Urbina
Cymopterus multinervatus (J. M. Coult. & Rose) Tidestrom

Eryngium beecheyanum Hook. & Arn.
Eryngium carlinae F.Delaroche
Eryngium deppeanum Schltdl. & Cham. (*E. painteri* Hemsl. & Rose)
Eryngium foetidum L.
Eryngium heterophyllum Engelm.
Lomatium dissectum (Nutt.) Mathias & Constance
Osmorhiza mexicana Griseb.
Rhodosciadium tuberosum Drude [unresolved name] (*Deanea tuberosa* J. M.Coult. & Rose)

Apocynaceae
Asclepias linaria Cav.
Asclepias notha W.D.Stevens
Asclepias oenotheroides Schltdl. & Cham. (*A. longicornu* Benth.)
Asclepias subulata Decne.
Asclepias tuberosa L.
Asclepias verticillata L.
Cascabela gaumeri (Hemsl.) Lippold (*Thevetia gaumeri* Hemsl.)
Cascabela thevetia (L.) Lippold [*Thevetia neriifolia* Juss. ex Steud., *T. peruviana* (Pers.) K.Schum.]
Cascabela thevetioides (L.) Lippold (*Thevetia yccotli* A.DC.)
Gonolobus chloranthus Schltdl.
Gonolobus erianthus Decne.
Gonolobus nummularius (Decne.) Hemsl.
Gonolobus pedunculatus (Decne.) Hemsl.
Gonolobus prostratus Baldwin [unresolved name]
Haplophyton cimicidum A.DC.
Mandevilla tubiflora (M. Martens & Galeotti) Woodson (*Echites tubiflorus* M.Martens & Galeotti)
Marsdenia mexicana Decne.
Marsdenia zimapanica Hemsl.
Matelea reticulata (Engelm. ex A.Gray) Woodson (*Gonolobus reticulatus* Engelm. ex A.Gray)
Plumeria alba L.
Plumeria rubra L. (*P. acutifolia* Poir., *P. bicolor* Ruiz & Pav., *P. mexicana* Lodd., *P. tricolor* Ruiz & Pav.)
Rauvolfia tetraphylla L. (*R. heterophylla* Willd. ex Roem. & Schult.)
Tabernaemontana amygdalifolia Jacq.

Araceae
Anthurium cordatum (L.) Schott
Monstera deliciosa Liebm.
Philodendron goeldii G. M. Barroso
Philodendron mexicanum Engl.
Philodendron radiatum Schott var. *pseudoradiatum* (Matuda) Croat
Philodendron sagittifolium Liebm. (*P. sanguineum* Regel)
Philodendron subincisum Schott

Philodendron tripartitum (Jacq.) Schott (*P. affine* Hemsl.)
Spathiphyllum cannifolium (Dryand. ex Sims) Schott
Syngonium podophyllum Schott
Xanthosoma robustum Schott
Xanthosoma sagittifolium (L.) Schott (*X. roseum* Schott)
Xanthosoma violaceum Schott

Araliaceae
Hydrocotyle ranunculoides L.f.
Polyscias guilfoylei (W. Bull) L.H.Bailey

Arecaceae
Acrocomia aculeata (Jacq.) Lodd. ex Mart. (*A. mexicana* Karw. ex Mart.)
Attalea sp. (*Maximiliana* sp.)
Bactris mexicana Mart.
Brahea dulcis (Kunth) Mart.
Chamaedorea elatior Mart.
Chamaedorea tepejilote Liebm.
Chamaedorea tuerckheimii (Dammer) Burret
Sabal yapa (C. Wright ex Becc. (*Inodes japa* Standl.)

Aristolochiaceae
Aristolochia anguicida Jacq. (*A. mexicana* Willd.)
Aristolochia odoratissima L.
Aristolochia subclausa S. Watson
Asparagaceae /Agavaceae
Agave americana L.
Agave atrovirens Karw. ex Salm-Dyck
Agave ellemeetiana Jacobi
Agave filifera Salm-Dyck
Agave fourcroydes Lam.
Agave horrida Lem. ex Jacobi
Agave lophantha Schiede (*A. univittata* Haw., *A. heteracantha* Zucc.)
Agave parryi Engelm.
Agave potatorum Zucc.
Agave salmiana Otto ex Salm-Dyck (*A. cochlearis* Jacobi)
Agave sisalana Perrine [*A. rigida* Mill. var. *sisalana* (Perrine) Engelm.]
Agave vera-cruz Mill. (*A. mexicana* Lam.)
Agave vivipara L.
Agave xylonacantha Salm-Dyck
Dasylirion sp.
Echeandia flavescens (Schult. & Schult. f.) Cruden [*Anthericum leptophyllum* (Benth.) Baker]
Echeandia mexicana Cruden
Manfreda guttata (Jacobi & C.D.Bouché) Rose (*Agave guttata* Jacobi & C.D.Bouché)

Manfreda scabra (Ortega) McVaugh (*Agave saponaria* Lindl.)
Polianthes geminiflora (Lex.) Rose (*Bravoa geminiflora* Lex.)
Polianthes tuberosa L. (*P. gracilis* Link)
Prochnyanthes mexicana (Zucc.) Rose (*P. viridescens* S.Watson)
Yucca aloifolia L.
Yucca filamentosa L.
Yucca gigantea Lem. (*Y. elephantipes* Regel ex Trel.)
Yucca faxoniana Sarg. (*Y. australis* Trel.)

Asteraceae
Achillea millefolium L.
Acourtia sp.
Acourtia cordata (Cerv.) B.L.Turner (*Perezia hebeclada* (DC.) A.Gray)
Acourtia cuernavacana (B.L.Rob. & Greenm.) Reveal & R.M.King (*Perezia cuernavacana* B.L.Rob. & Greenm.)
Acourtia fruticosa (La Llave & Lex.) B. L.Turner (*Perezia fruticosa* La Llave & Lex.)
Acourtia humboldtii (Less.) B.L.Turner (*Perezia adnata* A. Gray)
Acourtia moschata DC. (*Perezia moschata* Lex.)
Acourtia runcinata (Lag. ex D. Don) B.L.Turner
Acourtia thurberi (A. Gray) Reveal & King
Adenophyllum glandulosum (Cav.) Strother (*A. coccineum* Pers.)
Aganippea dentata DC. [unresolved name]
Ageratina adenophora R.M King & H.Rob.
Ageratina deltoidea (Jacq.) R.M.King & H.Rob. (*Eupatorium deltoideum* Jacq.)
Ageratina petiolaris (Moc. & Sessé ex DC.) R.M.King & H.Rob. (*Eupatorium petiolare* Moc. & Sessé ex DC.)
Ageratina ligustrina (DC.) R.M.King & H. Rob.
Ageratum littorale Kunth
Alomia alata Hemsl.
Ambrosia ambrosioides (Delpino) W.W.Payne
Artemisia longifolia Nutt.
Artemisia ludoviciana Nutt. subsp. *redolens* (A.Gray) D.D.Keck (*A. redolens* A.Gray)
Artemisia mexicana Willd.
Baccharis brachylaenoides DC. var. *ligustrina* (DC.) Maguire & Wurdack (*Baccharis saligna* Mart. ex Baker)
Baccharis conferta Kunth
Baccharis multiflora Kunth
Baccharis salicina Torr. & A.Gray (*B. glutinosa* Pers.)
Barkleyanthus salicifolius (Kunth) H.Rob. & Brettell
Bartlettina sordida (Less.) R.M.King & H.Rob. (*Eupatorium sordidum* Less.)
Bidens aurea (Aiton) Sherff [*B. tetragona* (Serv.) DC.]
Bidens laevis (L.) "Britton, Sterns & Poggenb." (*B. chrysanthemoides* Michx.)
Bidens odorata Cav. [unresolved name]
Bidens pilosa L. (*B. odorata* Cav.)

Blumea viscosa (Mill.) V.M.Badillo (*Conyza lyrata* Kunth)
Calea integrifolia (DC.) Hemsl.
Calea ternifolia Oliv. ex Thurn var. *hypoleuca* (B.L.Rob. & Greenm.) B.L.Turner (*C. hypoleuca* B.L.Rob. & Greenm.) [unresolved name]
Calea zacatechichi Schltdl.
Chaenactis stevioides Hook. & Arn.
Chaptalia tomentosa Ventenat
Chloracantha spinosa (Benth.) G.L.Nesom
Chromolaena collina (DC.) R.M.King & H.Rob. (*Eupatorium collinum* DC.)
Chromolaena odorata (L.) R.M.King & H.Rob. (*Eupatorium odoratum* L.)
Chromolaena pulchella (Kunth) R.M.King & H.Rob. (*Eupatorium pulchellum* Kunth)
Chrysactinia mexicana A.Gray
Cirsium acantholepis (Hemsl.) Petr.
Cirsium ehrenbergii Sch.Bip.
Cirsium mexicanum DC.
Conyza filaginoides (DC.) Hieron.
Corethrogyne filaginifolia (Hook. & Arn.) Nutt. [*Lessingia filaginifolia* (Hook. & Arn.) M.A.Lane].
Cosmos bipinnatus Cav.
Cosmos diversifolius Otto & Otto
Cosmos diversifolius Otto & Otto var. *dahlioides* (S.Watson) Melchert (*Bidens dahlioides* S.Watson)
Cosmos sulphureus Cav.
Critonia hemipteropoda (B.L.Rob.) R.M.King & H.Rob. (*Eupatorium hemipteropodum* B.L.Rob.)
Cyrtocymura scorpioides (Lam.) H. Rob.
Dahlia sp.
Dahlia coccinea Cav.
Dahlia lehmannii Hieron.
Dahlia pinnata Cav. [*D. variabilis* (Willd.) Desf.]
Dyssodia papposa (Vent.) Hitchc.
Elephantopus mollis Kunth
Emilia fosbergii Nicolson
Erigeron affinis DC.
Erigeron canadensis L. [*Conyza canadensis* (L.) Cronquist]
Erigeron karvinskianus DC.
Eupatorium sessilifolium L.
Flaveria angustifolia (Cav.) Pers.
Flourensia thurifera (Molina) DC. (*Helianthus thurifer* Molilna)
Galinsoga parviflora Cav.
Gymnosperma glutinosum (Spreng.) Less. [unresolved name] (*Selloa glutinosa* Spreng.)
Helenium mexicanum Kunth
Helianthus annuus L.
Heliopsis longipes (A.Gray) S.F.Blake
Heterotheca inuloides Cass.
Hieracium sp.

Hieracium mexicanum Less.
Hulsea heterochroma A. Gray
Iostephane heterophylla (Cav.) Benth.
Isocarpha oppositifolia (L.) Cass.
Jaegeria bellidiflora (Moc. & Sessé ex DC.) A.M.Torres & Beaman
Jaegeria pedunculata Hook. & Arn.
Jungia pringlei Greenm.
Lactuca graminifolia Michx.
Lygodesmia sp.
Melampodium divaricatum (Rich. ex Rich.) DC.
Mikania cordifolia (L.f.) Willd.
Mikania scandens (L.) Willd.
Montanoa floribunda (Kunth) DC. [unresolved name]
Montanoa frutescens (Mairet ex DC.) Hemsl.
Montanoa tomentosa Cerv. (*Eriocoma floribunda* Kunth)
Oclemena acuminata (Michx) Greene
Onoseris onoseroides (Kunth) B.L.Rob
Packera bellidifolia (Kunth) W.A.Weber & Á.Löve (*Senecio vulneraria* DC.)
Parthenium confertum A.Gray
Perymenium pringlei B.L.Rob. & Greenm.
Perymenium rude B.L.Rob. & Greenm. [unresolved name]
Petasites frigidus (L.) Fr. var. *palmatus* (Aiton) Cronquist
Plectocephalus americana (Nutt.) D. Don (*Centaurea americana* Nutt.)
Pinaropappus roseus (Les.) Less.
Pippenalia delphinifolia (Rydb.) McVaugh
Piqueria trinervia Cav.
Pluchea odorata (L.) Cass.
Pluchea salicifolia (Mill.) S.F.Blake (*P. adnata* (Humb. & Bonpl. ex Willd.) C.Mohr)
Podachaenium eminens (Lag.) Sch.Bip. ex Sch.Bip.
Porophyllum coloratum (Kunth) DC. (*P. seemannii* Sch.Bip.)
Porophyllum punctatum (Mill.) S.F.Blake
Porophyllum ruderale (Jacq.) Cass. subsp. *macrocephalum* (DC.) R.R.Johnson (*P. macrocephalum* DC.)
Porophyllum tagetoides (Kunth) DC.
Porophyllum viridiflorum (Kunth) DC.
Prenanthes trifoliata (Cass.) Fernald
Psacalium peltigerum (B.L.Rob. & Seaton) Rydb.
Pseudelephantopus spicatus (B.Juss. ex Aubl.) Rohr ex C.F.Baker
Pseudognaphalium canescens (DC.) Anderb. (*Gnaphalium canescens* DC.)
Pseudognaphalium conoideum (Kunth) Anderb. [*Gnaphalium conoideum* (ex Humb., Bonpl. & Kunth) Kunth]
Pseudognaphalium macounii (Greene) Kartesz
Pseudogynoxys chenopodioides (Kunth) Cabrera (*Senecio confusus* Burtt)
Rafinesquia neomexicana A. Gray (*R. californica* Nutt.)

Roldana ehrenbergiana (Klatt) H.Rob. & Brettell (*Senecio canicidus* Sessé & Moc.)
Schkuhria pinnata (Lam.) Kuntze ex Thell.
Senecio helodes Benth. (*S. purpurascens* Klatt)
Senecio ovalifolius Turcz. [unresolved name]
Senecio praecox (Cav.) DC.
Senecio salignus DC.
Senecio sessilifolius (Hook. & Arn.) Hemsl. (*S. cardiophyllus* Hemsl)
Senecio toluccanus DC.
Simsia amplexicaulis (Cav.) Pers.
Smallanthus sp.
Solidago paniculata DC.
Solidago sempervirens Michx. var. *mexicana* (L.) Fernald
Sonchus oleraceus (L.) L. (*Sonchus ciliatus* Lam.)
Stebbinsoseris heterocarpa (Nutt.) K.L.Chambers
Stenotus acaulis (Nutt.) Nutt.
Stevia jorullensis Kunth (*S. clinopodia* DC.)
Stevia linoides Sch. Bip. ex Klotzsch
Stevia salicifolia Cav.
Symphyotrichum laeve (L.) Á. Löve & D. Löve var. *geyeri* (A. Gray) G. L. Nesom
Tagetes coronopifolia Willd. (*T. multifida* DC.)
Tagetes erecta L.
Tagetes tenuifolia Cav. (*T. peduncularis* Cav.)
Tagetes lucida Cav.
Tagetes patula L.
Townsendia sp.
Tragopogon porrifolius L.
Verbesina crocata (Cav.) Less. (*Bidens crocata* Cav.)
Vernonanthura deppeana (Less.) H.Rob. [unresolved name] (*Vernonia deppeana* Less.)
Vernonia marginata (Torr.) Raf.,
Vigethia mexicana (S. Watson) W.A.Weber (*Wyethia mexicana* S.Watson
Viguiera excelsa (Willd.) Benth. & Hook.f.
Xylorhiza wrightii (A. Gray) Greeene
Zinnia elegans Jacq.
Zinnia pauciflora Phil.

Basellaceae
Anredera baselloides (Kunth) Baill. (*Boussingaultia baselloides* Kunth)

Begoniaceae
Begonia gracilis Kunth

Berberidaceae
Berberis sp.

Betulaceae
Alnus acuminata Kunth subsp. *arguta* (Schltdl.) Furlow (*Alnus arguta* (Schltdl.) Spach)
Alnus jorullensis Kunth (*A. firmifolia* Fernald)

Bignoniaceae
Amphilophium buccinatorium (DC.) L.G.Lohmann (*Distictis buccinatoria* (DC.) A.H.Gentry)
Amphilophium crucigerum (L.) L.G.Lohmann [*Pithecoctenium echinatum* (Jacq.) Baill.]
Amphitecna macrophylla (Seem.) Miers ex Baill. (*Crescentia macrophylla* Seem.)
Astianthus viminalis (Kunth) Baill.
Amphitecna macrophylla (Seem.) Miers ex Baill.
Chilopsis linearis (Cav.) Sweet
Crescentia alata Kunth
Crescentia cujete L.
Dolichandra unguis-cati (L.) L.G.Lohmann (*Bignonia unguis-cati* L.)
Parmentiera aculeata (Kunth) Seem. (*Crescentia aculeata* Kunth, *P. edulis* DC.)
Tabebuia heterophylla (DC.) Britton (*T. pentaphylla* Hemsl.)
Tabebuia rosea (Bertol.) Bertero ex A.DC.
Tecoma stans (L.) Juss. ex Kunth

Bixaceae
Bixa orellana L.
Cochlospermum vitifolium (Willd.) Spreng. [*Maximilianea vitifolia* (Willd.) Krug & Urb.]

Boraginaceae/Hydrophyllaceae + Namaceae
Bourreria huanita (Lex.) Hemsl. [*B. formosa* (A.DC.) Hemsl.]
Bourreria litoralis D.A.Sm.
Cordia alba (Jacq.) Roem. & Schult. (*C. dentata* Poir.)
Cordia alliodora (Ruiz & Pav.) Oken
Cordia boissieri A.DC.
Cordia bullata (L.) Roem. & Schult. var. *globosa* (Jacq.) Govaerts [*C. globosa* (Jacq.) Kunth]
Cordia elaeagnoides A.DC.
Cordia sebestena L.
Cordia sonorae N.E.Rose
Cordia tinifolia Willd. ex Roem. & Schult.
Cynoglossum grande Douglas ex Lehm.
Ehretia anacua (Terán & Berland.) I.M.Johnst. (*E. scabra* Kunth & C.D.Bouché)
Ehretia tinifolia L.
Heliotropium angiospermum Murray (*H. parviflorum* L.)
Heliotropium curassavicum L.
Lithospermum sp.
Nama pringlei B.L.Rob. & Greenm.
Nama undulata Kunth
Phacelia campanularia A.Gray
Phacelia crenulata Torr.

Phacelia integrifolia Torr.
Tournefortia capitata M. Martens & Galeotti
Tournefortia hirsutissima L.
Tournefortia mexicana Vatke
Tournefortia suffruticosa L. [unresolved name]
Tournefortia volubilis L.
Wigandia urens (Ruiz & Pav.) Kunth (*W. caracasana* Kunth)

Brassicaceae
Caulanthus heterophyllus (Nutt.) Payson
Lepidium armoracia Fisch. & C. A. Mey. (*L. intermedium* A. Rich.)
Lepidium sativum L.
Lepidium sordidum A.Gray (*L. granulare* Rose)
Lepidium virginicum L.
Raphanus raphanistrum L.
Rorippa palustris (L.) Besser

Bromeliaceae
Ananas comosus (L.) Merr. (*A. sativus* Schult. & Schult.f.)
Bromelia humilis Jacq.
Bromelia pinguin L.
Guzmania lingulata (L.) Mez.
Tillandsia chaetophylla Mez
Tillandsia prodigiosa (Lem.) Baker
Tillandsia usneoides (L.) L.

Burseraceae
Bursera ariensis (Kunth) McVaugh & Rzed. (*B. sessiliflora* Engl.)
Bursera bipinnata (Moc. & Sessé ex DC.) Engl. [*Elaphrium bipinnatum* (Moc. & Sessé ex DC.) Schltdl.]
Bursera copallifera (Sessé & Moc. ex DC.) Bullock [*B. jorullensis* (Kunth) Engl., *Elaphrium jorullense* Kunth)
Bursera fagaroides (Kunth) Engl. (*Elaphrium fagaroides* Kunth, *E. odoratum* (Brandegee) Rose]
Bursera glabrifolia (Kunth) Engl. (*Elaphrium aloexylon* Schied. ex Schltdl.)
Bursera graveolens (Kunth) Triana & Planch.
Bursera lancifolia (Schltdl.) Engl. (*B. trijuga* Ramirez)
Bursera linanoe (La Llave) Rzed., Calderón & Medina [*B. aloexylon* (Schiede ex Schltdl.) Engl.]
Bursera longipes (Rose) Standl. (*Terebinthus longipes* Rose)
Bursera microphylla A.Gray [*Elaphrium microphyllum* (A.Gray) Rose]
Bursera multijuga Engl. [*Elaphrium multijugum* (Engl.) Rose]
Bursera simaruba (L.) Sarg. (*B. gummifera* L.)
Bursera tecomaca (Sessé & Moc. ex DC.) Standl. [*Elaphrium tecomaca* (Moc. & Sessé ex DC.) Standl.]
Elaphrium pubescens Schltdl. [unresolved name] [*Bursera pubescens* (Schltdl.) Standl.]

Protium copal (Schltdl. & Cham.) Engl.
Protium heptaphyllum (Aubl.) Marchand

Cabombaceae
Brasenia schreberi G. F. Gmel.

Cactaceae
Carnegiea gigantea (Engelm.) Britton & Rose
Cylindropuntia sp.
Cylindropuntia imbricata (Haw.) F.M.Knuth [*Opuntia imbricata* (Haw.) DC.]
Disocactus ackermannii (Haw.) Ralf Bauer [*Phyllocactus ackmerannii* (Haw.) Salm-Dyck, *Epiphyllum ackermannii* Haw.]
Disocactus flagelliformis (L.) Barthlott [*Aporocactus flagelliformis* (L.) Lem., *Nyctocereus serpentinus* (Lag. & Rodr.) Britton & Rose]
Disocactus phyllanthoides (DC.) Barthlott [*Nopalxochia phyllanthoides* (DC.) Britton & Rose]
Ferocactus recurvus (Mill.) Borg [*F. latispinus* (Haw.) Britton & Rose]
Heliocereus speciosus (Cav.) Britton & Rose
Hylocereus undatus (Haw.) Britton & Rose
Isolatocereus dumortieri (Scheidw.) Backeb.
Lophophora williamsii (Lem. ex Salm-Dyck) J.M. Coult.
Mammillaria bicolor Lehm. [unresolved name]
Mammillaria dolichacantha Lem. ex C.F.Först [unresolved name]
Mammillaria polythele Mart. (*M. tetracantha* Salm-Dyck)
Nopalea cochenillifera (L.) Salm-Dyck
Nopalxochia sp.
Opuntia ficus-indica (L.) Mill.
Opuntia maxima Mill. (*O. amyclaea* Ten.)
Opuntia megacantha Salm-Dyck
Opuntia megarhiza Rose
Opuntia microdasys (Lehm.) Peiff.
Opuntia pseudo-tuna Salm-Dyck [unresolved name]
Opuntia streptacantha Lem.
Opuntia tomentosa Salm-Dyck
Opuntia tuna (L.) Mill.
Pachycereus weberi (J. M. Coult.) Backeb.
Pereskia portulacifolia (L.) DC.
Pereskiopsis sp.
Pseudorhipsalis ramulosa (Salm-Dyck) Bertholott)
Selenicereus grandiflorus (L.) Britton & Rose

Calophyllaceae
Mammea americana L.

Campanulaceae

Centropogon sp.
Lobelia cardinalis L. (*L. fulgens* Humb. & Bonpl. ex Willd.)
Lobelia laxiflora Kunth

Cannabaceae
Celtis caudata Planch.

Cannaceae
Canna indica L.

Caprifoliaceae
Lonicera pilosa (Kunth) Spreng.
Scabiosa fruticulosa C.Sm. ex DC. [unresolved name]

Caricaceae
Carica papaya L.

Caryophyllaceae
Drymaria ovata Willd. ex Schult.
Silene menziesii Hook.
Stellaria nemorum L.

Celastracaceae
Maytenus phyllanthoides Benth.

Chenopodiaceae
Chenopodium berlandieri Moq. [*C. berlandieri* Moq. subsp. *nuttalliae* (Saff.) H.D.Wilson & Heiser]
Chenopodium vulvaria L. (*C. foetidum* Lam., *C. graveolens* Lag. ex Rodr.)
Dysphania ambrosioides (L.) Mosyakin & Clemants (*Chenoposium ambrosioides* L.)

Chrysobalanaceae
Chrysobalanus icaco L.
Couepia polyandra (Kunth) Rose
Licania arborea Seem.

Combretaceae
Conocarpus erectus L.

Commelinaceae
Commelina coelestis Willd.
Commelina diffusa Burm.f.
Commelina erecta L.
Commelina pallida Willd.

Commelina tuberosa L.
Tradescantia crassifolia Cav.
Tripogandra disgrega (Kunth) Woodson (*Tradescantia disgrega* Kunth)

Convolvulaceae
Cuscuta americana L.
Cuscuta odontolepis Engelm.
Cuscuta tinctoria Mart. ex Engelm.
Cuscuta umbellata Kunth
Distimake tuberosus (L.) Simões & Staples (*Ipomoea tuberosa* L.)
Evolvulus alsinoides (L.) L.
Ipomoea arborescens (Humb. & Bonpl. ex Willd.) G. Don (*Convolvulus arborescens* Humb. & Bonpl. ex Willd.)
Ipomoea batatas (L.) Lam. (*Convolvulus batatas* L.)
Ipomoea bracteata Cav.
Ipomoea capillacea (Kunth) G.Don
Ipomoea conzatii Greenm. [*Exogonium conzatii* (Greenm.) House]
Ipomoea cristulata Hallier f.
Ipomoea emetica Choisy
Ipomoea hederifolia L. [*Quamoclit hederifolia* (L.) G.Don)
Ipomoea jalapa (L.) Pursh [*Batatas jalapa* (L.) Choisy]
Ipomoea jicama Brandegee
Ipomoea mexicana A. Gray [unresolved name]
Ipomoea murucoides Roem. & Schult.
Ipomoea nil (L.) Roth [*I. hederacea* (L.) Jacq.]
Ipomoea orizabensis (G. Pelletan) Ledeb. ex Steud. (*I. longipedunculata* M. Martens & Galeotti)
Ipomoea plummerae A.Gray var. *cuneifolia* (A. Gray) J.F.Mabr.
Ipomoea pubescens Lam.
Ipomoea purga (Wender.) Hayne
Ipomoea purpurea (L.) Roth
Ipomoea rosea Choisy
Ipomoea schaffneri S. Watson
Ipomoea stans Cav.
Ipomoea wolcottiana Rose
Merremia quinquefolia (L.) Hallier f. (*Ipomoea quinquefolia* L.)
Merremia tuberosa (L.) Rendle (*Ipomoea tuberosa* L.)
Operculina pteripes (G. Don) O'Donell [*O. alatipes* (Hook.f.) House]
Quamoclit bracteata (Cav.) Roberty (*Exogonium bracteatum* (Cav.) Choisy)
Turbina corymbosa (L.) Raf. [*Ipomoea burmanni* Choisy; *I. sidaefolia* (Kunth) Sweet; *Rivea corymbosa* (L.) Hallierf.]

Corariaceae
Coriaria ruscifolia L. subsp. *microphylla* (Poir.) J.E.Skog (*C. atropurpurea* Moc. & Sessé ex DC.)

Cornaceae
Cornus excelsa Kunth
Cornus florida L. var. urbiniana (Rose) Wangerin (*C. urbiniana* Rose)

Crassulaceae
Bryophyllum pinnatum (Lam.) Oken (*B. calycinum* Salisb.)
Dudleya ingens Johans.
Echeveria sp.
Echeveria fulgens Lem.
Echeveria gibbiflora DC.
Sedum bourgaei Hemsl.
Sedum dendroideum Moc. & Sessé ex DC.
Sedum oxypetalum Kunth
Sedum praealtum A. DC. subsp. *parvifolium* (R.T.Clausen) R.T.Clausen

Cucurbitaceae
Cayaponia laciniosa (L.) C.Jeffrey (*Bryonia variegata* Mill.)
Cayaponia racemosa (Mill.) Cogn.
Cucurbita argyrosperma C.Huber (*C. mixta* Pangalo)
Cucurbita ficifolia Bouché
Cucurbita foetidissima Kunth
Cucurbita maxima Duchesne
Cucurbita moschata Duchesne
Cucurbita pepo L.
Cucurbita radicans Naudin
Ibervillea sonorae (S. Watson) Greene
Lagenaria siceraria (Molina) Standl. (*L. vulgaris* Ser.)
Luffa acutangula (L.) Roxb.
Luffa cylindrica (L.) M.Roem.
Luffa operculata (L.) Cogn.
Melothria pendula L. [*M. guadalupensis* (Spreng.) Cogn.]
Melothria scabra Naudin
Microsechium palmatum (Ser.) Cogn. [*M. guatemalense* Hemsl., *M. helleri* (Peyr.) Cogn.]
Momordica charantia L.
Sechiopsis triqueter (Ser.) Naudin
Sechium edule (Jacq.) Sw.

Cyperaceae
Cyperus articulatus L.
Cyperus esculentus L.
Rhynchospora corymbosa (L.) Britton (*R. aurea* Vahl)
Schoenoplectus sp.

Dilleniaceae

Curatella americana L.

Dioscoreaceae
Dioscorea composita Hemsl.
Dioscorea convolvulacea Cham. & Schltdl.
Dioscorea floribunda M.Martens & Galeotti
Dioscorea galeottiana Kunth
Dioscorea mexicana Schweidw.
Dioscorea remotiflora Kunth

Ebenaceae
Diospyros nigra (J.F.Gmel.) Perrier (*D. digyna* Jacq., *D. ebenaster* Auct.)

Ericaceae
Arbutus sp.
Arctostaphylos pungens Kunth
Arctostaphylos tomentosa (Pursh) Lindl.
Chimaphila umbellata (L.) Nutt.
Gaultheria acuminata Schltdl. & Cham. (*G. nitida* Benth., *G. ovata* DC.)
Gaultheria myrsinoides Kunth [*Pernettya ciliata* (Cham. & Schltdl.) Small]
Gaultheria procumbens L.
Macleania insignis M. Martens & Galeotti
Vaccinium caespitosum Michx.

Euphorbiaceae
*Acalypha monostachy*a Cav. (*A.* hederácea Torr.)
Cnidoscolus multilobus (Pax) I.M.Johnst.
Cnidoscolus texanus Müll. Arg.) Small
Cnidoscolus urens (L.) Arthur (*Jatropha urens* L.)
Croton ciliatoglandulifer Ortega
Croton cortesianus Kunth
Croton dioicus Cav.
Croton draco Schltdl. & Cham
Croton eluteria (L.) W.Wright (*C. cascarilla* L.)
Croton gossypiifolius Vahl (*C. sanguifluus* Kunth)
Croton guatemalensis Lotsy
Croton niveus Jacq.
Croton salutaris Casar.
Croton suberosus Kunth (*C. cladotrichus* Müll. Arg.)
Croton vulpinus Sessé & Moc.
Euphorbia bracteata Jacq. [*Pedilanthus pavonis* (Klotzsch & Garcke) Boiss.]
Euphorbia calyculata Kunth
Euphorbia chamaesula Boiss.
Euphorbia glyptosperma Engelm. [*Chamaesyce glyptosperma* (Engelm.) Small]

Euphorbia helioscopia L.
Euphorbia hirta L.
Euphorbia hypericifolia L
Euphorbia macropus (Klotzsch & Garcke) Boiss. (*E. biformis* S. Watson)
Euphorbia maculata L.
Euphorbia nutans Lag. [*Chamaesyce nutans* (Lag.) Small]
Euphorbia parviflora L. (*E. pilulifera* L.)
Euphorbia prostrata Aiton
Euphorbia pulcherrima Willd. ex Klotzsch
Euphorbia radians Benth.
Euphorbia seguieriana Neck. (*E. campestris* S. Geuns)
Euphorbia serpens Kunth (*E. radicans* Moric. ex Klotzsch & Garcke)
Euphorbia thymifolia L. [*Chamaesyce thymifolia* (L.)
Euphorbia tithymaloides L. [*P. tithymaloides* (L.) Poit.]
Euphorbia tithymaloides L. subsp. *parasitica* (Boiss. ex Klotzsch) V.W.Steinm. (*Pedilanthus itzaeus* Milsp.)
Havardia albicans (Kunth) Britton & Rose [*Pithecellobium albicans* (Kunth) Benth.]
Hura crepitans L.
Hura polyandra Baill.
Jatropha cardiophylla (Torr.) Müll.Arg.
Jatropha cathartica Terán & Berland.
Jatropha gossypifolia L.
Jatropha cordata (Ortega) Müll.Arg.
Jatropha curcas L.
Jatropha dioica Sessé [*J. spathulata* (Ortega) Mull. Arg.]
Jatropha peltata Sessé (*J. platyphylla* Müll. Arg.)
Manihot esculenta Crantz [*M. aipi* Pohl, *M. dulcis* (J. F. Gmel.) Pax, *M. utilissima* Pohl]
Manihot foetida (Kunth) Pohl
Manihot pringlei S. Watson
Manihot rubricaulis I. M. Johnst.
Pithecellobium circinale (L.) Benth. [*Inga circinalis* (L.) Willd.; *Mimosa circinalis* L.]
Pithecellobium dulce (Roxb.) Benth.
Pithecellobium unguis-cati (L.) Benth. [*Inga unguis-cati* (L.) Willd.]
Ricinus communis L.

Fabaceae
Acacia angustissima (Mill.) Kuntze
Acacia cornigera (L.) Willd.
Acacia coulteri A.Gray
Acacia farnesiana (L.) Willd.
Andira inermis (Wright) DC.
Arachis hypogaea L.
Astragalus crassicarpus Nutt.
Bauhinia racemosa Lam.

Bauhinia subrotundifolia Cav. (*B. spathacea* DC.)
Caesalpinia bonduc (L.) Roxb. [*C. bonducella* (L.) Fleming]
Caesalpinia coriaria (Jacq.)Willd.
Caesalpinia crista L.
Caesalpinia echinata Lam.
Caesalpinia pulcherrima (L.) Sw.
Caesalpinia vesicaria L.
Calliandra houstoniana (Mill.) Standl. var. *anomala* (Kunth) Barneby [*C. anomala* (Kunth) J.F.Macbr., *C. grandiflora* (L'Her.) Benth.]
Canavalia villosa Benth.
Cassia fistula L.
Cassia grandis L.f. (*C. brasiliana* Lam.)
Chamaecrista nictitans (L.) Moench var. *mensalis* (Greenm.) H.S.Irwin & Barneby (*Cassia leptadenia* Greenm.)
Cochliasanthus caracalla (L.) Trew ex A. Delgado et al. (*Phaseolus caracalla* L.)
Crotalaria cajanifolia Kunth (*C. eriocarpa* Benth.)
Crotalaria pumila Ortega
Dalbergia brownei (Jacq.) Urb. (*D. amerimnum* Benth.)
Dalea exserta (Rydb.) Gentry (*D. lagopus* sensu auct.)
Dalea foliolosa (Aiton) Barneby
Dalea leporina (Aiton) Bullock
Dalea lutea (Cav.) Willd. [*Parosela plumosa* (S. Watson) Rose]
Desmodium amplifolium Hemsl.
Desmodium angustifolium (Kunth) DC.
Desmodium orbiculare Schltdl.
Desmodium parviflorum (Dalzell) H.Oshahi
Desmodium scoparium Wall. [unresolved name]
Desmodium triflorum (L.) DC. (*D. stipulaceum* Burm.f.)
Enterolobium cyclocarpum (Jacq.) Griseb.
Erythrina americana Mill.
Erythrina coralloides DC.
Erythrina flabelliformis Kearney
Erythrina mexicana Krukoff
Eysenhardtia polystachya (Ortega) Sarg.
Gliricidia sepium (Jacq.) Walp.
Guibourtia hymenaefolia (Moric.) J.Leonard [*Copaifera hymenaeifolia* (Moric.) Kuntze]
Haematoxylum brasiletto H. Karst.
Haematoxylum campechianum L.
Hoffmannseggia glauca (Ortega) Eifert (*H. densiflora* Benth.)
Hymenaea courbaril L.
Indigofera suffruticosa Mill. (*I. anil* L.)
Leucaena esculenta (DC.) Benth.
Leucaena leucocephala (Lam.) de Wit (*L. glauca* Benth.)
Leucaena pulverulenta (Schltdl.) Benth.

Lysiloma acapulcense (Kunth) Benth.
Lupinus elegans Kunth
Lupinus montanus Humb., Bonpl. & Kunth
Macroptilium atropurpureum (DC.) Urb. (*Phaseolus atropurpureus* DC.)
Macroptilium gibbosifolium (Ortega) A.Delgado (*Phaseolus heterophyllus* Willd.)
Marina diffusa (Moric.) Barneby (*Dalea diffusa* Moric.)
Marina nutans (Cav.) Barneby (*Dalea nutans* (Cav.) Willd.)
Mariosousa willardiana (Rose) Seigler & Ebinger (*Acacia willardiana* Rose)
Mimosa aculeaticarpa Ortega [*Mimosa aculeaticarpa* var. *biuncifera* (Benth.) Barneby]
Mimosa albida Willd.
Mimosa ervendbergii A.Gray
Mimosa sensitiva L.
Mucuna pruriens (L.) DC.
Myroxylon balsamum (L.) Harms var. *pereirae* (Royle) Harms (*Toluifera pereirae* Baill.)
Myroxylon sonsonatense Tonduz [unresolved name]
Pachyrhizus erosus (L.) Urb. (*P. angulatus* DC., *P. palmatilobus* (DC.) Benth. & Hook.f.]
Parkinsonia aculeata L.
Pediomelum pentaphyllum (L.) J.W.Grimes (*Psoralea pentaphylla* L.)
Phaseolus coccineus L. (*Phaseolus multiflorus* Willd.)
Phaseolus vulgaris L.
Piscidia piscipula (L.) Sarg.
Pithecellobium dulce (Roxb.) Benth.
Prosopis chilensis (Molina) Stuntz
Prosopis juliflora (Sw.) DC.
Prosopis laevigata (Willd.) M.C.Johnst. (*P. dulcis* Kunth)
Prosopis pubescens Benth.
Psoralea glandulosa L.
Pterocarpus sp.
Ramirezella strobilophora (Robinson) Rose
Rhynchosia phaseoloides (Sw.) DC. [*Dolicholus phaseoloides* (DC.) Kuntze]
Senna alata (L.) Willd. (*Cassia alata* L.)
Senna bacillaris (L.f.) H.S.Irwin & Barneby (*Cassia bacillaris* L.f.)
Senna occidentalis (L.) Link (*Cassia laevigata sensu auct., C. occidentalis* L.)
Senna pallida (Vahl) H.S.Irwin & Barneby (*Cassia acapulcensis* Kunth)
Senna septemtrionalis (Viv.) H.S.Irwin & Barneby (*Cassia laevigata* Willd.)
Senna sophera (L.) Roxb. (*Cassia esculenta* Roxb.)
Styphnolobium burseroides M. Sousa, Rudd & González Medrano, Francisco
Swartzia guttata (D. Don) Standl.
Vataireopsis araroba (Aguiar) Ducke (*Andira araroba* Aguiar)
Zapoteca formosa (Kunth) H.M.Hern. (*Calliandra gracilis* Baker)

Fagaceae
Quercus acutifolia Née
Quercus elliptica Née (*Q. lanceolata* M.Martens & Galeotti ex A.DC.)

Quercus fusiformis Small
Quercus laurina Bonpl. (*Q. barbinervis* Benth.)
Quercus peduncularis Née (*Q. tomentosa* Willd.)
Quercus polymorpha Schltdl. & Cham.
Quercus rugosa Née

Garryaceae
Garrya laurifolia Benth. (*G. oblonga* Benth.)
Garrya laurifolia Benth. subsp. *racemosa* (Ramírez) Dahling (*G. racemosa* Ramírez)
Garrya ovata Benth.

Gentianaceae
Frasera albicaulis Griseb. var. *nitida* (Benth.) C.L.Hitchc. [*Swertia nitida* (Benth.) Jeps.]
Zeltnera stricta (Schiede) G.Mans. (*Erythraea stricta* Schiede)

Geraniaceae
Geranium alchemilloides L. [unresolved name]
Geranium carolinianum L.
Geranium hernandesii Moc. & Sessé ex DC.
Geranium mexicanum Kunth
Geranium tenue Hanks

Gesneriaceae
Achimenes erecta (Lam.) H.P.Fuchs [*A. coccinea* (Scop.) Pers.]
Diastema hispidum (DC.) Fritsch.
Moussonia deppeana (Schltdl. & Cham.) Klotzsch ex Hanst. [*Isoloma deppeanum* (Schltdl. & Cham.) Hemsl., *Kohleria deppeana* (Schltdl. & Cham.) Fritsch]

Grossulariaceae
Ribes cereum Douglas
Ribes malvaceum Sm.
Ribes montigenum McClatchie

Hydrangeaceae
Philadelphus coulteri S.Watson

Iridaceae
Tigridia pavonia (L.f.) Redouté
Tigridia violacea Schiede ex Schltdl.

Juncaceae
Juncus balticus Willd. subsp. *mexicanus* (Willd. ex Schult. & Schult.f.) Snogerup (*J. mexicanus* Willd. ex Schult. & Schult.f.)

Lamiaceae
Agastache mexicana (Kunth) Lint & Epling [*Cedronella mexicana* (Kunth) Benth.]
Brazoria arenaria Lundell
Clinopodium macrostemum (Moc. & Sessé ex Benth.) Kuntze [*Calamintha macrostema* (Moc. & Sessé ex Benth.) Benth., *Satureja macrostema* (Moc. & Sessé ex Benth.) Briq.]
Cunila lythrifolia Benth.
Cunila polyantha Benth. (*C. longiflora* A. Gray ex S.Watson)
Hedeoma acinoides Scheele
Hedeoma costata Hemsl. var. *costata* (*H. quinquenervata* Bartlett)
Hedeoma piperita Benth.
Hyptis albida Kunth
Mesosphaerum suaveolens (L.) Kuntze (*Hyptis suaveolens* (L.) Poit.)
Ocimum campechianum Mill. (*O. micranthum* Willd.)
Salvia ballotiflora Benth.
Salvia cacaliifolia Benth.
Salvia divinorum Epling & Játiva
Salvia formosa L'Hér.
Salvia gesneriiflora Lindl. & Paxton
Salvia hispanica L.
Salvia longispicata M.Martens & Galeotti
Salvia microphylla Kunth
Salvia occidentalis Sw.
Salvia polystachya Cav.
Scutellaria caerulea Moc. & Sessé ex Benth.
Scutellaria mexicana (Torr.) A. J.Paton (*Salazaria mexicana* Torr.)
Teucrium laciniatum Torr.
Vitex mollis Kunth
Vitex pyramidata B.L.Rob.

Lauraceae
Litsea glaucescens Kunth
Litsea parvifolia (Hemsl.) Mez
Persea americana Mill. (*P. gratissima* C.F.Gaertn.)
Sassafras albidum (Nutt.) Nees (*S. officinalis* T.Nees & C.H.Eberm.)

Lemnaceae
Lemna minor L.

Liliaceae
Calochortus purpureus (Kunth) Baker (*C. bonplandianus* Schult.f.)
Clintonia andrewsiana Torr.

Loasaceae

Gronovia grandis L.
Gronovia scandens L.
Mentzelia aspera L.
Mentzelia hispida Willd.
Mentzelia lobata (Hook.) Hemsl.

Loganiaceae
Buddleja americana L.
Buddleja cordata Kunth
Buddleja sessiliflora Kunth (*B. verticillata* Kunth)
Spigelia anthelmia L.

Loranthaceae
Psittacanthus calyculatus (DC.) G.Don
Struthanthus interruptus (Klunth) G.Don [*S. venetus* (Kunth) Blume]

Lythraceae
Cuphea aequipetala Cav.
Cuphea angustifolia Jacq. ex Koehne
Cuphea jorullensis Kunth
Cuphea lanceolata W.T.Aiton
Heimia salicifolia (Kunth) Link
Lythrum alatum Pursh
Lythrum maritimum Kunth
Lythrum vulneraria Aiton ex Schrank
Nesaea syphilitica (DC.) Steud. (*Heimia syphilitica* DC.)

Magnoliaceae
Magnolia macrophylla Michx. var. *dealbata* (Zucc.) D.L.Johnson (*M. dealbata* Zucc.)
Magnolia mexicana DC. [*Talauma mexicana* (DC.) G.Don]
Magnolia schiedeana Schltl.
Magnolia virginiana L. [*M. glauca* (L.) L.]

Malpighiaceae
Byrsonima crassifolia (L.) Kunth (*B. cotinifolia* Steud.)
Galphimia glauca Cav. [Thryallis glauca (Cav.) Kuntze]
Galphimia paniculata Bartl. (*G. humboldtiana* Bartl.)
Gaudichaudia cynanchoides Kunth
Malpighia glabra L.
Malpighia souzae Miranda

Malvaceae
Anoda cristata (L.) Schltdl.
Apeiba tibourbou Aubl.

Bombax ceiba L.
Bombax palmeri S. Watson [unresolved name]
Ceiba pentandra (L.) Gaertn. [*Eriodendron occidentale* (Spreng.) G.Don]
Chiranthodendron pentadactylon Larreat. (*C. platanoides* Bonpl.)
Gossypium hirsutum L. (*G. mexicanum* Tod., *G. palmeri* G. Watt, *G. schottii* G.Watt)
Guazuma ulmifolia Lam. (*G. tomentosa* Kunth)
Helicteres baruensis Jacq.
Heliocarpus americanus L.
Heliocarpus terebinthinaceus (DC.) Hochr. (*Grewia terebinthinacea* DC., *Heliocarpus reticulatus* Rose)
Kosteletzkya pentacarpos (L.) Ledeb. (*Hibiscus pentacarpos* L.)
Malvastrum coromandelianum (L.) Garcke
Malvastrum spicatum (L.) A.Gray
Malvaviscus arboreus Cav.
Malvaviscus drummondii Torr. & A.Gray
Melochia tomentosa L.
Pachira aquatica Aubl. [*P. macrocarpa* (Schltdl. & Cham.) Walp.]
Pachira insignis (Sw.) Savigny
Pavonia schiedeana Steud. (*Malache rosea* Kuntze)
Pseudobombax ellipticum (Kunth) Dugand (*Bombax ellipticum* Kunth)
Quararibea funebris (La Llave) Vischer (*Lexarza funebris* La Llave)
Sphaeralcea angustifolia Cav.) G.Don (*Malva angustifolia* Cav.)
Theobroma angustifolium Sessé & Moc. ex DC.
Theobroma bicolor Humb. & Bonpl.
Theobroma cacao L.
Triumfetta semitriloba Jacq.

Marantaceae
Calathea loeseneri (J.F. Macbr.) Borschs. & Suárez
Maranta arundinacea L.

Melanthiaceae
Anticlea frigida (Schltdl. & Cham.) Zomlefer & Judd [*Stenanthium frigidum* (Schltdl. & Cham.) Kunth, *Veratrum frigidum* Schltdl. & Cham.]
Schoenocaulon caricifolium (Schltdl.) A.Gray [*S.coulteri* (Baker) Kuntze]
Schoenocaulon officinale (Schltdl. & Cham.) A.Gray (*Sabadilla officinarum* Brandt & Ratzeb., *Veratrum officinale* Schltdl. & Cham., *V. sabadilla* Retz.)

Melastomataceae
Conostegia xalapensis (Bonpl.) D. Don ex DC.
Leandra sp.
Melastoma sp.
Miconia laevigata (L.) D. Don
Tibouchina longifolia (Vahl) Baill. (*T. bourgaeana* Cogn.)

Meliaceae
Swietenia humilis Zucc. (*S. cirrhata* S.F.Blake)
Swietenia macrophylla King
Swietenia mahagoni (L.) Jacq.

Melianthaceae
Zigadenus sp.

Menispermaceae
Cissampelos pareira L.

Menyathaceae
Menyanthes trifoliata L.
Nymphoides aquatica (J.F.Gmel.) Kuntze

Molluginaceae
Mollugo verticillata L.

Moraceae
Brosimum alicastrum Sw.
Castilla elastica Cerv.
Dorstenia contrajerva L.
Dorstenia lindeniana Bureau in DC.
Ficus aurea Nutt. [*F. tecolutensis* (Liebm.) Miq.]
Ficus congesta Roxb. (*F. fasciculata* F.Muell. ex Benth.)
Ficus cotinifolia Kunth
Ficus insipida Willd. (*F. glabrata* Kunth)
Ficus maxima Mill. (*F. guadalajarana* S.Watson, *F. radula* Willd.)
Ficus nymphaeifolia Mill.
Ficus obtusifolia Kunth [*F. involuta* (Liebm.) Miq.]
Ficus petiolaris Kunth (*F. jaliscana* S.Watson)
Ficus pertusa L.f. (*F. arbutifolia* Pers., *F. complicata* Kunth, *F. padifolia* Kunth
Morus celtidifolia Kunth (*M. microphylla* Buckley)
Trophis racemosa (L.) Urb.

Myricaceae
Morella cerifera (L.) Small (*Myrica xalaplensis* Kunth)

Myrtaceae
Pimenta dioica (L.) Merr. (*P. officinalis* Lindl.)
Psidium guajava L. (*P. pomiferum* L.)

Nelumbonaceae
Nelumbo lutea (Willd.) Pers.

Nyctaginaceae
Allionia incarnata L.
Boerhavia coccinea Mill. (*B. caribaea* Jacq.)
Mirabilis jalapa L.

Nymphaeaceae
Cassytha filiformis L.
Nymphaea ampla (Salisb.) DC.
Nymphaea gracilis Zucc. (*Castalia gracilis* (Zucc.) Rose)
Nymphaea mexicana Zucc.

Onagraceae
Epilobium mexicanum Moc. & Sessé ex DC.
Fuchsia thymifolia Kunth
Lopezia racemosa Cav.
Ludwigia ascendens Hall in Eat. & Wr. [unresolved name] (*Jussiaea repens* L.)
Oenothera elata Kunth subsp. *hookeri* (Torr. & A. Gray) W. Dietr. & W. L. Wagner (*O. hookeri* Torr. & A. Gray)
Oenothera laciniata Hill (*O. sinuata* L.)
Oenothera pubescens Willd. ex Spreng.
Oenothera rosea L'Hér. ex Aiton (*O. rubra* Cav.)

Orchidaceae
Arpophyllum spicatum Lex.
Bletia campanulata Lex.
Bletia coccinea Lex.
Catasetum integerrimum Hook.
Catasetum maculatum Kunth
Cranichis speciosa Lex. [unresolved name]
Cranichis tubularis Lex. [unresolved name]
Euchile citrina (Lex.) Withner [*Cattleya citrina* (Lex) Lindl., *Prosthechea citrina* (Lex.) W. E. Higgins)
Galeoglossum tubulosum (Lindl.) Salazar & Soto Arenas (*Cranichis tubulosa* Lindl.)
Govenia liliacea (Lex.) Lindl.
Govenia superba (Lex.) Lindl.
Laelia anceps Lindl.
Laelia autumnalis (Lex.) Lindl.
Laelia speciosa (Kunth) Schltr.
Prosthechea pastoris (Lex.) Espejo & López-Ferr. [*Encyclia pastoris* (Lex.) Schltr., *Epidendrum pastoris* Lex.]
Prosthechea vitellina (Lindl.) W.E.Higgins (*Epidendrum vitellinum* Lindl.)
Sobralia macrantha Lindl.
Stanhopea hernandezii (Kunth) Schltr.
Stanhopea oculata (Lodd.) Lindl.

Stanhopea tigrina Bateman ex Lindl.
Trichocentrum cebolleta (Jacq.) M.W.Chase & N.H.Williams [*Oncidium cebolleta* (Jacq.) Sw.]
Vanilla mexicana Mill. (*Epidendrum vanilla* L.)
Vanilla planifolia Jacks. ex Andrews (*V. fragrans* Ames)

Orobanchaceae
Castilleja tenuiflora Benth. (*C. canescens* Benth.)
Escobedia sp.

Oxalidaceae
Biophytum dendroides (Kunth) DC. (*Oxalis dendroides* Kunth)
Oxalis acetosella L. (*O. americana* Bigelow)
Oxalis albicans Kunth
Oxalis corniculata L.
Oxalis frutescens subsp. *angustifolia* (Kunth) Lourteig (*Oxalis angustifolia* Kunth)
Oxalis hernandezii DC.
Oxalis latifolia Kunth
Oxalis stricta L.
Oxalis tetraphylla Cav.
Oxalis verticillata DC. [unresolved name]
Oxalis violacea L.

Papaveraceae
Argemone grandiflora Sweet
Argemone mexicana L.
Argemone ochroleuca Sweet
Bocconia arborea S.Watson
Bocconia frutescens L.
Bocconia integrifolia Bonpl.

Passifloraceae
Passiflora coriacea Juss.
Passiflora jorullensis Kunth
Passiflora morifolia Mast.
Passiflora sicyoides Schltdl. & Cham.
Passiflora suberosa L.

Pentaphylacaceae
Ternstroemia tepezapote Cham. & Schltdl.

Penthoraceae
Penthorum sedoides L.

Phrymaceae

Mimulus cardinalis Douglas ex Benth.
Mimulus glabratus Kunth
Mimulus luteus L.
Mimulus rupestris Greene

Phytolaccaceae
Phytolacca americana L. (*P. decandra* L.)
Phytolacca dioica L.
Phytolacca icosandra L.
Phytolacca longespica Moq. [unresolved name]
Phytolacca octandra L.
Phytolacca rugosa A.Braun & C.D.Bouché
Rivina humilis L.

Picramniaceae
Alvaradoa amorphoides Liebm.

Piperaceae
Peperomia galioides Kunth
Peperomia umbilicata Ruiz & Pav.
Piper amalago L.
Piper auritum Kunth
Piper palmeri C.DC. [unresolved name]
Piper sanctum (Miq.) Schltdl. ex C.DC.
Piper umbellatum L. (*P. cuernavacanum* C. DC.)

Plantaginaceae
Herpestis sp.
Penstemon barbatus (Cav.) Roth
Penstemon campanulatus (Cav.) Willd.
Penstemon roseus (Cerv. ex Sweet) G.Don
Plantago australis Lam. (*Plantago australis* Lam. subsp. *hirtella* (Kunth) Rahn)
Plantago mexicana Link
Russelia polyedra Zucc.
Tetranema roseum (M.Martens & Galeotti) Standl. & Steyerm. (*T. mexicanum* Benth.)
Veronica americana Schwein. ex Benth.

Plumbaginaceae
Plumbago pulchella Boiss.
Plumbago zeylanica L. (*P. scandens* L.)

Poaceae
Andropogon sp.
Arundinella hispida (Willd.) Kuntze

Bambusa sp.
Cenchrus echinatus L.
Cenchrus tribuloides L.
Cynodon dactylon (L.) Pers. [*Capriola dactylon* (L.) Kuntze]
Distichlis spicata (L.) Greene
Muhlenbergia macroura (Humb., Bonpl. & Kunth) Hitchc. (*Epicampes macroura* Benth.)
Phragmites australis (Cav.) Trin. ex Steud. (*P. communis* Trin.)
Saccharum officinarum L.
Sporobolus sp.
Zea sp.
Zea mexicana (Schrad.) Kuntze (*Euchlaena mexicana* Schrad.)

Polemoniaceae
Cobaea biaurita Standl.
Ipomopsis pinnata (Cav.) V.E.Grant
Loeselia mexicana (Lam.) Brand [*L. coccinea* (Cav.) G.Don]
Loeselia scariosa (M. Martens & Galeotti) Walp.
Polemonium grandiflorum Benth.

Polygalaceae
Monnina salicifolia Ruiz & Pav.
Polygala alba Nutt.
Polygala scoparia Kunth
Polygala verticillata L.

Polygonaceae
Antigonon flavescens S. Watson
Coccoloba uvifera (L.) L.
Neomillspaughia emarginata (H.Gross) S.F.Blake
Oxyria digyna (l.) Hill
Persicaria hydropiper (L.) Delarbe (*Polygonum hydropiper* L.)
Persicaria hydropiperoides (Michx.) Small (*Polygonum hydropiperoides* Michx.)
Persicaria punctata (Elliott) Small (*Polygonum acre* Lam.)
Polygonum aviculare L.
Rumex mexicanus Meisn.
Rumex patientia L.
Rumex pulcher L.
Rumex pulcher subsp. *woodsii* (De Not.) Arcang. (R. *brevipes* Meisn.)

Portulacaceae
Portulaca oleracea L.
Portulaca pilosa L.

Primulaceae
Ardisia sp.

Bonellia macrocarpa (Cav.) B.Ståhl & Källersjö subsp. *macrocarpa* (*Jacquinia aurantiaca* W.T.Aiton)

Ranunculaceae
Actaea rubra (Aiton) Willd. f. *neglecta* (Gillman) B.L.Rob.
Anemone narcissiflora L.
Anemone patens L.
Anemone tuberosa Rydb.
Ranunculus dichotomus Moc. & Sessé ex DC.
Ranunculus hydrocharoides A. Gray (*R. stolonifer* Hemsl.)
Ranunculus petiolaris Humb., Bonpl. & Kunth ex DC. (*Ranunculus geoides* Humb., Bonpl. & Kunth ex DC.)
Thalictrum fendleri Engelm. ex A.Gray
Thalictrum hernandezii Tausch ex J. Presl
Thalictrum mexicanum DC. [unresolved name]

Rhamnaceae
Ceanothus caeruleus Lag.
Colubrina triflora Brongn. ex Sweet (*C. guatemalensis* Standl.)
Condalia hookeri M.C.Johnst. (*C. obovata* Hook.)
Rhamnus humboldtiana Willd. ex Schult. [*Karwinskia humboldtiana* (Schult.) Zucc.]
Rhamnus serrata Willd. ex Schult.

Rosaceae
Crataegus gracilior J.B.Phipps [*C. pubescens* (Kunth) Steud.]
Crataegus laevigata (Poir.) DC. (*C. oxyacantha* auct.)
Crataegus mexicana Moc. & Sessé ex DC.
Potentilla candicans Willd. ex Schltdl.
Prunus serotina Ehrh. var. *salicifolia* (Kunth) Koehne [*P. serotina* Ehrh. subsp. *capuli* (Cav. ex Spreng.) McVaugh, *P.capollin* (Cav. ex Spreng.) Zucc., *P.capuli* Cav.]
Rubus pumilus Focke [unresolved name]
Rubus trivialis Michx.

Rubiaceae
Bouvardia bouvardioides (Seem.) Standl.)
Bouvardia erecta (DC.) Standl.
Bouvardia ternifolia (Cav.) Schltdl.
Carapichea ipecacuanha (Brot.) L. Andersson [*Cephaelis ipecacuanha* (Brot.) Stokes]
Cephalanthus occidentalis L.
Didymaea mexicana Hook.f.
Exostema caribaeum (Jacq.) Schult.
Ferdinandusa sp.
Galium mexicanum Kunth
Galium trifidum L.

Hamelia calycosa Donn.Sm.
Hamelia patens Jacq. (*H. erecta* Jacq.)
Hintonia latiflora (Sessé & Moc. ex DC.) Bullock (*Coutarea latiflora* Sessé & Moc. ex DC.)
Randia aculeata L. (*R. mitis* L.)
Spermacoce tenella Kunth

Rutaceae
Casimiroa edulis La Llave
Casimiroa greggii (S. Watson) F. Chiang (*Sargentia greggii* S. Watson)
Casimiroa pubescens Ramírez
Zanthoxylum fagara (L.) Sarg.

Salicaceae
Populus sp.
Salix sp.
Salix lasiolepis Benth.
Salix paradoxa Kunth
Salix pentandra L.
Xylosma flexuosa (Kunth) Hemsl.

Sapindaceae
Acer negundo L.
Cardiospermum halicacabum L.
Dodonaea viscosa (L.) Jacq.
Paullinia fuscescens Kunth
Sapindus marginatus Willd.
Sapindus saponaria L.
Serjania mexicana (L.) Willd. (*Paullinia mexicana* L.)
Urvillea ulmacea Kunth

Sapotaceae
Manilkara zapota (L.) P.Royen [*Achras sapota* L., *Calocarpum mammosum* (L.) Pierre, *Lucuma mammosa* (L.) C.F.Gaertn.]
Pouteria campechiana (Kunth) Baehni [*Lucuma salicifolia* Kunth, *Vitellaria sphaerocarpa* (A.DC.) Radlk.]
Pouteria dominigensis (C.F.Gaertn.) Baehni (*Lucuma dominigensis* C.F.Gaertn.)
Pouteria sapota (Jacq.) H.E.Moore & Stearn
Sideroxylon palmeri (Rose) T.D.Penn. (*Bumelia laetevirens* Hemsl.)
Sideroxylon persimile (Hemsl.) T.D.Penn. subsp. *subsessiliflorum* (Hemsl.) T.D.Penn. (*Bumelia subsessiliflora* Hemsl.)
Sideroxylon tepicense (Standl.) T.D.Penn. (*S. mexicanum* Hemsl.)

Saxifragaceae
Lithophragma affine A.Gray

Smilacaceae
Smilax sp.
Smilax aristolochiifolia Mill.
Smilax cordifolia Humb. & Bonpl. ex Willd.
Smilax moranensis M.Martens & Galeotti
Smilax pseudochina L.
Smilax rotundifolia L.
Smilax spinosa Mill. (*S. mexicana* Griseb. ex Kunth)

Solanaceae
Brugmansia arborea (L.) Steud. (*Datura arborea* L.)
Capsicum annuum L.
Capsicum annuum L. var. *glabriusculum* (Dunal) Heiser & Pickersgill
Capsicum baccatum L. (*C. microcarpum* Cav.)
Capsicum frutescens L. (*C. minimum* Blanco)
Cestrum dumetorum Schltdl.
Cestrum nocturnum L.
Datura ceratocaula Ortega
Datura stramonium L.
Datura wrightii Regel (*D. meteloides* auct.)
Lycianthes moziniana (Dun.) Bitter.
Lycianthes somniculenta (Kuntze ex Schltdl.) Bitter (*Solanum somniculentum* Kuntze ex Schltdl.)
Margaranthus solanaceus A.Gray
Nicandra physalodes (L.) Gaertn.
Nicotiana rustica L.
Nicotiana tabacum L. (*N. mexicana* Schltdl.)
Physalis aequata J. Jacq. ex Nees
Physalis angulata L.
Physalis coztomatl Dunal
Physalis ixocarpa Brot. ex Hornem.
Physalis mollis Nutt.
Physalis nicandroides Schltdl.
Physalis peruviana L.
Physalis pubescens L.
Solandra brevicalyx Standl.
Solandra grandiflora Sw.
Solandra guerrerensis Martinez
Solandra maxima (Moc. & Sessé ex Dunal) P.S.Green (*Datura maxima* Moc. & Sessé ex Dunal, *S. hartwegii* C.F.Ball, *S. nitida* Zuccagni)
Solanum americanum Mill. (*S. nigrum* L.)

Solanum angustifolium Mill. (*S. cornutum* Lam.)
Solanum donianum Walp. (*S. verbascifolium* L.) [unresolved name]
Solanum fontanesianum Dunal [unresolved name]
Solanum hernandesii Dunal
Solanum lycopersicum L. (*Lycopersicon esculentum* Mill.)
Solanum scandens Mill.
Swartzia nitida (Zuccagni) Standl.

Stegnospermataceae
Stegnosperma halimifolium Benth.

Tropaeolaceae
Tropaeolum majus L.

Typhaceae
Typha angustifolia L.
Typha latifolia L.

Urticaceae
Cecropia obtusifolia Bertol. (*C. mexicana* Hemsl.)
Discocnide mexicana (Liebm.) Chew (*Urticastrum mexicanum* (Liebm.) Kuntze)
Urera baccifera (L.) Gaudich. ex Wedd.
Urera caracasana (Jacq.) Gaudich. ex Griseb.
Urtica chamaedryoides Pursh
Urtica chichiaztli Sessé & Moc. [unresolved name]
Urtica dioica L.
Urtica urens L.

Valerianaceae/Caprifoliaceae
Valeriana albonervata B.L.Rob.
Valeriana ceratophylla Kunth
Valeriana edulis Nutt. subsp. *procera* (Kunth) F.G. Mey.
Valeriana mexicana DC.
Valeriana sorbifolia Kunth (*V. toluccana* DC.)

Verbenaceae
Citharexylum affine D.Don
Citharexylum flabellifolium S.Watson
Duranta erecta L. (*D. repens* L.)
Lantana camara L.
Lantana hirsuta M.Martens & Galeotti
Lippia alba (Mill.) N.E.Br. ex Britton & P.Wilson [*L. geminata* (Kunth) Spreng.]
Lippia graveolens Kunth (*L. berlandieri* M.Martens & Galeotti)
Lippia mexicana G.L.Nesom

Lippia umbellata Cav.
Petrea volubilis L.
Phyla scaberrima (Juss. ex Pers.) Moldenke (*Lippia dulcis* Trevir.)
Priva grandiflora (Ortega) Moldenke (*P. tuberosa* S.Watson)
Priva lappulacea (L.) Pers.
Priva mexicana (L.) Pers. (*P. hispida* Juss.)
Stachytarpheta sp.
Verbena carolina L.
Verbena litoralis Kunth
Verbena officinalis L.

Violaceae
Viola bicolor Pursh (*V. rafinesquei* Greene)

Vitaceae
Cayratia japonica (Thunb.) Gagnep. (*Vitis mollis* Wall. ex M.A.Lawson)
Cissus tiliacea Kunth [*Vitis tiliacea* (Kunth) Hemsl.]
Cissus verticillata (L.) Nicolson & C.E.Jarvis (*C. sicyoides* L.)
Vitis tiliifolia Humb. & Bonpl. ex Schult. (*V. caribaea* DC.

Xanthorrhoeaceae
Asphodelus ramosus L.

Zingiberaceae
Renealmia alpinia (Rottb.) Maas
Zingiber officinale Roscoe

Zygophyllaceae
Bulnesia arborea (Jacq.) Engl. [*Guaiacum arboreum* (Jacq.) DC.]
Guaiacum sanctum L.

Index

A. O. Tucker, J. Janick, *Flora of the Voynich Codex*,
https://doi.org/10.1007/978-3-030-19377-5

MIX
Papier aus verantwortungsvollen Quellen
Paper from responsible sources
FSC® C105338

If you have any concerns about our products,
you can contact us on
ProductSafety@springernature.com

In case Publisher is established outside the EU,
the EU authorized representative is:
Springer Nature Customer Service Center GmbH
Europaplatz 3, 69115 Heidelberg, Germany

Printed by Libri Plureos GmbH
in Hamburg, Germany